Joint Source-Channel Decoding

EURASIP and Academic Press Series in Signal and Image Processing

Series Advisor

Sergios Theodoridis

Titles in the series:

Duhamel and Kieffer

Joint Source-Channel Decoding, ISBN 9780123744494

Thiran, Marqués and Bourlard

Multimodal Signal Processing, ISBN 9780123748256

Joint Source-Channel Decoding

A Cross-Layer Perspective with Applications in Video Broadcasting over Mobile and Wireless Networks

Pierre Duhamel
Michel Kieffer

ELSEVIER

AMSTERDAM • BOSTON • HEIDELBERG • LONDON
NEW YORK • OXFORD • PARIS • SAN DIEGO
SAN FRANCISCO • SINGAPORE • SYDNEY • TOKYO
Academic Press is an imprint of Elsevier

AP

Academic Press is an imprint of Elsevier
The Boulevard, Langford Lane, Kidlington, Oxford OX5 1GB, UK
30 Corporate Drive, Suite 400, Burlington, MA 01803, USA

First edition 2010

British Library Cataloguing in Publication Data
A catalogue record for this book is available from the British Library.

Library of Congress Cataloging-in-Publication Data
A catalog record for this book is available from the Library of Congress.

ISBN-13: 978-0-12-374449-4

For information on all Academic Press publications
visit our Web site at *www.elsevierdirect.com*

Typeset by: diacriTech, India

Printed and bound in the UK
10 11 12 13 14   10 9 8 7 6 5 4 3 2 1

Contents

Preface and Acknowledgements

Il ne faut pas toujours tellement épuiser un sujet qu'on ne laisse rien à faire au lecteur. Il ne s'agit pas de faire lire mais de faire penser.
Montesquieu (L'esprit des lois)

(approximate) English translation (and in short):
One should not always exhaust a subject in such a way that this leaves nothing to do to the reader. This is not to read but to think.

Joint source-channel coding/decoding is a rather specialized term for designating techniques allowing to transmit efficiently audiovisual contents through wireless channels by means of possibly unconventional methods. Such a simple goal turns out to be quite complex since it involves many different techniques (namely, source coding, networking, and transmission techniques) usually mastered by different teams.

Therefore, this book does not intend to address with full accuracy all these aspects, but rather tries to give a precise idea of the basics of each and to provide a state of the art of the "joint source/network/channel decoding" results obtained on video signals. Many examples illustrate the corresponding concepts and algorithms. It is essentially based on a tutorial by the same authors, which was presented at the IEEE International Conference on Acoustics, Speech, and Signal Processing in 2007. The joint source-channel coding (JSCC) aspect is outlined only at the end of the book.

Being a topic with such a wide scope, we were obviously helped and stimulated by many people. Our only hope is that we will not forget too many of them in the list below.

Very special thanks are due to several colleagues: Olivier Rioul, who introduced us to the topic of JSCC, before we even thought of JSCD, Pierre Siohan, who always showed a strong interest in this topic (and led one of the corresponding projects), and Christine Guillemot and her team, with whom the neologism "coopetition" seems to be the most appropriate for characterizing our stimulating cooperation/competition. Catherine Lamy-Bergot and Kave Salamatian also took part in many fruitful discussions on the topic. Parts of this book were proofread by colleagues who did a very detailed job, many thanks to Aurelia Fraysse, Jean-Benoist Leger, Pablo Piantanida, Phil Schniter, and Claudio Weidmann.

Obviously, when working on this topic, we were helped by many PhD students, namely, Barham Zahirazami, Avi Gabay, Hang Nguyen, Rachel Chiang, Chang-Ming Lee, Galina Sabeva, Salma Ben-Jamaa, Cedric Marin, Usman Ali, and Zied Jaoua who worked directly on the topics covered by this book. All of them were funded by either industry grants or collaborative contracts funded by the French government. Other PhD students, namely, José Saavedra and Olivier Derrien, were also of great help for some parts of the book.

In fact, we were very lucky to obtain funding for such a long-term research topic, almost continuously since 1997. Special thanks are thus due to Gerard Eude, from France Telecom R&D, who initiated these fundings, followed by CNES (the French Space Agency, with Catherine Lambert-Nebout as an interface), and Alcatel-Lucent, a long-term contributor to this work, with an initial strong commitment of Denis Rouffet. CNES (with Gaël Scot as contact person, and Jerôme Lacan as coworker) also funded a cross-layer study on JSCD, which demonstrated to us that both aspects were two sides of the same coin.

Part of the material described in this book was funded by the French Ministry of Research and Industry by means of collaborative contracts named COSOCATI, DIVINE, VIP, DITEMOI, and TVMSL.

Finally, most of this work benefited from contacts established by means of two Networks of Excellence of the European Commission, NEWCOM and NEWCOM++, scientifically led in a friendly and efficient manner by Sergio Benedetto. We also benefited from fundings through these NoEs and helpful collaborative work with Lajos Hanzo, Michael Ng, and Morten Holm Larsen.

The final thanks will go to the people who showed the highest patience during the writing of this book: our families, who did not sign to write the book, but had to undergo the consequences… and our editor, Tim Pitts, whose patience and sense of humor were combined with sufficient pressure so that the book finally came to an end without too much delay.

More materials (Appendices X, Y and Z) can be found at website: http://www.elsevierdirect.com/companions/9780123744494.

Chapter 1

Introduction: Context

This book addresses a question of growing importance: How can one improve the perceived quality of video communicated over a wireless network? For successful communication of multimedia data, especially video, the network must behave as if it is predictable and stable. Wireless networks are not known for predictability and stability, but rather variability and scarcity of resources. Oversizing, the usual way of coping with variability, is not practical in wireless systems due to the scarcity of resources. Yet, the number of applications for video broadcasting over wireless channels keeps increasing. Therefore, there is a growing need for techniques that provide sufficient received video quality without overspending the scarce resources of the wireless network.

In the classical approach to network design, the networking task is partitioned into distinct layers, where each layer, assuming that the lower layers behave perfectly, attempts to provide perfect information to the upper layers. In recent years, the so-called joint approaches have been proposed for improved performance and better use of resources. In joint approaches, the network layers become more transparent so that the information previously available at only one layer is accessible to other layers. The downside to this ambitious approach is a loss of the architectural coherence that was the primary driving force behind the use of decoupled layers.

Wireless network designs, thus, fall between two extreme points: a generic layered architecture, which would be safe but not very efficient at resource utilization, and an optimized architecture (i.e., optimal in both resource utilization and performance), which would be dedicated to a single application. Finding joint mechanisms that are compatible with the use of layered, generic architectures is a serious challenge.

Given the complexity of the task, the joint mechanisms usually proposed are quite limited. Most techniques adapt the parameters used in source coding, channel coding, and other procedures based on a limited set of information provided by other layers and thus, do not reach their full potential. While these so-called cross-layer approaches have made some impact on wireless network performance, they are not truly joint schemes and constitute a first step toward a global optimization of the network. In fact, there seems to be a large difficulty in obtaining fully joint techniques compatible with generic network architectures.

This book intends to fill the gap by developing – in some detail – tools that allow the implementation of fully joint techniques for robust reception of multimedia transmitted over wireless/Internet Protocol (IP) networks. Since these techniques require the cooperation of many layers, each with its own peculiar characteristics; this book covers a wide variety of topics from source coding to channel coding to networking.

While there are many books that focus on any one of these topics (e.g., source coding, channel coding, wireless communications, or wireless networking), the readerships of these books are often very different. For example, the community that works on source coding does not overlap too much with the one that works on the so-called physical layer. So, years ago, a tool known as joint source-channel coding (JSCC) was proposed to address optimal source transmission for applications in which Shannon's separation principle does not hold. However, the problem was considered as a curiosity by many practitioners, and even if a (relatively small) number of articles have appeared regularly on this topic, the impacts have not been widespread. The main drawback of JSCC is that, since it is not compatible with any existing standards, widespread implementation would require significant time and effort, even if the performance improvements were known to be potentially worthwhile. Motivated by the practical challenges of JSCC, an interest in joint source-channel decoding (JSCD) emerged, whereby the goal is to extract the best performance out of the received signal without considering changes to the transmitted signal. Clearly, the ability to use JSCD with existing standards makes it potentially very practical. Therefore, there is a good possibility that JSCD tools could become useful in the near future, provided that all the different layers can be incorporated. For example, many JSCD articles treat the network layers as transparent, which would clearly be a drawback for practical implementation. At the same time, cross-layer communication protocols have become a great interest to some who, motivated by multimedia applications, have worked on jointly adapting the characteristics of each layer to the type of signal that was transmitted. The aim of this book is precisely to provide a cross-layer perspective on JSCC/JSCD tools so that their applicability is increased; on one hand, JSCD tools become more practical when network layers are not treated as fully transparent, and on the other hand, the cross-layer approach benefits from additional efficient tools.

Though this book necessarily treats topics from a wide range of fields (e.g., source coding, channel coding, and networking), it attempts to be as self-contained as possible.

1.1 MULTIMEDIA WIRELESS: THE NEED FOR NEW TOOLS

1.1.1 The Classical Approach to Multimedia Wireless

Many years ago, when communication systems were first designed, they were targeted toward a specific application. For example, the wired phone system was principally designed to carry speech signals, as were the first generations of wireless phone systems, while the Internet was designed to carry data files. From that starting point, these systems have evolved greatly and now support a much wider variety of applications, using (almost) the same channels and protocols. For example, fax was added to the wired phone system, short message services (SMS) and multimedia services (MMS) were added to wireless phone systems, and Voice over IP (VoIP) was added to the Internet. Furthermore, the systems have been integrated to the point that, now, many phone calls go (at least partially) through the Internet without users being notified or even able to tell the difference, and many e-mails are typed and/or read on wireless mobile phones.

Moreover, the bitrate available to most systems has grown so high that they now support very demanding signals. For example, high-definition television (HDTV) goes through wired phone lines, via asymmetric digital subscriber line (ADSL), and video is routinely transmitted through wireless channels. As discussed below, these changes have led to pronounced shifts in both the usage and the design of these systems.

Changes in System Design

The dominant approach to the design of wireless multimedia systems is the layered approach, where the network layers are designed to function independently from each other. Each layer is dedicated to a specific networking task, and the layers are designed regardless of the signal types carried by the network.

This layered approach is motivated to some extent by Shannon's separation principle for source-channel coding, which (loosely speaking) established that, under certain conditions, the best joint source/channel coder is the concatenation of the best source coder with the best channel coder. In such situations, the source coder generates the lowest-rate bitstream given an allowed distortion – free of redundancy, whereas the channel coder generates the lowest-rate bitstream enabling the correction of "almost" all channel-induced errors – via judicious use of redundancy. Thus, the optimal source coding is invariant to the channel characteristics, whereas optimal channel coding is invariant to the source characteristics. However, this separation principle (described in more detail in Chapter 2) holds only under certain not-so-practical conditions, such as in the limit of infinite code-word length. Implementations that attempt to satisfy these conditions directly would involve very high complexity and possibly very large transmission delays.

Changes in System Usage

Predicting the way that wireless multimedia systems will be used turns out to be very difficult. In the early 1970s, during the time of analog telephones, there were already proposals for videophones in various parts of the world. At that time, cost was a significant problem. Several years later, in the early 1990s, digital techniques had evolved, and the first studies of video compression (again targeted at videophones, i.e., visiophony) resulted in the widely accepted H.263 standard, published in 1996. It took some time for this technology to be viable on widely deployed terminals, but at that time, visiophony did not seem to be a driving need. Only recently did some "killer applications" for mobile networks emerge: video database consultation and video broadcasting. While these uses of wireless networks appear to be linked primarily to the ways that mobile phones are now being used (e.g., more like a PC for surfing the Internet and downloading files, images, and videos), they show a clear shift in the way that networks are being used: the resources required for transmitting a 3-minute video clip are far beyond those required to support voice communication for the same duration. Even downloading audio files is about 10 times more costly (in terms of bitrate) than communicating speech. Since multimedia traffic is increasing at a much faster rate than speech traffic, one is well justified to concentrate on the multimedia aspects of mobile devices.

As a result of these trends, it may well happen that demands for increased bandwidth, in conjunction with the use of limited-efficiency layered networks, will overload the global network. While many actions have been taken to prevent this occurrence (e.g., the addition of broadcast mode in some standards), new tools as JSCC/JSCD and cross-layer optimization could be of great help.

1.1.2 A New Approach to Multimedia Wireless

The claim that current networks are implemented using fully independent layers is really an oversimplification. For example, the second generation European mobile telephone system (GSM) residual bit error rate (BER) is tuned so that it is compatible with the BER tolerated by GSM's speech coder. This, in fact, is already a form of cross-layer optimization in which the behavior of the physical (PHY) layer is matched to the needs of the application layer. Clearly, the tuning would be different for other applications and signal types (e.g., video).

The primary factors motivating the shift from fully independent layers to closely coupled layers are the high variability of wireless channels and the need for increased robustness and graceful degradation. If a network must support a wide variety of channels and users, then designing the network to work perfectly on all of them amounts (for the most part) to designing the network to satisfy the worst channel/user. In this case, if one user experiences a very bad channel, then that user would consume a disproportionate amount of system resources.

In addition, it should be remembered that the wireless multimedia networks are composed of many links, some wireless (e.g., to the terminals) and some wired. The means of protecting the signal is somewhat different over these two links due to the differing natures of their impairments. For the wired portion, the network adds headers (containing, e.g., the MAC address of the user and the number assigned to each packet) and cyclic redundancy checks (CRCs), which are used to check the integrity of the received packet. Because of the reliability of the wired channel, it is feasible to implement protection using an acknowledgment (ACK/NACK)-based procedure known as Automatic Repeat reQuest (ARQ): a packet is retransmitted if, upon reception, the CRC does not match the part that it protects (e.g., the payload, header, or both). Clearly, ARQ-based error correction requires feedback and induces large delays (though hopefully on only a few packets). If feedback was somehow unavailable (e.g., in the broadcast scenario), then the CRC would at least indicate which packets were not correctly received.

The amount of delay that is tolerable, as well as the channel conditions themselves, can vary widely across different applications.

Finally, practical communication links are rarely dedicated entirely to a single user, and a given user rarely communicates through only a single link. Rather, each user's signal is segmented into packets which are scheduled, along with the packets of many other users, through available links. This segmentation of packets has often been overlooked in the design of robust reception schemes, even though it can have a large (positive or negative) impact on those strategies. This topic will be addressed in the second part of the book, i.e., Chapters 7–9, describing the corresponding tools that are similar to the ones used in more traditional JSCD (explained in the first part).

To make these notions more concrete, a few examples of applications that might benefit from the tools discussed in this book are described, and their essential characteristics are analyzed.

1.2 EXAMPLE APPLICATIONS

Over the past several years, the authors have worked with engineers, network operators, and academic colleagues (all of them gratefully acknowledged) on various wireless multimedia system design projects. Some of these projects are described here and linked to later chapters of the book. These examples are useful in bridging the abstract models used later in the book with practical real-world problems.

1.2.1 Example 1: Visiophony Over IP (RNRT VIP)

This project was concerned with the classical scenario of a video communication through wireless terminal links that are connected by a wired central link that

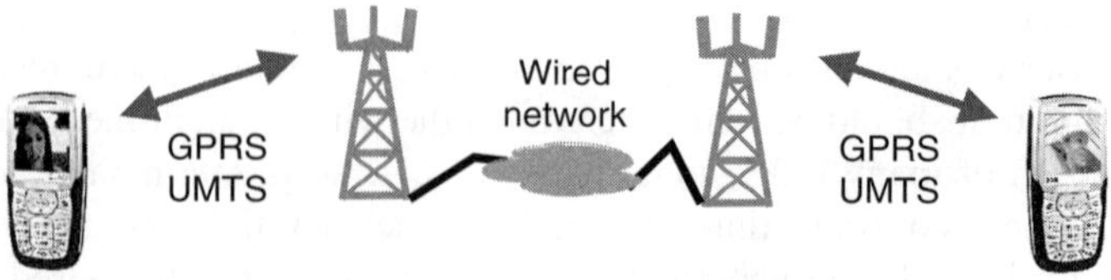

FIGURE 1.1 Wireless visiophony over IP.

uses the IP, see Figure 1.1. Admittedly, the model assumed in the project was abstract, and many implementation details were never considered. However, the essential constraints resulting from the visiophony application were (1) that the maximum delay induced by coding and transmission was at most ≈ 0.5s and (2) that the reconstructed speech/video signal was of sufficiently high quality.

The proposed solution leveraged the fact that the error mechanism on the wired link is different than that on the wireless link. For example, the wired link can introduce packet losses, which – in terms of transmission – are equivalent to erasures. While the delay constraint makes the retransmission of lost packets difficult, the knowledge of which packets are lost can be used for error concealment; portions of the stream received without error can be interpolated to "fill the holes." In contrast, the wireless link must contend with intersymbol interference (ISI) and noise, the mitigation of which can be accomplished using JSCD. Since it was assumed that the wireless link does not support retransmission, the decoder must be able to display an image on the screen, whatever the output of the channel. But even in this situation, CRC checks can be exploited by the PHY layer: the packets that can be correctly received using a simple "hard decision" procedure (as indicated by the CRC) do not need the application of a more complex JSCD procedure. Packets that do not pass this check can be processed by a robust decoder, after which CRC-based checks would be infeasible.

While the mechanisms described above are somewhat simple and ignore many practical aspects of implementation, they already go beyond what has been proposed in many JSCD studies, where, e.g., erasures were not taken into account. The primary aim of this study was to quantify the potential improvements of "classical" JSCD. (See Figure 1.2.)

Example 1: Summary

- Mixed wired/wireless links facilitate error concealment.
- Strong delay constraints limit the possibility for ARQ.
- CRCs can be used to determine which packets require complex JSCD.

1.2.2 Example 2: Video Broadcasting (CNES SDMB, Alcatel TVMSL)

The second example pertains to a project launched by Alcatel-Lucent in 2006, partially funded by the "Agence de l'Innovation Industrielle" (AII) of the French

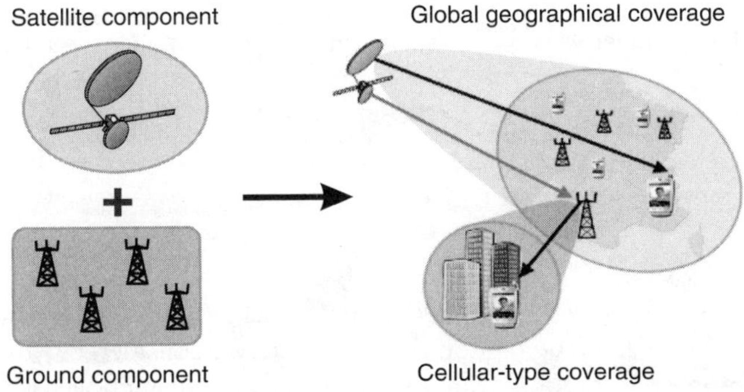

FIGURE 1.2 Video broadcasting using a mixture of satellite and terrestrial links.

government, where the goal was to provide (relatively) low-cost video broadcasting to mobiles. The project was motivated by the observation that the first users of 3G were using video consultation much more than anticipated. In the proposed solution, mobiles were addressed by satellites when possible or terrestrial relays when satellite reception was poor, see Figure 1.2. While, in this project, most of the effort was devoted to implementation, some effort was made to evaluate the impact of JSCD.

At first glance, this application does not seem to impose a delay constraint; since there is no need to support a comfortable conversation between two users, there does not seem to be a drawback to a several-second delay between source and destination. However, it turns out that users do want the ability to switch from one channel to another without waiting too long. Thus, the application does in fact impose a delay constraint, though perhaps not as strict as the one in the previous example. For this project, the DVS-S/H standard was assumed, which allowed the introduction of MAC-layer redundancy but required cross-layer tuning. Since a broadcast scenario was considered, feedback was not feasible, and the system was designed to maximize robustness and coverage. Finally, a wide range of terminal mobility (i.e., slow to fast) was allowed. Since mobility greatly impacts the fading characteristics of the wireless channel, it can greatly impact performance in terms of video quality.

Example 2: Summary

- Primarily, wireless links
- Mild delay constraints, e.g., up to several seconds
- No possibility of feedback
- Wide range of terminal mobility

1.2.3 Example 3: Multimedia Multicasting (RNRT DIVINE)

The application envisioned here is a museum in which supplemental information on each exhibit, in the form of multimedia files, is served to visitors' mobile

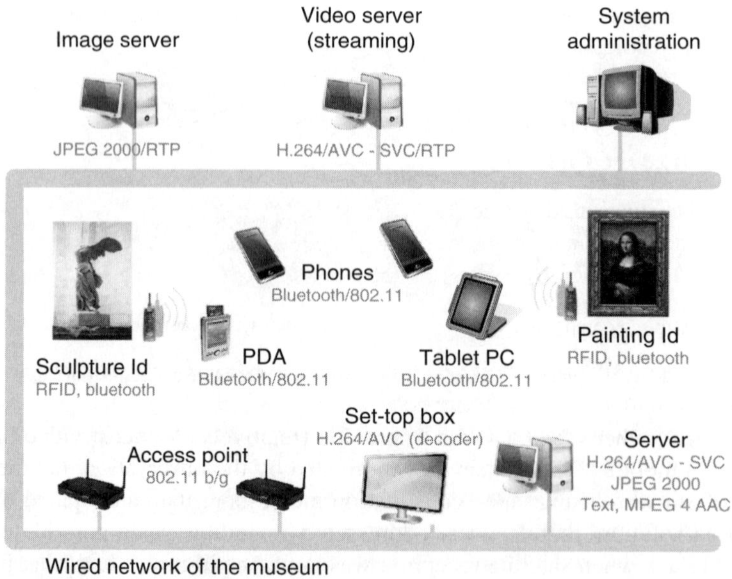

FIGURE 1.3 Multicasting on wireless links.

phones or Personal digital assistants (PDAs) via Wi-Fi. By using their own devices, visitors can avoid the need to rent dedicated devices, which often have fixed content (e.g., prerecorded audio) and offer little flexibility. Figure 1.3 illustrates some possibilities for such a situation, e.g., the reference number of the exhibit could be obtained wirelessly, allowing access to audio, images, or video, which provide useful information on the exhibit.

Besides providing a practical service, this project provides an interesting example of a scenario that falls somewhere between point-to-point and broadcast communication, because of the fact that several groups of visitors might want to download the same files at about the same time. In fact, by grouping these users and assigning a single channel to each group, the problem can be considered as a multicasting one. Here, a judicious use of delay can be used to synchronize users within a group. Relative to point-to-point communication, multicasting uses less channels, and relative to full broadcast, multicasting allows the transmission of only necessary files. Furthermore, the use of feedback (e.g., ARQ) is feasible, since the number of users per channel should not be very large (and, furthermore, can be controlled). In fact, choosing the best number of users per group is an interesting problem, considering the effects of power allocation and ARQ within a group.

Example 3: Summary

- Mixed wired wireless channels (though essentially wireless)

- Mild delay constraints
- ARQ possible, depending on multicast group size

1.3 JOINT SOURCE-CHANNEL CODING AND DECODING

Clearly, the situations described above can be addressed by the classical approach that uses independent network layers (e.g., decoupled source, network, and physical layers). However, by exploiting the finer details of each layer, video output quality can be greatly improved.

To better understand the classical motivation for the decoupled approach to network design and to support the notion that a joint approach can often be more efficient, Shannon's theorems are recalled in Chapter 2.

It can be seen in Chapter 2 that, in systems designed according to the principles of Shannon's separation theorem, both source and channel coders should be optimal. In this case, the bitstream output by the source coder would be totally free of redundancy. However, it is shown in Chapter 4 that this is not the case for practical video coders, especially if block segmentation and side information are taken into account.

For this case, Shannon (1948) explained:

> ...However, any redundancy in the source will usually help if it is utilized at the receiving point. In particular, if the source already has a certain redundancy and no attempt is made to eliminate it in matching to the channel, this redundancy will help combat noise. For example, in a noiseless telegraph channel one could save about 50% in time by proper encoding of the messages. This is not done and most of the redundancy of English remains in the channel symbols. This has the advantage, however, of allowing considerable noise in the channel. A sizable fraction of the letters can be received incorrectly and still reconstructed by the context. In fact this is probably not a bad approximation to the ideal in many cases...

The previous quote underscores the role of JSCD. For the receiver to obtain the best possible estimate of the source signal, it should exploit all information that it can obtain about the source-coded bitstream, the channel-coded bitstream, and the channel itself.

1.3.1 Practical JSCC and JSCD

In all practical situations, including those described in Section 1.2, the following occurs:

1. Some redundancy remains in the source-coded bitstream
2. Side information exists in the form of packet headers
3. Additional information is present, such as channel outputs corresponding to CRC bits

Thus, if JSCC and JSCD are to be made practical, the solution should encompass the source characteristics, the network – through its protocol, and the physical layer characteristics.

It turns out that all these aspects have been addressed to some extent in the literature, and in some cases, very early. For example, JSCD, which has been studied for many years, initially focused on exploiting the properties of the source more than those of the channel.

Early work on JSCD (e.g., Alajaji et al., 1996; Phamdo and Farvardin, 1994; Sayood and Borkenhagen, 1991) addressed the robust reception of quantized source indices, making explicit use of their correlation. The solution was obtained using a Bayesian framework, which allowed the formulation of maximum a posteriori (MAP) and minimum mean-square error (MMSE) estimates. While initially simplified by the assumption of fixed-length codes (FLCs), variable-length codes (VLCs) were eventually considered. Through the use of soft information, the decoder was able to capitalize on the suboptimality of the source coder, ultimately exploiting both the intersymbol correlation of nonwhite sources and the code-word redundancy that results from the suboptimality of entropy coding (Park and Miller, 1998, 1999; Wen and Villasenor, 1999, 2002). Entropy coding was first addressed for tree-based codes, such as Huffman codes (Park and Miller, 1998), and later extended to arithmetic codes (Guionnet and Guillemot, 2003; Pettijohn et al., 2001). When the communication chain includes error-correcting codes (ECCs), as would be the case when some links are wireless, optimal decoding requires the joint exploitation of both source redundancy and ECC redundancy (Garcia-Frias and Villasenor, 1997; Gortz, 2001). Such decoding can be implemented using an iterative approach reminiscent of turbo decoding (Bauer and Hagenauer, 2000a; Gortz, 2001).

In parallel with the aforementioned work on JSCD, another line of work attempted to either directly solve the problem of JSCC (Farvardin and Vaishampayan, 1991b) or to control the amount of redundancy left after source (or entropy) encoding (Buttigieg and Farrell, 1994; Buttigieg, 1995) (which, in terms of source encoding, can be regarded as an "imperfection").

While a rather large set of studies followed along these lines, few addressed the impact of the network on signal reception. An exception among the early literature is the study by Elmasry (1999), which involved ARQ in the decoding of arithmetic codes.

Many additional references on JSCD and JSCC are provided in later chapters.

1.3.2 System Models and Notation

The full system model is summarized by the block diagram of Figure 1.4, for which each block contains many subcomponents (e.g., subblocks) that are described in subsequent chapters.

The notation used throughout the book follows these conventions:

- X : a random variable
- $\mathbf{X}$: a vector of random variables
- x : a realization of random variable X

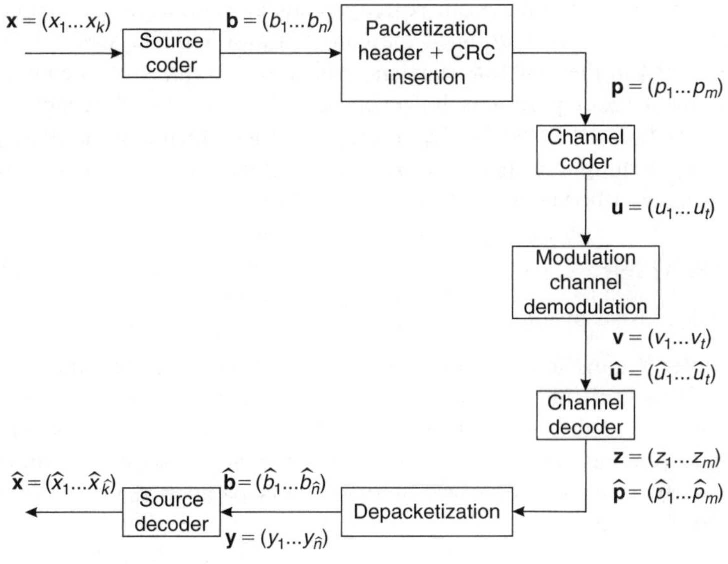

FIGURE 1.4 Summary block diagram of the multimedia network.

- $\mathbf{x}$: a realization of random vector $\mathbf{X}$
- $\mathbf{x}_{1:k}$: components 1 through k of the vector $\mathbf{x}$

Figure 1.4 denotes the variables used to describe each step of the communication chain:

- $\mathbf{X}_{1:k}$: the source signal, represented by its samples $\mathbf{x}_{1:k}$
- $\mathbf{b}_{1:n}$: the bitstream after source encoding
- $\mathbf{p}_{1:m}$: the packetized source bits, which may include the effects of several successive packetizations, with a time index to denote successive packets, we write $\mathbf{p}^N_{1:m}$
- $\mathbf{u}_{1:t}$: the channel-coded packets

At the receiver, a distinction is made between output quantities and estimated input quantities:

- $\mathbf{v}_{1:t}$: the demodulator output, from which one may obtain $\widehat{\mathbf{u}}_{1:t}$ the estimates of $\mathbf{u}_{1:t}$
- $\mathbf{z}_{1:m}$: the decoder output, from which one may obtain $\widehat{\mathbf{p}}_{1:m}$ the estimates of $\mathbf{p}_{1:m}$
- $\mathbf{y}_{1:\hat{n}}$: the depacketizer output, note that the length is $\hat{n}$ because the length is sometimes unknown and thus needs to be estimated
- $\widehat{\mathbf{b}}_{1:\hat{n}}$: the estimated source bitstream
- $\widehat{\mathbf{x}}_{1:\hat{k}}$: the estimated source samples

Clearly, the (soft) outputs of each block can be used as inputs by the next block in the chain. This notion should be clear in the corresponding chapters.

Though fully optimal system design eventually requires a detailed consideration of the network portion of the communication chain, these details are deemphasized in the first few chapters, which focus on source encoding. The effect of the network portion is, however, treated in detail in later chapters. This approach in fact reflects the pedagogy of the book, which is initially aided by simplifying assumptions that are gradually relaxed. A more complete outline of the book is described next.

1.4 OUTLINE

This book consists of four main parts:

1. Chapter 2 aims to make precise the benefits that can be expected from JSCC/JSCD. Using information-theoretic principles, Shannon's separation theorem is clearly stated to understand the conditions under which optimality requires joint processing; by making the underlying assumptions explicit, it can be ascertained whether the separation theorem applies to a given problem.
2. The second part considers the impact of source-coder imperfections on the joint decoding process. To provide essential background, Chapter 3 describes the fundamentals of source encoding (here, video coding). Going deeper, Chapter 4 evaluates the redundancy remaining in the bitstream after various processing stages or as a result of side information that could be used by the receiver. Tools for exploiting this redundancy are then discussed in Chapter 5. Finally, Chapter 6 discusses simplified practical algorithms. This part of the book offers a more-or-less traditional view on JSCD that is supplemented by practical information on simplified algorithms.
3. In the third part, the book considers the impact of the network protocol on end-to-end transmission. Chapter 7 gives a general description of the network layers and describes, in some detail, the network protocols of several modern systems, e.g., Wi-Fi, DVB-S/H, and WiMAX. These descriptions emphasize the fact that redundancy, which can be exploited to improve reception, is present at many levels. Chapter 8 then outlines the tools that can be used to harness two forms of redundancy: robust header reception and robust block segmentation. Here, the discussion is facilitated by a brief discussion of soft-decoding for block codes. Finally, Chapter 9 shows that, with the aid of cross-layer tuning, these two joint decoders can work together in a way that reduces complexity and improves performance. Examples are provided for certain applications.
4. The final part of the book describes open challenges. JSCC is addressed in Chapter 10. There it is shown that the main challenges are allocation of redundancy over the various stages of processing and the search for the best joint tools. Chapter 11 addresses some additional situations for which the tools provided in this book could prove useful and provides some preliminary results along these lines.

Chapter 2

Why Joint Source and Channel Decoding?

This chapter introduces theoretical justifications for joint source-channel coding (JSCC) and identifies certain expectations that one may have regarding JSCC and joint source-channel decoding (JSCD). While we mainly focus on a justification for JSCC, we also discuss JSCD toward the end of the chapter.

While JSCC was, initially, of interest to only a small group of researchers, it has, more recently, gained widespread attention. For some of the readers just learning about this field, the main concept may seem somewhat awkward: Is it really possible to forget Shannon's separation theorem? For other readers, Shannon's work may be remembered only vaguely. For these reasons, we start with a brief review of Shannon theory, we carefully explain the origin of the separation theorem, we outline limitations on source coding and on channel coding, and we define an important quantity – the *optimum performance theoretically attainable (OPTA)* – for the communication of a particular source over a particular channel. From these foundations, the assumptions under which the separation theorem holds become clear, and the expectations that one has when working on JSCC can be made more explicit.

These expectations are illustrated in two ways: In a first step, we describe some simple scenarios where, even though the separation theorem holds, separation is not the only approach to optimal system design. In a second step, we summarize key ways in which practical situations differ from the abstract model that was used to justify the separation theorem. We concentrate on the broadcast scenario, which, given modern applications, is more prevalent than the point-to-point scenario, and which, by its nature, results in imperfectly known channel parameters. In this case, it is shown that the separation theorem does not hold.

Our presentation closely follows the studies of Zahir Azami et al. (1996) and Gatspar et al. (2003).

2.1 INFORMATION THEORETIC PRELIMINARIES

This section provides the basic information theoretic tools that are needed to understand the Shannon bounds in their various incarnations. While, with channel coding, it is quite common to check the performance of a proposed system against the best attainable performance, it is not as common in source coding, even when one knows the rate-distortion curves. One of the reasons for this is that the corresponding computations are likely to be intractable, as a result of the high correlation found in practical source signals like video. In addition, it is less well known that a similar bound exists for JSCC, which is the real target of this book. In fact, one can evaluate, for simple sources, the minimum rate that is required to communicate the given source over a particular channel with a specified maximum end-to-end distortion. This point is clarified later in this section.

2.1.1 The Situation of Interest

When deriving bounds, one typically requires a simple model for the situation of interest. We use here the classical transmission model, as depicted in Figure 2.1. We now carefully analyze this situation, while keeping in mind the constraints that are required for the theoretical results to have practical utility.

We consider only block processing; the inputs and all intermediate quantities are vector valued, as defined below:

- The initial and reconstructed source words have k components (i.e., the "source symbols"), $\mathbf{X} = (X_1, X_2, \ldots, X_k)$, and $\widehat{\mathbf{X}} = (\widehat{X}_1, \widehat{X}_2, \ldots, \widehat{X}_k)$.
- The channel input and output words have n components (i.e., the "channel symbols"), $\mathbf{B} = (B_1, B_2, \ldots, B_n)$, and $\mathbf{Y} = (Y_1, Y_2, \ldots, Y_n)$.

All variables are modeled as random variables and, for generality, are not assumed to be either discrete or continuous. For this reason, following the

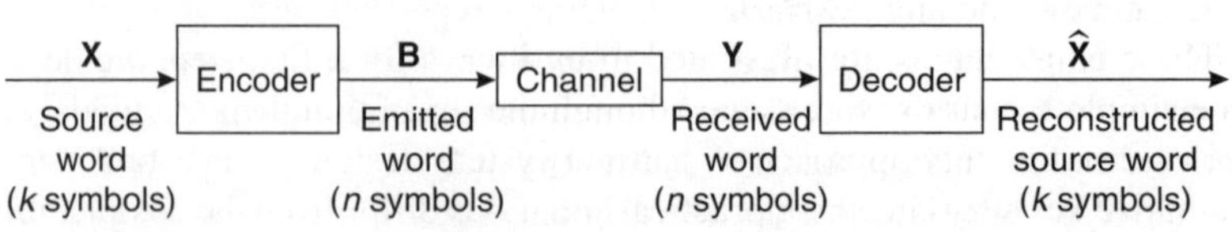

FIGURE 2.1 Shannon's paradigm.

notations in the studies by Rioul (2007) and Zahir Azami et al. (1996), we use $\sum\!\!\!\!\!\!\int$ to simultaneously denote summation, for discrete variables, and integration, for continuous variables (where we omit the measure symbol $d(.)$ for simplicity). With this notation, for example, expectation reads

$$E[\mathbf{X}] = \sum\!\!\!\!\!\!\int_{\mathbf{x}} \mathbf{x}p(\mathbf{x}). \tag{2.1}$$

This way, we can address a wide variety of sources and channels using a common notation.

The source is characterized by its probability distribution $p(\mathbf{x}) = p(x_1, \ldots, x_k)$. For example, a zero-mean i.i.d. Gaussian source is described by

$$p(\mathbf{x}) = \frac{1}{(2\pi\sigma_x^2)^{k/2}} \exp\left(-\frac{\|\mathbf{x}\|^2}{2\sigma_x^2}\right), \tag{2.2}$$

where σ_x^2 denotes the source variance, and a memoryless binary symmetric source (BSS) is described by the uniform distribution

$$p(\mathbf{x}) = \frac{1}{2^k}. \tag{2.3}$$

The channel is described by the transition probabilities $p(\mathbf{y}|\mathbf{b})$ that relate the input $\mathbf{b}$ to the output $\mathbf{y}$, i.e., the probability distribution of the output variable $\mathbf{Y}$ for a given realization of the input $\mathbf{b}$. For example, an additive white Gaussian noise (AWGN) channel is characterized by

$$p(\mathbf{y}|\mathbf{b}) = \frac{1}{(2\pi\sigma_n^2)^{n/2}} \exp\left(-\frac{||\mathbf{y}-\mathbf{b}||^2}{2\sigma_n^2}\right), \tag{2.4}$$

where σ_n^2 denotes the noise variance. A binary symmetric channel (BSC) is modeled by its crossover probability p such that

$$p(\mathbf{y}|\mathbf{b}) = p^d(1-p)^{n-d}, \tag{2.5}$$

where $d = d_H(\mathbf{b}, \mathbf{y})$ denotes the Hamming distance between the channel input and output.

We allow any combination of input and channel types, e.g., a Gaussian signal over a Gaussian channel, a Gaussian signal over a BSC, a binary signal over a Gaussian channel, a binary signal over a Gaussian channel, or even multivalued signals. Unless stated otherwise, the equations we provide hold in all of these cases.

Of course, when mixing input and channel types, one must adapt the source to the channel; this is the role of the encoder. The encoder is described by the mapping $\mathbf{b} = \mathbf{C}(\mathbf{x})$. Specifying $\mathbf{C}$ amounts to designing channel code words characterized by their probability distribution $p(\mathbf{b})$.

Conversely, one needs a decoder to convert the channel output to a reconstructed word in the source domain. The decoder is described by the mapping

$\widehat{\mathbf{x}} = \mathbf{D}(\mathbf{y})$. Specifying $\mathbf{D}$ amounts to designing source code words according to the distribution $p(\widehat{\mathbf{x}}|\mathbf{x})$.

The model is not yet complete since important degrees of freedom still remain. For example, when designing the transmission system, one needs to choose the channel input power. While increased channel input power generally yields improved performance (except in a few pathological cases), it comes with practical costs. For this reason, a power constraint is typically adopted. In a similar manner, one needs to decide on an acceptable level of end-to-end reconstruction error $\widehat{\mathbf{X}} - \mathbf{X}$. Various measures can be used to quantify the error, including error probability (when the source is discrete), mean square error (MSE), and so on. The best choice depends on the situation.

In summary, there are two fundamental constraints to consider:

(C1) Average Cost Per Channel Input

$$\frac{1}{n}E[N(\mathbf{B})] \leqslant P, \tag{2.6}$$

where $N(\mathbf{b})$ is a cost function chosen so that

$$N(\mathbf{b}) = \sum_{i=1}^{n} N(b_i) \geqslant 0. \tag{2.7}$$

Notice that this constraint operates on $p(\mathbf{b})$. For example, $N(\mathbf{b})$ can be chosen to yield the power constraint $\frac{1}{n}E[\|\mathbf{B}\|^2] \leqslant P$.

(C2) Reliability Criterion

$$\frac{1}{k}E[d(\mathbf{X}, \widehat{\mathbf{X}})] \leqslant D, \tag{2.8}$$

where $d(\mathbf{x}, \widehat{\mathbf{x}})$ is a distance metric chosen so that

$$d(\mathbf{x}, \widehat{\mathbf{x}}) = \sum_{i=1}^{k} d(x_i, \widehat{x}_i) \geqslant 0. \tag{2.9}$$

In terms of model parameters, one can consider this constraint to operate on $p(\widehat{\mathbf{x}}|\mathbf{x})$ because $Ed(\mathbf{X}, \widehat{\mathbf{X}}) = \sum\!\!\!\!\!\!\int_{\mathbf{x}, \widehat{\mathbf{x}}} p(\mathbf{x}) p(\widehat{\mathbf{x}}|\mathbf{x}) d(\mathbf{x}, \widehat{\mathbf{x}})$. Examples include the following:

- Lossless source coding, for which $D = 0$
- Mean square error (lossy source coding): $\frac{1}{k}E[\|\mathbf{X} - \widehat{\mathbf{X}}\|^2] \leqslant D$
- Bit error rate (BER): $P_e = \frac{1}{k}E[d_H(\mathbf{X}, \widehat{\mathbf{X}})] \leqslant D$, where $d_H(.,.)$ denotes Hamming distance

Note that, in this treatment, the distortion metric is constrained rather than minimized. We do this to stay consistent with the conventions of information theory, which usually maximizes the so-called *global rate* parameter

$$\rho = \frac{k}{n} \quad \text{source symbol/channel symbol,}$$

under the constraints *(C1)* and *(C2)*. In practice, however, one sometimes minimizes D for a given ρ under constraint *(C1)*. Note that the global rate, as defined above, takes the form similar to that of a channel rate, while at other places it is defined in a similar way as a source rate (Rioul, 2007), in which case it has to be minimized.

2.1.2 Optimum Performance Theoretically Attainable

Deriving a source- or channel-coding performance bound is often much easier than proving its attainability. Bounds can be obtained using the so-called *mutual information* between random vectors (or random variables) **B** and **Y**, which may be discrete or continuous.

The mutual information between **B** and **Y** is defined as

$$I(\mathbf{B},\mathbf{Y}) = \sum\!\!\!\!\!\!\int_{\mathbf{b},\mathbf{y}} p(\mathbf{b},\mathbf{y}) \log_2 \frac{p(\mathbf{b},\mathbf{y})}{p(\mathbf{b})p(\mathbf{y})}, \tag{2.10}$$

with units of (binary) bits.

It can be interpreted as a *dependency measure* between **B** and **Y**. When applied to vectors (e.g., $I(\mathbf{B},\mathbf{Y})$ and $I(\mathbf{X},\widehat{\mathbf{X}})$), mutual information is measured in terms of bits per word, and when applied to individual source or channel symbols (e.g., $\frac{1}{k}I(\mathbf{X},\widehat{\mathbf{X}})$ and $\frac{1}{n}I(\mathbf{B},\mathbf{Y})$), it is measured in terms of bits per symbol.

The generic definition given above is often rewritten for channel quantities as follows

$$I(\mathbf{B},\mathbf{Y}) = \sum\!\!\!\!\!\!\int_{\mathbf{b},\mathbf{y}} p(\mathbf{y}|\mathbf{b})p(\mathbf{b}) \log_2 \frac{p(\mathbf{y}|\mathbf{b})}{\sum_{\mathbf{b}'} p(\mathbf{y}|\mathbf{b}')p(\mathbf{b}')}. \tag{2.11}$$

When written as such, $I(\mathbf{B},\mathbf{Y})$ depends on (1) the probability density of the input: $p(\mathbf{b})$ and (2) the transition probability $p(\mathbf{y}|\mathbf{b})$ that describes the channel. It can also be rewritten in terms of the Kullback–Leibler (KL) divergence between $p(\mathbf{Y}|\mathbf{b})$ and $p(\mathbf{Y})$.

$$\begin{aligned} D(p(\mathbf{Y})|\mathbf{b})||p(\mathbf{Y})) &= \sum\!\!\!\!\!\!\int_{\mathbf{y}} p(\mathbf{y}|\mathbf{b}) \log_2 \frac{p(\mathbf{y}|\mathbf{b})}{p(\mathbf{y})} \\ &= \sum\!\!\!\!\!\!\int_{\mathbf{y}} p(\mathbf{y}|\mathbf{b}) \log_2 \frac{p(\mathbf{y}|\mathbf{b})}{\sum_{\mathbf{b}'} p(\mathbf{y}|\mathbf{b}')p(\mathbf{b}')} \end{aligned} \tag{2.12}$$

The KL divergence is a nonsymmetric distance measure between two pdfs (Kullback and Leibler, 1951). Thus, (2.12) measures the "distance" between the probability of the channel output given input **b** and the corresponding probability of the channel output.

As a dependency measure, the mutual information thus corresponds to the average amount of "information" that the knowledge of one realization of **B** brings to the knowledge of **Y** since it can be written as the expectation of the distance between the *actual output pdf and the output pdf given the actual input* pdfs, as shown below:

$$I(\mathbf{B},\mathbf{Y}) = \sum\!\!\!\!\!\!\int_{\mathbf{b}} p(\mathbf{b}) D(p(\mathbf{Y})|\mathbf{b}))||p(\mathbf{Y})). \tag{2.13}$$

For source-related quantities, mutual information is often rewritten as

$$I(\mathbf{X},\widehat{\mathbf{X}}) = \sum\!\!\!\!\!\!\int_{\mathbf{x},\widehat{\mathbf{x}}} p(\widehat{\mathbf{x}}|\mathbf{x})p(\mathbf{x}) \log_2 \frac{p(\widehat{\mathbf{x}}|\mathbf{x})}{\sum_{\mathbf{x}'} p(\widehat{\mathbf{x}}|\mathbf{x}')p(\mathbf{x}')}, \tag{2.14}$$

which involves (1) the probability density of the source, $p(\mathbf{x})$, and (2) the end-to-end transition probabilities $p(\widehat{\mathbf{x}}|\mathbf{x})$.

From the flow graph in Figure 2.1 and the comments above, it should seem intuitive (and can be rigorously proven) that

$$I(\mathbf{X},\widehat{\mathbf{X}}) \leqslant I(\mathbf{B},\mathbf{Y}), \tag{2.15}$$

i.e., that mutual information cannot be increased through the processing of the data. This is widely known in information theory as the *data processing inequality* (Blahut, 1987). Note that this inequality may be rewritten as

$$\rho \frac{1}{k} I(\mathbf{X},\widehat{\mathbf{X}}) \leqslant \frac{1}{n} I(\mathbf{B},\mathbf{Y}). \tag{2.16}$$

Following the conventions of information theory, we intend to maximize the global rate ρ. From the previous inequality, it is clear that an upper bound on ρ can be obtained by independently minimizing $I(\mathbf{X},\widehat{\mathbf{X}})$ and maximizing $I(\mathbf{B},\mathbf{Y})$. The minimum of $I(\mathbf{X},\widehat{\mathbf{X}})$ and the maximum of $I(\mathbf{B},\mathbf{Y})$ are precisely the quantities that are most important in information theory.

Capacity

The capacity of the channel characterized by $p(\mathbf{y}|\mathbf{b})$ is obtained by maximizing $I(\mathbf{B},\mathbf{Y})$ over all possible $p(\mathbf{b})$ under constraint *(C1)* and in the supremum on n:

$$C(P) = \sup_n \max_{p(\mathbf{b})} \left\{ \frac{1}{n} I(\mathbf{B},\mathbf{Y}) \mid \frac{1}{n} EN(\mathbf{B}) \leqslant P \right\}. \tag{2.17}$$

Rate Distortion

The rate-distortion curve $R(D)$ of the source characterized by $p(\mathbf{X})$ is obtained by minimizing $I(\mathbf{X},\widehat{\mathbf{X}})$ over all possible $p(\widehat{\mathbf{X}}|\mathbf{X})$ under constraint $C2$ and in the infimum of k:

$$R(D) = \inf_k \min_{p(\widehat{\mathbf{X}}|\mathbf{X})} \left\{ \frac{1}{k} I(\mathbf{X},\widehat{\mathbf{X}}) \mid \frac{1}{k} Ed(\mathbf{X},\widehat{\mathbf{X}}) \leqslant D \right\}. \tag{2.18}$$

Note that, in the lossless case (i.e., $D=0$), the $R(D)$ limit is exactly the entropy of the source (i.e., $H(\mathbf{X}) = I(\mathbf{X},\mathbf{X})$). In practical situations, this quantity is bounded (at least for **X** taking finitely many values).

OPTA

By taking both of these information-theoretic limits, (2.16) becomes

$$\rho \leqslant \frac{C(P)}{R(D)}. \tag{2.19}$$

Since this limit is indeed attainable (McEliece, 2002), it represents the ultimate performance of this simple communication system.

Example 2.1 (OPTA for a Gaussian source sent over a Gaussian channel): In the case of a memoryless Gaussian source communicated over a memoryless Gaussian channel, the channel capacity is

$$C(P) = \frac{1}{2}\log_2\left(1+\frac{P}{\sigma_n^2}\right), \tag{2.20}$$

the rate-distortion curve of the source is

$$R(D) = \frac{1}{2}\log_2\left(\frac{\sigma_x^2}{D}\right), \tag{2.21}$$

and the OPTA is thus

$$\rho \leqslant \frac{C(P)}{R(D)} = \frac{\log_2\left(1+\frac{P}{\sigma_n^2}\right)}{\log_2\left(\frac{\sigma_x^2}{D}\right)}, \tag{2.22}$$

which depends on the channel signal-to-noise ratio (SNR) $\frac{P}{\sigma_n^2}$ and on the source-SNR $\frac{\sigma_x^2}{D}$.

2.2 TO SEPARATE OR NOT TO SEPARATE?

2.2.1 Three Shannon Theorems

The properties mentioned previously are very general and apply to many types of signal (e.g., continuous, discrete-valued, correlated). We note that Shannon's theorems were initially proven under the memoryless assumption, but have since been proven under more general assumptions (Han, 2002; Vembu et al., 1995).

In practice, the techniques used to protect a signal from channel degradations usually assume more specific properties, to obtain simpler algorithms of wider application. Practical systems are thus implemented using more special-purpose blocks than those depicted in Figure 2.1. For example, the encoder can be separated into a transformation stage, a vector quantizer stage, an index assignment stage, and so on. Since practical systems involve many processing blocks, one objective of system design is to force the inputs/outputs of various blocks to meet certain desired properties. For example, it is typically desired that the input of the channel-coding block is binary, symmetric, and memoryless (which makes sense if the source coder is close to optimum, as will be shown in Chapter 3). The capacity can thus be obtained as an application of the OPTA above, when the input meets the above-mentioned properties and assuming lossless transmission.

The Channel-Coding Theorem

For the channel-coding theorem, the source is assumed to be discrete, and the "information word" is assumed to take on K different values with equal probability, which corresponds to the binary, symmetric, and memoryless properties mentioned above. The information word $\mathbf{X}$ thus corresponds to $\log_2 K$ information bits. In practice, though, some sort of preprocessing may be required to guarantee that this assumption (reasonably) holds.

The encoder outputs the code word $\mathbf{B}$, spread over n symbols. The channel-coding rate is defined as $R_c = \frac{\log_2 K}{n}$, which can be interpreted as the number of information bits per coded channel symbol.

Assuming lossless transmission of these equiprobable binary words, the OPTA inequality reads:

$$R_c = \frac{\log_2 K}{n} \leqslant C(P), \tag{2.23}$$

and Shannon's channel-coding theorem states that, for any arbitrarily small $\varepsilon > 0$ and for any $R_c < C(P)$, there exists a channel code of rate larger than R_c such that the error probability satisfies $P_e < \varepsilon$.

The Source-Coding Theorem

When considering the source-coding theorem, the channel is assumed to be perfect, that is, $\mathbf{Y} = \mathbf{B}$ (i.e., no errors). The source vector has length k, and the output of the source encoder is an information word that takes on K different values. Hence, the source-coding rate is $R_s = \frac{\log_2 K}{k}$ encoded bits per source symbol.

When applied to this situation, the OPTA inequality reads

$$R_s = \frac{\log_2 K}{k} \geqslant R(D). \tag{2.24}$$

Shannon's source-coding theorem states that, for any arbitrarily small $\varepsilon > 0$ and for any $R_s > R(D)$, there exists a source code of rate less than R_s such that the distortion introduced by this code is less than $D + \varepsilon$.

The Separation Theorem: OPTA

Thanks to the first two Shannon theorems; it should be clear that, at least in principle, the source/channel bound can be approached as closely as desired using the following strategy:

1. The source-coding procedure yields an information bitstream whose rate R_s is as close as possible to the bound $R(D)$ for a maximum specified distortion of D.

2. The channel-coding procedure generates, from the information bitstream, a coded bitstream capable of passing through the channel with "almost no" errors (i.e., P_e arbitrarily small) and whose rate R_c is as close as desired to the bound $C(P)$.

Using these two steps, one indeed can obtain a global rate ρ just below Shannon's limit $C(P)/R(D)$ by incurring a total distortion as close as desired to D.

This is the essence of Shannon's source and channel-coding theorem. Note that, using this method, the channel makes a negligible contribution to the overall distortion; distortion is entirely the result of source compression.

Note also that this result that separation is an optimal strategy was obtained under the implicit assumption of stationary and ergodic channels and in a point-to-point situation.

2.2.2 Simple (Counter) Examples

It turns out that, in Shannon's source and channel-coding strategy (which, as described above, exploits the separation principle), both source compression and channel coding are incredibly difficult tasks. For example, the implementation of each task may require the use of huge data blocks and huge amounts of processing complexity. These challenges have forced researchers to consider JSCC schemes, where the hope is that the elegance of separation can be exchanged for a much simplified implementation while keeping the system near-optimal.

While, at first glance, JSCC may not seem to be compatible with Shannon's methods, the two well-known examples we recall below show instances where joint approaches are in fact optimal. Moreover, in these two examples, the encoder and decoder are greatly simplified.

Example 2.2 (BSS sent over a BSC): Consider a memoryless BSS to be transmitted over a memoryless BSC with crossover probability p, at the global rate $\rho = 1$. We choose BER as the distortion measure. For a maximum BER of D, it can be shown that

$$R_s^* = R(D) = 1 - H_2(D),$$

where $H_2(x) = x\log_2 \frac{1}{x} + (1-x)\log_2 \frac{1}{1-x}$ denotes the binary entropy function. Since the channel capacity is

$$R_c^* = 1 - H_2(p),$$

Shannon's bound becomes

$$\frac{C}{R(D)} = \frac{1-H_2(p)}{1-H_2(D)} \geqslant \rho = 1,$$

which simplifies to

$$D \geqslant p.$$

Thus, Shannon's separation principle states that the optimal attainable performance in this situation for $\rho = 1$ is $D = p$. Moreover, it suggests that this optimal performance can be obtained by building a source coder with rate $R_s \approx 1 - H_2(D)$ for $D = p$ and a channel coder with rate $R_c \approx 1 - H_2(p)$ that generates almost no bit errors.

However, it can also be seen (and is rather trivial) that by transmitting the source bits directly over the channel (without any source or channel coding) we also obtain an end-to-end BER of $D = p$ and thus a global rate of $\rho = 1$. As a result, the simplest source and channel-coding strategy, i.e., "don't do any coding," is also optimal. Note that, with this simpler optimal solution, all errors are introduced by the channel and none through source coding; this is the opposite of what happens with Shannon's separation-based approach.

Example 2.3 (Gaussian source sent over an AWGN channel): A similar situation occurs when a memoryless Gaussian source, with variance σ_x^2, is to be transmitted through a memoryless Gaussian channel, with variance σ_n^2, at global rate $\rho = 1$. The natural distortion measure here is the MSE.

In this case, the optimal rate-distortion function is

$$R(D) = \frac{1}{2} \log_2 \gamma_s,$$

where $\gamma_s = \sigma_x^2 / D$ is the source-to-noise ratio, and the channel capacity is

$$C(P) = \frac{1}{2} \log_2 (1 + \gamma_c),$$

where $\gamma_c = P / \sigma_n^2$ is the channel-to-noise ratio. In this case, Shannon's bound becomes

$$\frac{C(P)}{R(D)} = \frac{\log_2(1 + \gamma_c)}{\log_2 \gamma_s} \geqslant \rho = 1,$$

which simplifies to

$$\gamma_s \leqslant 1 + \gamma_c.$$

Since optimal systems are those which attain equality above, a simple optimal solution is the one illustrated in Figure 2.2b, where the gains α and β obey $\frac{1}{\beta} = \alpha + \frac{1}{\alpha} \frac{\sigma_n^2}{\sigma_x^2}$. It can be verified that, here again, the OPTA is attained.

2.2.3 Possible Benefits

The previous two examples, although somewhat trivial once the problem is well understood, are not incompatible with Shannon's principles. They simply show

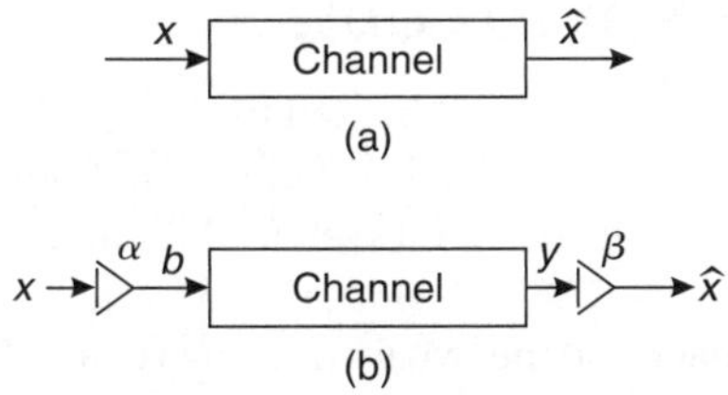

FIGURE 2.2 Two simple yet optimal systems: (a) binary source and channel and (b) Gaussian source and channel.

that the concatenation of optimal source coding with optimal channel coding is not necessary for end-to-end optimality. Furthermore, they suggest that, by looking beyond the separation principle, one may be able to obtain near-optimal performance at greatly reduced complexity.

However, another situation commonly arises: in the presence of constraints (on, e.g., complexity, processing delay, etc.), it may be impossible to implement unconstrained-optimal source and/or channel coders. The optimal system design, in this case, is generally not the cascade of subcomponents that were individually optimized under the constraints. In fact, this principle is familiar from optimization theory and is visible across many fields. For example, from the signal processing area, we recall the design of filter banks. Although brick-wall analysis and synthesis filters (i.e., the unconstrained-optimal ones) do lead to perfect reconstruction (PR) filter banks, the best finite-length approximation to these brick-wall filters does not yield PR filter banks. However, it is possible to construct a PR filter bank out of jointly designed finite-length analysis and synthesis filters.

The implication is that, when practical constraints prevent the implementation of optimal source coders and/or optimal channel coders, JSCC designs can yield performance gains over separately designed source and channel coders. The gains, however, may be small when the constraints are mild in this point-to-point situation, i.e., when it is possible to implement near-optimum source and channel coders. However, when the system is heavily constrained (in, e.g., complexity or delay), one can expect significant gains from the use of a jointly optimized source/channel coder, even in the very situation where the separation theorem was demonstrated.

Why It Is Not Always So Simple

The previous simple examples, while illuminating, do not translate well to practice, due to their reliance on the following properties:

- The channel and source are ideally "matched" to each other.
- When $\rho = 1$, the end-to-end error performance is limited by the noise level of the channel.

2.3 TO CODE OR NOT TO CODE?

A much more general setting was addressed by Gatspar et al. (2003), where the authors derived a set of necessary and sufficient conditions on the optimality of any discrete memoryless point-to-point system. Although some of the results hold only for discrete alphabets, Gatspar et al. (2003) generalizes both the classical separation-based approach and the two well-known examples of optimal uncoded communication recalled above (including the Gaussian one).

The main theorem in the study by Gatspar et al. (2003) gives simple conditions that can be used to check whether a given code of rate $\rho=1$ performs optimally for a given source/channel pair; they are paraphrased below. (Note that the result in the study by Gatspar et al. (2003) is much more complete.) In what follows, the theorem is restricted to single-letter codes, meaning that the encoder is restricted to be a function $b=f(x)$ and the decoder restricted to $\widehat{x}=g(y)$. Moreover, we exclude degenerate cases where (1) there would not be any tradeoff between channel-related cost and distortion, (2) the constraints have no impact on the involved source-related or channel-related mutual information, or (3) the optimal system would result in a mutual information of zero.

Theorem 2.4: *Consider source X with pdf $p(x)$ and distortion measure $\mathrm{d}(x,\widehat{x})$, and channel $p(y|b)$ with cost $N(b)$. Let C_0 be the unconstrained capacity of the channel (i.e., the capacity of the system for an infinite value of the channel constraint).*[1] *For transmission using a single-letter code, the following statements hold.*

- *If $I(B,Y)\neq I(X,\widehat{X})$, then the system does not perform optimally.*
- *If $0\leqslant I(B,Y)=I(X,\widehat{X})<C_0$, then the system is optimal if and only if the source distortion $d(x,\widehat{x})$ and the channel cost $N(b)$ satisfy*
  - *$N(b)=c_1D(p(Y|b)||p(Y)))+N_0$ if $p(x)>0$, and $N(b)\geq c_1D(p(Y|b)||p(Y)))+N_0$ otherwise*
  - *If $0<I(X,\widehat{X})$, then $d(x,\widehat{x})=-c_2\log_2(p(x|\widehat{x}))+\mathrm{d}_0(x)$*

for some N_0, some $d_0(x)$, and some constants c_1 and c_2.

Note that the result is not expressed in terms of adaptation of the source to the channel, as could be intuitive. Instead, the theorem relates the involved measures (cost, distortion) to the corresponding problem-related quantities : the cost must be adapted to channel-dependent quantities, while the distortion measure must be adapted to source-related quantities.

Example 2.5: As an example, it is shown below that Theorem 2.4 covers the Gaussian example discussed earlier. Suppose that the source is Gaussian with variance σ_x^2 and the channel is AWGN with noise variance σ_n^2. Suppose

---

1. In the Gaussian case, obviously, C_0 is infinite, which may not be the case in other situations.

also that uncoded transmission is used and that the decoder is $\widehat{x} = \frac{P}{P+\sigma_n^2}y$. Then, the conditions in Theorem 2.4 are satisfied via

$$N(b) = b^2 c_1 + N_0 \tag{2.25}$$

$$d(x, \widehat{x}) = (x - \widehat{x})^2 c_2 + d_0(x). \tag{2.26}$$

In other words, if the channel cost is power and the distortion is MSE, then uncoded transmission is optimal. An important observation is that this optimality holds for any noise level. This differs from the case of separately optimized source and channel coding, where the channel code rate is chosen in accordance with the channel noise level.

2.4 BACK TO THE SEPARATION PARADIGM

Until now, we have considered the (classical) case of a unidirectional point-to-point communication link. However, it is also important to consider other scenarios, where the problem of interest has quite different characteristics. Two important examples are *broadcast channels* and *channels with feedback*. Both examples are of practical importance since, in many of today's applications, a signal must be communicated to several users (registered or not) and/or feedback is provided by the network layers (e.g., via Automatic Repeat request [ARQ]).

The case of feedback has been studied for some time in the context of channel coding. Though it was shown, by Shannon, that feedback does not increase capacity, feedback does make it much easier to get close to channel capacity. Thus, for JSCC, feedback should make it easier to obtain performance close to the OPTA. However, we do not consider feedback here, even if the connection with ARQ is certainly worth considering for the future.

2.4.1 Broadcast Channels

Broadcast communication refers to the one-to-many scenario. In this case, the coder cannot be tuned to each user, and so, the goal is to facilitate that each user recovers as much as he or she can from the received signal.

The separation theorem is not directly applicable to the broadcast scenario, since there the channel is assumed to follow an "all or nothing" behavior: the channel either perfectly passes the encoded signal (when channel capacity is not exceeded) or completely destroys it. In contrast, the classical solution to broadcast scenario involves progressive encoding also denoted as successive refinement (SR) (Equitz and Cover, 1991).

The broadcast situation was also studied by Gatspar et al. (2003), which compared optimal usage of progressive transmission and uncoded transmission in the following simple setting. Consider a Gaussian i.i.d. source X of variance σ_x^2 transmitted to two users through Gaussian channels of variance σ_{n1}^2 and σ_{n2}^2, respectively, with $\sigma_{n1}^2 > \sigma_{n2}^2$, which ensures that this is a degraded channel.

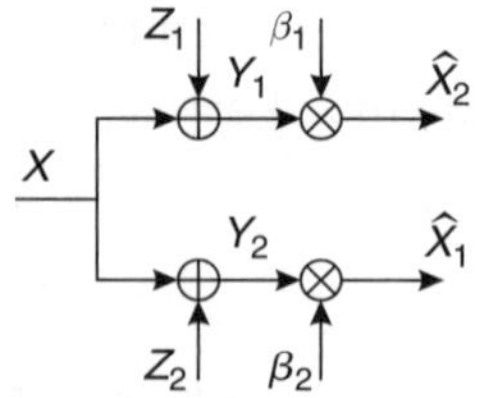

FIGURE 2.3 Sending a single source over a Gaussian broadcast channel using uncoded transmission.

According to the results above, uncoded transmission is optimal on each of the channels individually, assuming that the received signals are scaled by $\beta_1 = P/(P+\sigma_{n1}^2)$ and $\beta_2 = P/(P+\sigma_{n2}^2)$. In this case, the resulting distortion pair is

$$D_1 = P\sigma_{n1}^2/(P+\sigma_{n1}^2) \tag{2.27}$$

and

$$D_2 = P\sigma_{n2}^2/(P+\sigma_{n2}^2) \tag{2.28}$$

This scenario is illustrated in Figure 2.3.

If one chooses separately optimized source and channel coding, the source must be progressively encoded, i.e., encoded using a coarse version that can be successfully decoded by User 2, and a refinement can be successfully decoded (together with the coarse version) by User 1, whose channel has a higher capacity (Equitz and Cover, 1991; Rimoldi, 1994). The use of progressive encoding, as it turns out, incurs no loss. In other words, if $D(.)$ denotes the distortion-rate function of the source (Berger, 1971), and if R_2 bits are used for the coarse version and R_1 for the refinement, then the distortion achievable from the coarse part is $D(R_2)$, while the distortion achievable from the total bitstream is $D(R_1+R_2)$. This progressive encoding strategy is of great current interest. The rates available for these two regions are the pairs (R_1, R_2) in the capacity region of the corresponding Gaussian broadcast channel (GBC). Since it is a degraded channel, the property that the better channel can also decode the coarse information holds (Cover and Thomas, 1991). Therefore, the smallest distortion pair achievable by separately optimal source and channel coding is related to the capacity region of the degraded broadcast channel and equals $(D(R_1+R_2), D(R_2))$, where R_1+R_2 and R_2 lie on the boundary region of the GBC. This is illustrated in Figure 2.4 (for $P=1$, $\sigma_{n1}^2=0.1$, and $\sigma_{n2}^2=0.2$), where the distortion achievable by any separation-based system is denoted by the shaded region, and the distortion achievable by the uncoded system is denoted by the circle. One can observe that the distortion pair achieved by the nonseparation-based transmission scheme (i.e., the uncoded one) has superior distortion values than those achievable by the separation-based approach.

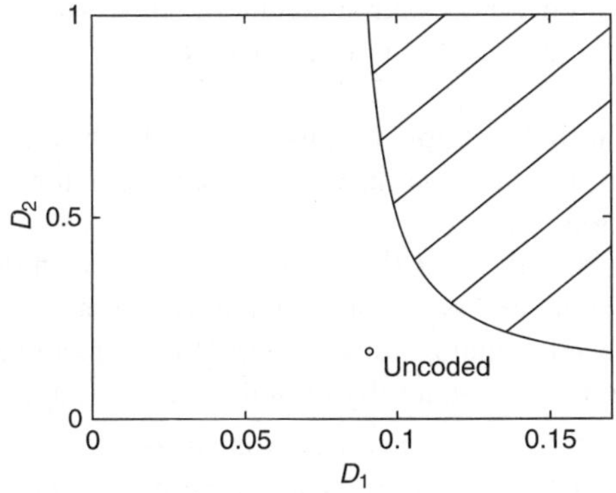

FIGURE 2.4 Distortion pairs achievable by uncoded transmission (circle) and by transmission schemes designed in accordance with the separation principle (shaded region).

2.4.2 The Encoder

The simple two-user example provides intuition about more general scenarii, since it is a special case of a number of situations that are interesting on their own and have been studied as such. The case of the so-called compound channel – where one does not know the exact parameter values of a channel, but only that they belong to some set – is one of these, and in particular a special case where the set contains two elements. As a consequence, the separation theorem does not apply to this case. A related scenario is that of Rayleigh slow fading channels, which also shares some properties with this simple two-user situation. For this case, the separation-based approach has been studied by Tian et al. (2008), in combination with SR. In this reference, the broadcast strategy coupled with SR source coding can be considered as a separation-based approach with optimized cross-layer resource allocation. More explicit JSCC approaches (e.g., those in the studies of Caire and Narayanan (2007) and Mittal and Phamdo (2002)) may provide better performance.

The possibility of increased robustness provides another strong argument in favor of JSCC, as argued in the sequel. The so-called *universality* of a code was introduced by Gatspar et al. (2003) as the property that some source-channel codes remain optimal for a given source when transmitted over a given class of channels. Universality can be used to define optimality for a compound channel, as well for the following two practical scenarios.

- For channels that are time-varying, where the channel variations define the class of channels
- When the same source is transmitted over several channels that, together, define the class of channels

In all these cases, a universal source-channel code performs optimally. Thus, universal-coding schemes can be considered perfectly robust to variations within the class.

Finally, note that in the simple two-user example above, since the (identity) code is universal at this rate among single-letter codes for Gaussian channels, the noncoded solution is also optimal.

Note, however, that in most of the situations encountered in this chapter, the joint source-channel codes that are efficient (i.e., optimal) are continuous in nature; they correspond to analog transmission. Whether similar properties can be (at least partially) preserved in a fully digital system remains an open question.

2.4.3 The Decoder

This chapter has so far focused on JSC coding rather than decoding. However, as pointed out above, our examples of optimal nonseparation-based schemes have mostly been analog ones. Also, it is not clear that analog transmission will be used heavily in the predictable future, even though analog processing is a topic of increasing interest in communications. (Consider, e.g., analog decoding of turbo codes using the tools provided by Loeliger et al. (1998) for probability propagation and decoding.) For these reasons, JSCD is an important topic. Furthermore, JSCD may lead to new ways of implementing JSCC that are more compatible with fully digital communication systems.

The Role of JSCD

Increasing the robustness of multimedia transmission can be likened to enforcing analog-like properties in a digital system, since the goal is to go from an "all or nothing" channel behavior to one where channel fading results in a graceful degradation of the channel output. If the source coder was perfect, i.e., if it exhibited the $R(D)$ curve of an optimal encoder, and if the side information carried by network layers was totally "hidden," this would be a hopeless task. However, as will be explained in later chapters, practical source coders are often slightly suboptimal and therefore leave some redundancy in the bitstream, see Chapter 4. Furthermore, the bitstream is not likely to be i.i.d., as required for the separation theorem to hold. The network layers themselves introduce redundancy which, although intended for proper operation of the network, contains useful side information. These aspects are described and evaluated in Chapter 7.

The presence of these various redundancies can be exploited by the decoder and, when done properly, can lead to significant performance increase. When the decoding exploits properties of both the source and channel, it can be considered as JSCD. When, in addition, the decoding exploits network-level redundancies, it can be considered as joint source, protocol, and channel decoding (JSPCD). All these aspects are addressed in this book. It is important to remember that JSCD is advantageous for point-to-point communication only when either the source coder or the channel coder is sufficiently far from optimality. In practice, although modern channel coders are sometimes nearly optimal (e.g., turbo or Low-density parity-check codes (LDPC) codes), it is rarely the case that the source coder is close to optimal. In fact, for most practical sources, nonstationarity and long correlation times make the $R(D)$ characteristic difficult to compute.

Joint Transmitter/Receiver Design

As argued above, improved performance results from the joint design and operation of receiver subfunctions like channel decoding and source decoding. Typically, however, these joint reception schemes exploit redundancies in the received bitstream that are unintentional or that at least were not included to help the decoder. Naturally, then, one might ask whether the source/channel coding blocks can be designed in such a way that the source/channel decoder can make the most of the redundancy that they contribute. In fact, many aspects of this problem have already been studied; see Chapter 10 for a review of these methods.

Finally, a global perspective may be taken; for a given bitstream length, one may wonder where the redundancy is most efficient. One may then ask: What is the best allocation of redundancy? This remains an open question which is discussed in Chapter 11.

Feasibility Issues

At first sight, it is not clear whether these advanced system designs are practical, especially if they require making drastic changes to the transmitted signal structure. In general, one should keep the following goals in mind when proposing new techniques that are in competition with existing ones.

- Backward compatibility: It is very often important that the new system is able to coexist with already deployed ones. Note that this is not a problem when the changes relative to existing systems occur only at the receiver, as in JCSD.
- Complexity: The methods described in this book sometimes require significantly more computation than existing ones. However, it will be shown that, in some cases, it is enough to apply the more complex processing to only a small fraction of the received signal. Furthermore, reduced-complexity algorithms are also presented in Chapters 6 and 8.
- Broad applicability: The classical system architecture, based on independent network layers, was designed to keep the operation of the network decoupled from the signal types carried by the network. This attractive property is likely to remain important in the foreseeable future. Therefore, the system should not be designed to operate in a completely application-specific manner, and some independence in the functioning of layers should be retained.

JSCD seems better suited to meeting these goals than JSCC, unless JSCC is implemented in the same spirit. These issues are discussed in Chapter 10.

2.5 CONCLUSION

This chapter provided the core motivation for JSCC/JSCD. In particular, the separation theorem (justifying separate design of the source and channel coders) was shown to be strongly dependent on the assumption of point-to-point communication of stationary sources without constraints on encoder/decoder complexity. It was then argued that practical applications (with complexity constraints, one-to-many architectures, and nonstationary sources and channels) can be better served by joint source/channel decoders. Furthermore, it was argued that JSCD, compared to separate channel decoding and source decoding, is better able to exploit imperfections (i.e., redundancy) in the received bitstream, especially when side information introduced by network protocols are considered.

Chapter 3

Source-Coding Primer

As pointed out in Chapter 2, joint source-channel decoding (JSCD) aims at exploiting, at receiver side, the redundancy remaining in the bitstream generated by source coders and transmitted over noisy channels. This obviously requires a precise knowledge of the bitstream provided by the source coder and the properties of the involved building blocks. In particular, one should understand the ultimate characterization of each constituent of the coder and to what extent the corresponding properties are met or not in actual standards. This chapter, thus, provides a short description of the building blocks commonly used within source coders with emphasis on the properties, which will help in expressing the residual redundancy.

Source coders may be lossless, as usual, e.g., for text compression, the price to be paid for not losing any information being a poor compression performance. When some distortion is acceptable from an applicative point of view, lossy source coders are used, e.g., for audio, speech, still image, and video signals. In this case, a much higher compression performance is obtained at the cost of a controlled amount of distortion.

Video coders are taken as examples in this chapter because on one side, video is quite a demanding signal in terms of raw bitrate, therefore high compression ratios are important, and on another side, their structure allows to describe most building blocks found in source coders for other multimedia contents.

Due to lack of space, only basic concepts are presented. Details about the theoretical foundations of source coding may be found, e.g., in a study by Cover and Thomas (1991). A more practical presentation of source coding is provided, e.g., in the books by Hanzo et al. (2007) and Sayood (2005). Finally, let us point out that the design of efficient joint source-channel decoders requires a good understanding of the source-coding standards and their limitations, as well as the precise characteristics of the generated bitstream. Ultimately, a careful reading of these standards becomes unavoidable.

3.1 COMPONENTS OF SOURCE CODERS

3.1.1 General Structure of Source Coders

Audiovisual signals, as well as many other signals to be compressed, e.g., text files, contain a lot of redundancy, which is intrinsically linked to their perception and meaning. This redundancy takes the form of correlation of various kinds. To generate compressed bitstreams, source coders take advantage of the *temporal* or *spatial correlation* between samples of the source. In monophonic music records, temporal correlation is exploited: when the sampling frequency is large enough, the value of a sample is closely linked to the values taken by the previous samples. For stereophonic records, the samples of both channels are usually fairly correlated. For still images, the spatial correlation is taken into account, provided that the spatial resolution is high enough, and neighboring pixels have usually intensities that vary smoothly. Video sequences present both types of correlation: temporal correlation between successive pictures of the sequence and spatial correlation within pictures.

Therefore, source coders first have to take into account the correlation of the initial signal, either by some kind of prediction (temporal, spatial, inter-channel...) or by some kind of "transform" (from discrete cosine transforms [DCTs] of any type, 1D, 2D DCTs, to filter banks, or wavelets), which approximately turns the initial signal into a set of decorrelated signals.

Prediction

Correlation can be used to *predict* data from previously encoded data. Data that may be perfectly predicted are very efficiently encoded; the coder has only to indicate which data have to be used to perform the prediction. When the prediction is not perfect, only the unpredictable part, the *innovation*, or *residue* has to be encoded. This idea has led to the popular *predictive coding* schemes used in many video-coding schemes, whose structure is detailed in Section 3.1.

Transforms or subbands

Another method for taking advantage of the signal correlation is based on transform or subband (Section 3.5) coding schemes, the role of which is to represent the data by a set of signals that have some kind of hierarchy, in such a way that one can determine those which are really worth keeping and those which may be more coarsely represented (for transform- and subband-coding schemes).

Quantization

The tools used for making use of correlation usually do not perform any bitrate reduction, they can be understood as a preparation step. In contrast, quantizers, described in Section 3.3, are responsible for the compression efficiency and for the distortion of lossy source-coding schemes. It is their interplay with the decorrelation tools that makes the main contribution to the bitrate reduction.

Entropy codes

Entropy coders, presented in Section 3.2, are the corner stones of lossless source coders and form the last stage of almost all lossy source coders. Their task is to take advantage of the source statistics to further reduce the bitrate.

Packetization

Finally, Section 3.7 provides some issues linked to the packetization of compressed data. As will become clear in Chapter 4, if all preceding building blocks of a source coder are usually quite efficient, the packetization of data results in most of the residual redundancy present in the compressed bitstream.

To summarize

The general architecture of a source coder is made of several parts:

- The (usually invertible) representation of the initial signal in terms of decorrelated samples or signals
- The quantization of the resulting representation, which is the place where the loss of information takes place
- The lossless entropy coding of the resulting discrete data
- Some packetization usually takes place since most transmission standards are now based on packets

3.1.2 Structure of Video Coders

Video coders take advantage of many properties of the signal. First, the color signal is represented in a domain that tolerates some kind of downsampling, thus already reducing the number of samples to be encoded. Then, some prediction is used to take advantage of the correlation between successive images, and a transform is used to exploit the correlation inside an image. Finally, these predictions should be performed in such a way that the decoder and the encoder perform the computation based on the same quantities, which introduces some constraints on the encoder structure.

Signal Representation

Video coders are designed to encode sequences of color pictures, each of which is represented by three color matrices (red, green, and blue [RGB]) or more frequently by a *luminance* (Y plane) matrix and two possibly subsampled *chrominance* matrices (U and V planes), which may be deduced by linear transformation of the RGB matrices. Video sequences in black and white consist only of a sequence of Y matrices.

In what follows, most of the illustrations are with the H.264/AVC coder applied to a video sequence where each picture is represented by a Y matrix and a pair (U,V) of matrices for the two-time subsampled chrominance.

Global Architecture

Figure 3.1 describes the architecture used by most standardized video coders from H.261 (ITU-T, 1990), MPEG 1 (ISO/IEC JTC 1, 1993), H.263 (ITU-T, 2005), and MPEG 4 (ISO/IEC JTC 1, 1999) to H.264/AVC (ITU-T and ISO/IEC JTC 1, 2003; Sullivan and Wiegand, 2005), see also Hanzo et al. (2007).

For H.264/AVC, the basic processing unit is the macroblock consisting of four blocks of 8×8 pixels in the Y plane and two blocks of 8×8 pixels in each component of the chrominance planes. To encode the *n*th picture, two modes may be considered: *intra* and *inter*. In the intra mode, the block may be predicted by other parts of the same image, which were already processed by the encoder. This mode is used to avoid too much error propagation. In the inter mode, blocks are allowed to be predicted by other images, either forward or backward. Obviously, for the first picture, only intra is allowed.

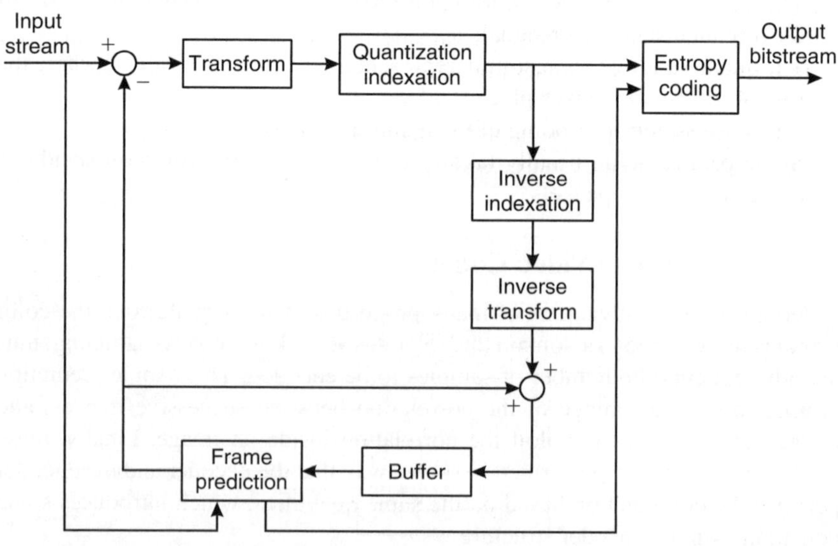

FIGURE 3.1 Structure of a video coder.

Inter and Intra Frames

Using the intra mode, the content of each block of a picture is predicted from surrounding blocks that have been already encoded, see Section 3.4. The *prediction residuals*, i.e., the difference between the predicted and actual blocks, are then transformed (Section 3.5), quantized (Section 3.3), and finally entropy-coded (Section 3.2).

Using the inter mode, coding starts with building a predicted picture from previously encoded pictures, usually on a block by block basis. The motion between pictures (or blocks) is estimated to build a better prediction. *Motion vectors* are used to describe this motion information. The prediction residuals between the actual and predicted pictures are then transformed, quantized, and entropy-coded as for intra pictures. Spatial correlation is exploited between motion vectors to code them efficiently using predictive coding.

Decoder Inside the Encoder

The definition of prediction residuals as the difference between the "predicted" and actual blocks has to be emphasized. In fact, if the prediction was performed based on the original image sequence, the decoder could not work exactly with the same quantities, unless using lossless coding, which is not a realistic situation. Therefore, the encoder has the task of reconstructing the same quantities as those which are to be computed at the decoder so that there is no drift between the predicted sequences. Figure 3.2 shows the structure of the corresponding decoder. The buffer stores already decoded blocks and pictures to build predicted blocks (when decoding intra-encoded pictures) or predicted pictures (for inter pictures). The encoded prediction residuals are then entropy-decoded, inverse transformed, and added to the predicted information to get an estimate of the initial data. This decoder is also clearly seen inside the encoder of Figure 3.1 in which the frame prediction outputs the same values as the corresponding part of the decoder.

The remainder of this chapter is dedicated to the description of the various building blocks used in source coders.

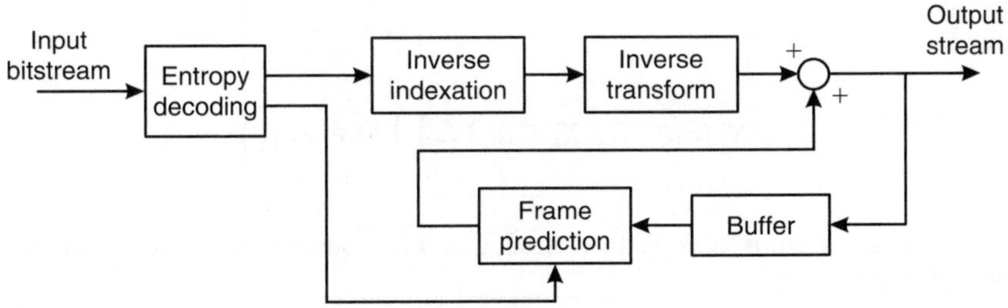

FIGURE 3.2 Structure of a video decoder.

3.2 ENTROPY CODING

This is usually the last step involved in a source coder as illustrated, e.g., in Figure 3.1. The aim is to generate a bitstream from the succession of symbols produced by the preceding parts of the source coder, as in Figure 3.1, or by the source itself, when dealing, e.g., with text compression. Entropy coding is a lossless step in which the statistical properties of the symbols to be encoded (assumed to be realizations of random variables) are taken into account to produce a bitstream that is as compact as possible. Typically, symbols or sequences of symbols, which have a high probability of occurrence, are encoded with a short code, whereas symbols or sequences, which are less likely to occur, are represented by longer codes. Probabilistic models to describe the source outcomes are thus very important and are introduced first before describing some entropy-coding schemes.

3.2.1 Source Models

A source emits symbols to be encoded, which are considered as realizations of a sequence of k random variables X_κ, $\kappa = 1, \ldots, k$, with values belonging to some set (alphabet) $\mathcal{A} = \{a_1, \ldots, a_J\}$. Here and in what follows, the source is assumed to be stationary, i.e., the probability that X_κ generates some a_i is independent of κ,

$$p(X_\kappa = a_j) = p(X_1 = a_j), \quad \kappa = 2, \ldots, k. \tag{3.1}$$

The *stationary probability* model (3.1) does not account for the dependence with time between symbols. The source is modeled as memoryless when each symbol is assumed independent from the others. When there is some dependency, it is usually considered limited in range (time or space) and represented with a *Markov* model.

Memoryless Model

For a memoryless source model, one has

$$p(X_\kappa = a_{j_\kappa} | X_{\kappa-1} = a_{j_{\kappa-1}}, X_{\kappa-2} = a_{j_{\kappa-2}}, \ldots) = p(X_\kappa = a_{j_\kappa});$$

thus,

$$p(X_1 = a_{j_1}, \ldots, X_k = a_{j_k}) = \prod_{\kappa=1}^{k} p(X_\kappa = a_{j_\kappa}).$$

A stationary memoryless source model is thus described by its stationary probability vector,

$$\mathbf{p} = (p_1, \ldots, p_J). \tag{3.2}$$

Markov Model

A (first-order) Markov model of a source is such that the probability that X_κ takes a_{j_κ} depends only on the value taken by $X_{\kappa-1}$, i.e.,

$$p\left(X_\kappa = a_{j_\kappa} | X_{\kappa-1} = a_{j_{\kappa-1}}, X_{\kappa-2} = a_{j_{\kappa-2}}, \ldots\right) = p\left(X_\kappa = a_{j_\kappa} | X_{\kappa-1} = a_{j_{\kappa-1}}\right). \tag{3.3}$$

Moreover, the model is time-homogeneous when

$$\begin{aligned} p\left(X_\kappa = a_j | X_{\kappa-1} = a_i\right) &= p\left(X_{\kappa-1} = a_j | X_{\kappa-2} = a_i\right) \\ &= p_{ij} \end{aligned} \tag{3.4}$$

for all κ and $(a_i, a_j) \in \mathcal{A}^2$.

The state of a Markov model at a given time instant κ is the information that determines the outcome probability at κ. In the case of a first-order Markov source, the state at time κ is the symbol emitted at time $\kappa - 1$ since the probability of the symbol at time κ is entirely determined by the symbol at time $\kappa - 1$. The probabilities p_{ij} may then be interpreted as state-transition probabilities. Provided that the state space is finite, transition probabilities for time-homogeneous Markov models are compactly described by a transition probability matrix $\underline{\mathbf{P}}$, whose (i, j)-entry is p_{ij}.

The stationary probability vector $\mathbf{p}$ has to satisfy

$$\mathbf{p} = \mathbf{p}\underline{\mathbf{P}}, \tag{3.5}$$

It is the left eigenvector of $\underline{\mathbf{P}}$ associated to the eigenvalue 1. It is easily seen that the stationary probability corresponds to the probabilities of the states (being taken from the eigenvector of the transition matrix, their value does not change by running the Markov model). Therefore, in some sense, it is the probability vector of a memoryless model of the first-order Markov source.

Example 3.1: Consider a binary-valued source described by the Markov model with transition probability matrix

$$\underline{\mathbf{P}} = \begin{pmatrix} 0.8 & 0.2 \\ 0.3 & 0.7 \end{pmatrix} = \begin{pmatrix} p_{00} & p_{01} \\ p_{10} & p_{11} \end{pmatrix}.$$

One has thus, for example, $p(X_\kappa = 1 | X_{\kappa-1} = 0) = p_{01} = 0.2$ or $p(X_\kappa = 1 | X_{\kappa-1} = 1) = p_{11} = 0.7$. The transitions between states of this Markov source are described by the *directed graph* or *finite-state machine* of Figure 3.3. Each circle corresponds to a *state* (the previously emitted symbol), arrows indicate transitions between states. The stationary probability vector of this model is $\mathbf{p} = (0.6, 0.4)$.

A Markov model of order m for a source is a model such that the probability that X_κ takes the value a_{j_κ} depends only on the m preceding realizations of the source, i.e.,

$$\begin{aligned} & p\left(X_\kappa = a_{j_\kappa} | X_{\kappa-1} = a_{j_{\kappa-1}}, X_{\kappa-2} = a_{j_{\kappa-2}}, \ldots\right) \\ & \quad = p\left(X_\kappa = a_{j_\kappa} | X_{\kappa-1} = a_{j_{\kappa-1}}, \ldots, X_{\kappa-m} = a_{j_{\kappa-m}}\right). \end{aligned} \tag{3.6}$$

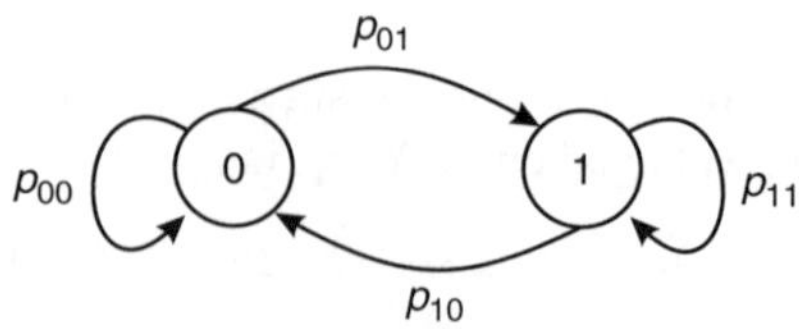

FIGURE 3.3 Directed graph or automaton describing the transition of the Markov source of Example 3.1.

Nevertheless, one may transform a Markov model of order m in a Markov model (of order 1) by considering the random vector $\mathbf{X}_\kappa = (X_\kappa, \ldots, X_{\kappa-m+1})$ taking values in $\mathcal{A}^m$. Then from (3.6), one may easily deduce

$$p\left(\mathbf{X}_\kappa = \mathbf{a}_{j_\kappa} | \mathbf{X}_{\kappa-1} = \mathbf{a}_{j_{\kappa-1}}, \mathbf{X}_{\kappa-2} = \mathbf{a}_{j_{\kappa-2}} \ldots, \right) = p\left(\mathbf{X}_\kappa = \mathbf{a}_{j_\kappa} | \mathbf{X}_{\kappa-1} = \mathbf{a}_{j_{\kappa-1}}\right), \tag{3.6'}$$

for any $\mathbf{a}_{j_{\kappa-1}} \in \mathcal{A}^m$ and $\mathbf{a}_{j_\kappa} \in \mathcal{A}^m$. The state of the Markov model is now a vector of dimension m belonging to $\mathcal{A}^m$. Since this transformation is always possible, in the remainder of this book, only memoryless and Markov models of order 1 are considered.

3.2.2 Entropy

This notion has been introduced in Shannon's original paper (Shannon, 1948) to quantify the average amount of information provided by the knowledge of a source outcome. Entropy represents an absolute limit on the best possible lossless compression of a given message. Assuming that the messages to be encoded are outcomes of a sequence of independent and identically distributed (i.i.d.) random variables, Shannon's source-coding theorem shows that, in the limit, the average length of the shortest possible representation of the messages is their entropy.

Definitions and basic properties useful in the remainder of this book are provided here. Many more details are found in information-theory textbooks (Cover and Thomas, 1991; Yeung, 2002).

Consider a source X with values in $\mathcal{A}$ and described by a probability vector $\mathbf{p}$.

Definition 3.2: *The* self information *associated to the outcome $X = a_j \in \mathcal{A}$ is*

$$I(a_j) = -\log_2(p_j). \tag{3.7}$$

This self information is measured in bits *per source symbol.* ■

Symbols that have a small probability have a high self information, whereas symbols that have a high probability have a small self information.

Definition 3.3: *The* entropy *of X is the expected value of the self information*

$$H(X)=E[I(X)]=-\sum_{i=1}^{J} p_i \log_2(p_i), \tag{3.8}$$

again in bits per source symbol. ■

Self information and entropy are always *positive* quantities. For a set of possible source outcomes $\mathcal{A}$ of J symbols, using Jensen's inequality (Cover and Thomas, 1991), one may show that

$$H(X)=E\left[\log_2 \frac{1}{p(X)}\right] \leqslant \log_2 E\left[\frac{1}{p(X)}\right]=\log_2 J, \tag{3.9}$$

with equality when all symbols are equally likely.

The extension of order k of the source X is a stationary and memoryless source model consisting of k i.i.d. replicas $(X_1, X_2, \ldots, X_k)$ of X. The entropy of the extension of order k of a memoryless source X satisfies

$$H(X_1, \ldots, X_k)=kH(X). \tag{3.10}$$

Example 3.4: A binary source X is described by $\mathbf{p}_1=(0.1, 0.9)$. Thus

$$H(X)=-0.1\log 0.1-0.9\log 0.9=0.469 \text{ bit/symbol}.$$

The extension (X_1, X_2) of order 2 of the source X generates symbols belonging to $\{(0,0), (0,1), (1,0), (1,1)\}$ with an associated probability vector $\mathbf{p}_2 = (0.01, 0.09, 0.09, 0.81)$. Its entropy is

$$\begin{aligned} H(X_1, X_2) &= 0.838 \text{ bit/symbol} \\ &= 2H(X), \end{aligned}$$

as predicted by (3.10).

The entropy of a source described by a Markov model with finite alphabet is more difficult to evaluate. One may show that the average amount of information provided by each outcome of the semi-infinite sequence $X_1, X_2, \ldots$ is the conditional entropy of the source (Cover and Thomas, 1991).

Definition 3.5: *The* conditional entropy *of a source described by a Markov model with finite alphabet $\mathcal{A}$, stationary probability vector $\mathbf{p}$, and transition matrix $\underline{\mathbf{P}}$ is*

$$\begin{aligned} H(X_\kappa|X_{\kappa-1}) &= \sum_i p(X_{\kappa-1}=a_i) H(X_\kappa|X_{\kappa-1}=a_i) \\ &= -\sum_i p_i \sum_j p_{ij} \log_2 p_{ij}. \end{aligned} \tag{3.11}$$

■

Entropy is reduced by conditioning since for any pair (X, Y) of random variables,

$$H(X,Y) = H(X|Y) + H(Y), \tag{3.12}$$

which implies

$$H(X,Y) - H(X|Y) = H(Y) \geqslant 0. \tag{3.13}$$

This is illustrated by Example 3.6.

Example 3.6: Consider again the source of Example 3.1. This source is described by a Markov model, and its entropy is

$$H(X_\kappa|X_{\kappa-1}) = 0.80\,\text{bit/symbol}.$$

Assume now that the Markov property is not known, and that the source is described by a memoryless model with probability, the stationary probability vector **p**. Now, the entropy of the source is

$$H(X_\kappa) = 0.97\,\text{bit/symbol}.$$

One has $H(X_\kappa) > H(X_\kappa|X_{\kappa-1})$.

Using a model of the source, one evaluates the entropy of the model, which is an approximation of the entropy of the source. The better the model, the closer the evaluated entropy to the actual entropy of the source. This aspect is further elaborated in Chapter 4.

3.2.3 Source Codes

This section provides definitions and basic properties of source codes. Only binary source codes are considered here since, at the time this book has been written, non binary-valued codes have not yet been used in any compression standard.

Definitions

Consider a source X with values in the finite set $\mathcal{A}$. A *single-letter* binary source code for X is defined by an *encoding function* C from $\mathcal{A}$ with values in $\{0,1\}^*$, the set of finite-length binary sequences, i.e.,

$$\{0,1\}^* = \{0, 1, (0,0), (0,1), \ldots, (1,1), (0,0,0), \ldots\}.$$

The set of code words $\mathcal{C} = C(\mathcal{A})$ contains all encoded symbols of $\mathcal{A}$. A source code has to be *decodable*, i.e., one has to be able to retrieve the symbols generated by the source from the code words. For that purpose, the code has to satisfy some properties.

Depending on the context, a source code will be denoted by its encoding function (which is important for determining its performance) or by its set of code words (which is enough to test whether it is decodable).

Definition 3.7: *A code is* nonsingular *if for any* $a, a' \in \mathcal{A}$,

$$a \neq a' \Longrightarrow C(a) \neq C(a'),$$

the encoding function has to be injective. ■

This property ensures that one is able to decode a single encoded symbol, but it is not sufficient to ensure the decodability of a sequence of bits consisting in the concatenation of several code words.

Definition 3.8: *The* extension $C^* : \mathcal{A}^* \to \{0, 1\}^*$ *of a code C is a function that takes any finite sequence of source symbols and associates to it the sequence of code words*

$$C^*(a_1, a_2, \ldots a_k) = C(a_1)\, C(a_2) \ldots C(a_k),$$

where $C(a_1)C(a_2)\ldots C(a_k)$ *denotes the concatenation of the code words* $C(a_1)$, $C(a_2), \ldots,$ *and* $C(a_k)$. *The set of code words of the extension is* $C^* \subset \{0, 1\}^*$. ■

Definition 3.9: *A code is* uniquely decodable *if its extension is nonsingular.* ■

This definition is straightforward, but it does not provide constructive means to determine whether a code C is uniquely decodable. For that purpose, one may use the Sardinas–Patterson (Rodeh, 1982; Sardinas and Patterson, 1953) criterion. Moreover, unique decodability does not necessarily mean easy decodability. To decode a sequence of code words on the fly, *prefix* or *instantaneous* codes are very useful.

Definition 3.10: *A code is* prefix *or* instantaneous *if no code word is the prefix of another code word.* ■

This property allows to isolate each code word in a bitstream without having to consider the whole coded sequence. Example 3.11 illustrates the previous definitions.

Example 3.11: Consider a source X with alphabet $\mathcal{A} = \{1, 2, 3, 4\}$. Then the code

- $C_1 = \{0, 0, 0, 0\}$ is singular.
- $C_2 = \{0, 010, 01, 10\}$ is nonsingular, but not uniquely decodable, since when considering its extension, 010 may be interpreted as the concatenation of 01 and 0.
- $C_3 = \{10, 00, 11, 110\}$ is uniquely decodable.
- $C_3 = \{0, 10, 110, 111\}$ is prefix.

The efficiency of a source code is mainly determined by the average length of the code words associated to the source outcomes.

Definition 3.12: *The average length of a code C is*

$$\ell(C) = \sum_{a \in \mathcal{A}} p(X=a)\, \ell(C(a)), \tag{3.14}$$

where $\ell(C(a))$ is the length of the code word associated to a. ■

Kraft and Kraft–Mc Millan Inequalities

The length of the code words has to be made as small as possible to minimize (3.14). The Kraft and Kraft–Mc Millan inequalities provide a necessary and sufficient condition on the length of the code words to allow the existence of prefix codes (Kraft) or uniquely decodable codes (Kraft–Mc Millan) with these lengths. Both theorems provide the same condition; thus, only the Kraft inequality is given.

Theorem 3.13 (Kraft): *Consider a set $\mathcal{C}$ of J binary sequences of length $\ell_1, \ldots, \ell_J$. There exists a prefix code with these code word length if and only if*

$$K(\mathcal{C}) = \sum_{j=1}^{J} 2^{-\ell_j} \leqslant 1. \tag{3.15}$$

A proof of Theorem 3.13 may be found, e.g., in the book by Cover and Thomas (1991). If (3.15) is an equality, the code is *complete*, otherwise, it is *redundant*, see Chapter 4.

Optimal Prefix Code

Consider a source X with probability vector $\mathbf{p} = (p_1, \ldots, p_J)$. One aims at determining the optimal lengths $\ell_1, \ldots, \ell_J$ of the code words of a single-letter binary prefix code $\mathcal{C}$ for X. For that purpose, one has to minimize

$$\ell(\mathcal{C}) = \sum_{j=1}^{J} p_j \ell_j, \tag{3.16}$$

under the constraint (3.15).

This is a constrained optimization problem. Relaxing the constraint on the integer length of the code words and introducing the Lagrange multiplier λ, one gets an unconstrained cost function

$$c(\ell_1, \ldots, \ell_J, \lambda) = \sum_{j=1}^{J} p_j \ell_j + \lambda \left(\sum_{j=1}^{J} 2^{-\ell_j} \right).$$

The minimum of which is obtained when

$$\frac{\partial c}{\partial \ell_j} = p_j - \lambda 2^{-\ell_j} \log_e 2 = 0, \quad j = 1, \ldots, J,$$

which leads to

$$2^{-\ell_j} = \frac{p_j}{\lambda \log_e 2}, \quad j = 1, \ldots, J. \tag{3.17}$$

Substituting (3.17) into (3.15), one gets

$$\lambda \geqslant \frac{1}{\log_e 2},$$

which introduced back in (3.17) leads to

$$\widehat{\ell}_j \geqslant \ell_j^* = -\log_2 p_j, \quad j = 1, \ldots, J \tag{3.18}$$

and to the optimal average code word length

$$\widehat{\ell}\,(\mathcal{C}) = \sum_{j=1}^{J} p_j \widehat{\ell}_j \geqslant -\sum_{j=1}^{J} p_j \log_2 p_j = H(X), \tag{3.19}$$

with equality only if $\mathbf{p}$ is a *dyadic* probability vector, i.e., a vector in which each probability is a negative integer power of 2.

Example 3.14: Consider a source X with probability vector $\mathbf{p} = (1/2, 1/4, 1/8, 1/16, 1/16)$, then the optimum code word lengths for this source are given by (3.18) and are $\widehat{\ell} = (1, 2, 3, 4, 4)$. The average length of the code words is

$$\widehat{\ell}\,(\mathcal{C}) = 1.875\,\text{bits}$$

and is equal to the entropy of X.

Example 3.15: Consider now a source X with probability vector $\mathbf{p} = (3/4, 3/16, 1/16)$. In this case, since the probability vector is no more dyadic, the optimum code word lengths $\ell^* = (0.415, 2.415, 4)$ are not integers. One may take $\widehat{\ell}_1 = (1, 3, 4)$, which leads to $\widehat{\ell}_1\,(\mathcal{C}) = 1.5625\,\text{bits}$ or $\widehat{\ell}_2 = (1, 2, 2)$, which also satisfies the Kraft inequality and leads to $\widehat{\ell}_2\,(\mathcal{C}) = 1.25\,\text{bits}$. In both cases, the entropy of the source $H(X) = 1.014$ bits/sample is not reached.

A natural choice that allows to meet (3.18) and Kraft inequality is to round $-\log_2 p_j$ upwards. This allows to get a code, whose average code word length is at most 1 bit/sample away from entropy.

$$\begin{aligned} \sum_{j=1}^{J} p_j \lceil -\log_2 p_j \rceil &\leqslant \sum_{j=1}^{J} p_j \left(-\log_2 p_j + 1\right) \\ &\leqslant H(X) + 1. \end{aligned}$$

Nevertheless, as shown by Example 3.15, this procedure does not lead to the optimal code in general. The Huffman coding procedure provides an algorithm for obtaining the code with minimum average code word length, even for nondyadic probability vectors.

3.2.4 Huffman Code

The Huffman coding procedure provides a way to build an optimal variable-length binary prefix code for a source X described by a given probability vector $\mathbf{p}$. It is based on the following theorem providing some necessary conditions to be satisfied by an optimal variable-length prefix code.

Theorem 3.16: *A binary prefix code with minimum, average code word length for a probability vector* $\mathbf{p}$ *has to be such that*

1. *if* $p_j \geqslant p_{j'}$, *then* $\ell_j \leqslant \ell_{j'}$;
2. *the two code words associated to the least probable source symbols have the same length;*
3. *if, among the least probable symbols, at least two code words have the same length, then two of these code words share the same bits, except the least significant one.*

See Cover and Thomas (1991) for the proofs. From Theorem 3.16, one may easily deduce a recursive algorithm for building an optimum variable-length code.

Consider a source X with J symbols and probability vector $\mathbf{p}$. A Huffman code $\mathcal{C}$ for this source is built recursively as follows.

1. Select the two least probable source outcomes a_j and a_k, and merge them into a virtual symbol a_{jk} with probability $p_j + p_k$ to get a new source $X^{(1)}$ with $J-1$ symbols and probability vector $\mathbf{p}^{(1)}$.
2. Determine the optimum code $\mathcal{C}^{(1)}$ for the source $X^{(1)}$.
3. Once $\mathcal{C}^{(1)}$ has been obtained, the code words for a_j and a_k are obtained by concatenating 0 and 1 at the end of the code words for the virtual symbol a_{jk}. $\mathcal{C}$ is thus simply deduced from $\mathcal{C}^{(1)}$.

For a source with only two symbols, the variable-length code is $\{0, 1\}$.

Example 3.17: Consider a source X with alphabet $\mathcal{A} = \{A, B, C\}$ and probability vector $\mathbf{p} = \left(\frac{3}{4}, \frac{3}{16}, \frac{1}{16}\right)$. Figure 3.4 illustrates the steps of the Huffman code construction. The two least probable symbols are first grouped to get the virtual symbol BC with probability 1/4. The new source X' has only two symbols A and BC, with code words 0 and 1. The codes for B and C are deduced by appending 0 and 1 to the code for BC to get the set of code words $\mathcal{C}_H^1 = \{0, 10, 11\}$ for the source X. The average length $\widehat{\ell}(\mathcal{C}_H^1) = 1.25$ bits/symbol, which is still larger than $H(X) = 1.014$ bits/symbol.

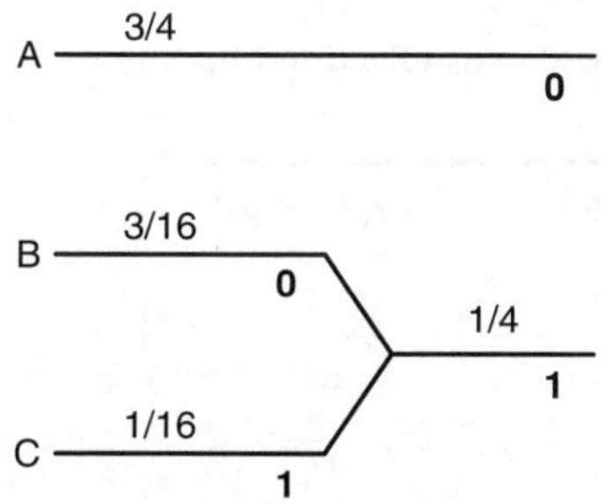

FIGURE 3.4 Single-letter Huffman code.

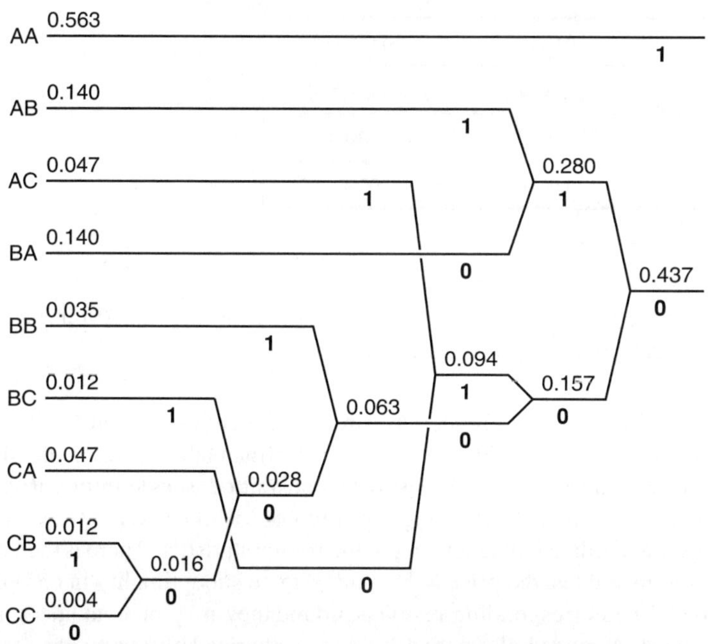

FIGURE 3.5 Huffman code for the extension of order 2 of the source of Example 3.17.

For nondyadic probabilities, designing a Huffman code for the order 2 or 3 extension of the source is an efficient way to get closer to the entropy of the source, as shown by Example 3.18.

Example 3.18: Consider the extension of order 2 of the source X. The set of code words belongs to $\mathcal{A}^2$. A Huffman code $\mathcal{C}_H^2$ may be built for this extension, as shown in Figure 3.5. The set of code words $\mathcal{C}_H^2$ is provided in Table 3.1. The average length of the code words is now $\widehat{\ell}\left(\mathcal{C}_H^2\right) = 2.070$ bits/

TABLE 3.1 Set of Code Words Obtained for Example 3.18

Symbol	Probability	Code word
AA	0.5625	1
AB	0.1406	011
AC	0.0469	0011
BA	0.1406	010
BB	0.0352	0001
BC	0.0117	00001
CA	0.0469	0010
CB	0.0117	000001
CC	0.0039	000000

symbol of the order 2 extension, which corresponds to 1.035 bits/symbol of X, which is now much closer to $H(X)$.

The obtained Huffman code, thus, strongly depends on the probability vector of the source. To perform the decoding, the Huffman table used for encoding has to be known at decoder side. It has to be transmitted as side information to the receiver. Alternatively, as done, e.g., in JPEG or MPEG 4/AAC, *static* Huffman tables are standardized, independent of the source statistics. There is then no need to transmit the tables, the price to be paid is some suboptimality in compression efficiency. The corresponding residual redundancy may be evaluated, as done in Chapter 4. A second alternative is to use *adaptive* Huffman codes, see, e.g., Gallager (1978) and Vitter (1987) for more details.

Instead of using a table, a Huffman coder or decoder may also be implemented using a *Mealy transducer*, i.e., a finite-state machine consisting of states linked by transitions. For a coder, these transitions are fired by symbols generated by the source and produce bits, whereas for decoders, transitions are fired by the bits to decode and produce decoded source symbols. More details on Mealy transducers are provided in Section 4.2.2, page 97.

3.2.5 Run-Length Codes

For a source with small alphabet and very skewed probability vector (resulting in an entropy that may be much smaller than one), the Huffman coding algorithm may produce a variable-length code whose performance is far from entropy, unless considering high-order extensions of the source, and thus complex Huffman tables. Moreover, getting Huffman codes that adapt to time-varying source statistics is far from being trivial. This is the main reason for the proposal of run-length codes (Zehavi and Wolf, 1988) and arithmetic coding, which can easily handle adaptation.

A typical example of skewed source is the one that would model the black-and-white pixels of a scanned version of this book. If 0 represents white and 1 black, pages containing mainly text would consist of many more 0s than 1s. Moreover, long periods of 0s would separate short series of 1s. Such source is typically very difficult to encode with a Huffman code.

The idea of *run-length coding* for a binary source is to output only the number of 0s separating two consecutive 1s. One gets then a nonbinary source, which is easier to encode with a Huffman code, see Example 3.19.

Example 3.19: Consider a binary source generating the message

$$00000101100001001100001000000000.$$

This message may be run-length encoded to get

$$5, 1, 0, 4, 2, 0, 4.$$

To recover the initial sequence, its length has to be transmitted as side information; since when the sequence does not end with 1, there is no way to know whether there are one or several thousands of 0s at the end of the sequence.

When the source is not binary, but still generates a large proportion of 0s, one may encode the sequence with a set of pairs containing the number of 0s before a non-zero component followed by the value of this non-zero component, see Example 3.20.

Example 3.20: Consider a source generating the message

$$00000401-30000100640000-3000000000.$$

This message may be run-level encoded to get

$$(5, 4), (1, 1), (0, -3), (4, 1), (2, 6), (0, 4), (4, -3).$$

Again, the length of the initial sequence is needed to perform decoding.

Run-level encoding has been used in the H263+ video standard as a source transformation technique used in combination with entropy coding, see Section 4.3.2, page 112, for more details.

3.2.6 Arithmetic Code

As already mentioned, designing Huffman codes that adapt to time-varying source statistics is far from being trivial. This is the main reason why arithmetic coding, which can easily handle adaptation, has gained a lot of interest in recent image and video coding standards (ISO/IEC JTC 1, 2000; Wiegand et al., 2003).

Consider a source generating a message $\mathbf{x}=(x_1,\ldots,x_k)$ assumed to be the realization of a sequence of k random variables X_κ, $\kappa=1,\ldots,k$, with values belonging to some set (alphabet) $\mathcal{A}=\{a_1,\ldots,a_J\}$. Arithmetic coding (Howard and Vitter, 1992, 1994; Rissanen, 1976; Witten et al., 1987) encodes directly a sequence of source outcomes by associating a unique identifier or tag to $\mathbf{x}$. This tag may be an integer, or a rational number belonging to the interval [0, 1[. The tag has to be such that its knowledge allows an unambiguous decoding of $\mathbf{x}$.

The next section describes the way this tag is obtained in the case numbers are represented with infinite precision (infinite-precision arithmetic coding [IP-AC]). A more practical arithmetic coder is presented in a second part. The presentation focuses on binary arithmetic codes, the variables X_κ, $\kappa=1,\ldots,k$, are assumed to be k iid replicas of a variable X with

$$p(X=0)=p_0 \quad \text{and} \quad p(X=1)=p_1=1-p_0.$$

Extension to the nonbinary case may be found in a study by Sayood (2005).

Infinite-Precision Arithmetic Coder

In this case, the tag $C(\mathbf{x})$ is a rational number in the interval [0, 1[. Starting from the *coding interval* $[\ell_0,h_0[=[0,1[$, the encoding of the first symbol x_1 is done by first bisecting $[\ell_0,h_0[$ into two subintervals, the width of which is proportional to p_0 and $1-p_0$. Then, one of these two subintervals is selected depending on the value of x_1 according to

$$[\ell_1,h_1[=\begin{cases}[0,p_0[& \text{if } x_1=0,\\ [p_0,1[& \text{if } x_1=1.\end{cases}$$

This procedure is repeated for x_2 using $[\ell_1,h_1[$ as starting interval. For x_κ, the bisection and selection rules are the same, but starting from $[\ell_{\kappa-1},h_{\kappa-1}[$, to get

$$[\ell_\kappa,h_\kappa[=\begin{cases}[\ell_{\kappa-1},\ell_{\kappa-1}+p_0(h_{\kappa-1}-\ell_{\kappa-1})[& \text{if } x_\kappa=0,\\ [\ell_{\kappa-1}+p_0(h_{\kappa-1}-\ell_{\kappa-1}),h_{\kappa-1}[& \text{if } x_\kappa=1,\end{cases} \tag{3.20}$$

see also Figure 3.6.

Once the kth symbol has been encoded, the interval $[\ell_k,h_k[$ is obtained with

$$h_k-\ell_k=p(\mathbf{x}). \tag{3.21}$$

Now consider

$$\lambda(\mathbf{x})=\lceil -\log_2 p(\mathbf{x})\rceil+1, \tag{3.22}$$

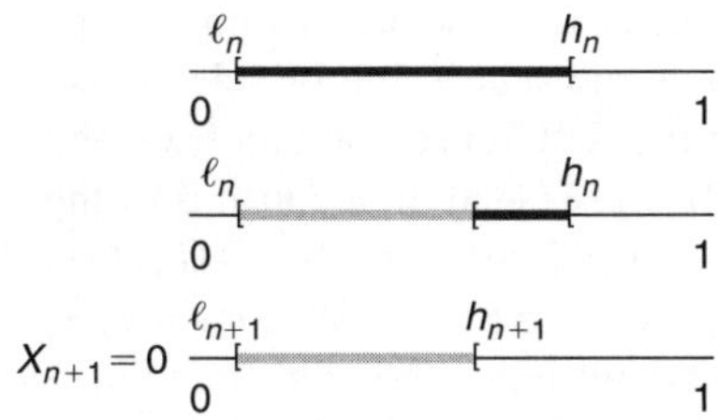

FIGURE 3.6 Division and selection performed during arithmetic encoding.

where $\lceil\cdot\rceil$ stand for rounding upwards. Let $\mu(\mathbf{x}) = (h_k + \ell_k)/2$ be the middle of $[\ell_k, h_k[$ and define the tag

$$C(\mathbf{x}) = \lfloor \mu(\mathbf{x}) \rfloor_{\lambda(\mathbf{x})} \tag{3.23}$$

as being $\mu(\mathbf{x})$ represented in base 2 and truncated to the $\lambda(\mathbf{x})$-th bit. The decimal part of $C(\mathbf{x})$ is then considered as the arithmetic code word for $\mathbf{x}$.

One may prove that

$$C(\mathbf{x}) \in [l_k, h_k[, \tag{3.24}$$

which ensures that $C(\mathbf{x})$ is associated to a *unique* sequence $\mathbf{x}$ of k bits. Moreover, the average length of $C(\mathbf{x})$ for source sequences of k bits is such that

$$E(\lambda(X_1, \ldots, X_k)) = \sum_{\mathbf{x} \in \{0,1\}^k} p(\mathbf{x}) \left(\lceil -\log_2 p(\mathbf{x}) \rceil + 1 \right) \tag{3.25}$$

$$\leqslant kH(X) + 2, \tag{3.26}$$

since the source is memoryless. Provided that k is sufficiently large, $C(\mathbf{x})$ is thus an efficient code for $\mathbf{x}$.

Example 3.21: Assume that $p_0 = 1/5$ and that the sequence $\mathbf{x} = (10111)$ has to be encoded. Starting from $[\ell_0, h_0[= [0, 1[$, the succession of coding intervals that would be obtained in this case are

$$[\ell_1, h_1[= [1, 5[/5$$
$$[\ell_2, h_2[= [5, 9[/25$$
$$[\ell_3, h_3[= [29, 45[/125$$
$$[\ell_4, h_4[= [161, 225[/625$$
$$[\ell_5, h_5[= [869, 1125[/3125.$$

One may then get $\lambda(\mathbf{x}) = 5$ and $C(\mathbf{x}) = 0.01010$. The code for $\mathbf{x} = (10111)$ is thus 01010.

To perform decoding, assume that an arithmetic code word has been received. This code is used to build a tag $c \in [0, 1[$ by considering a null integer part and a base-2 decimal part consisting of the arithmetic code word. The arithmetic decoder has to determine the succession of selections the arithmetic coder has performed. To decode the first source outcome x_1, the decoder starts with the interval $[\ell_0, h_0[=[0, 1[$ and bisects it according to the source probability into $[0, p_0[$ and $[p_0, 1[$. If $c \in [0, p_0[$, then $x_1 = 0$ and $[\ell_1, h_1[=[0, p_0[$, else $x_1 = 1$ and $[\ell_1, h_1[=[p_0, 1[$. The procedure is repeated for x_2, x_3, etc. The main difficulty is that it is not possible to determine when the decoding has to be stopped, unless the length of the sequence to be decoded has been sent as side information or a special sequence of bits (end of sequence) has been identified in **x**.

Example 3.22: Assume again that $p_0 = 1/5$ and that a sequence 01010 has been received. This sequence leads to the tag $c = (0.01010)_2 = (0.3125)_{10}$. To perform the decoding, one notices that $c \in [1, 5[/5$, thus, $x_1 = 1$; $c \in [5, 9[/25$, thus $x_2 = 0$ etc.

This arithmetic coder is not suited for practical implementations since it requires the representation of ℓ_κ and h_κ with a number of bits that is usually not compatible with the accuracy of standard processors. Finite-precision versions of the arithmetic coder have been proposed to solve this problem.

Finite-Precision Arithmetic Coder

In this version of the arithmetic coder, the tag is an integer and all computations are done on integers starting from the integer interval $[\ell'_0, h'_0[=[0, 2^p[$, where $p \in \mathbb{N}$ is the finite precision arithmetic coder (FP-AC) (Howard and Vitter, 1992, 1994; Witten et al., 1987). This integer interval may be viewed as a representation on p-bits of the interval $[\ell_0, h_0[=[0, 1[$ of the IP-AC.

Bisections and selections are done as in (3.20), with possible rounding to keep the integer bounds (encoder and decoder have to agree on the rounding direction). The tag c is built iteratively, starting from an empty tag. During encoding, as soon as some bits of the tag may be determined unambiguously, they are appended to it. Obviously, it is unambiguous determination when both binary representations of the lower and upper bound agree on their most significant bits since any further segmentation of this interval does not change their value. Moreover, to prevent $[\ell'_\kappa, h'_\kappa[$ from getting too small, its size is doubled as soon as it becomes less than half of that of the initial coding interval. This rescaling may be quiet, i.e., without modifying the tag, in which case, a variable f (for bits to *follow*) is incremented. It may also be done with a modification of the tag. The rescaling operations are performed when the following conditions are satisfied.

- If $[\ell'_\kappa, h'_\kappa[\subset [0, 2^{p-1}[$, one 0 and f 1s are appended to c, $[\ell'_\kappa, h'_\kappa[=[2\ell'_\kappa, 2h'_\kappa[$, and f is set to 0
- If $[\ell'_\kappa, h'_\kappa[\subset [2^{p-1}, 2^p[$, one 1 and f 0s are appended to c, $[\ell'_\kappa, h'_\kappa[= [2\ell'_\kappa - 2^{p-1}, 2h'_\kappa - 2^{p-1}[$, and f is set to 0
- If $[\ell'_\kappa, h'_\kappa[\subset [2^{p-2}, 3 \times 2^{p-2}[$, $f = f + 1$, and $[\ell'_\kappa, h'_\kappa[=[2\ell'_\kappa - 2^p, 2h'_\kappa - 2^p[$

Rescaling is done until none of the above conditions is satisfied. Consider, for example, the first case, the situation equivalent to $\left[\ell'_{\kappa}, h'_{\kappa}\right[\subset \left[0, 2^{p-1}\right[$ for the IP-AC is $[\ell_{\kappa}, h_{\kappa}[\subset [0, 0.5[$. If the IP-AC encoding would stop, the most significant bit of the resulting code word would be 0. This bit is appended on-the-fly to the FP-AC tag, and the size of the coding interval is doubled. In the second case, similarly, the most significant bit would be 1, which is appended to the tag before doubling the size of the coding interval. The third case corresponds to a situation where the most significant bit is not fully determined, at least in the FP-AC. Thus, no bit is output yet, but to prevent the coding interval to get too small, its size is doubled, and the variable f stores the number of consecutive quiet rescaling. These quiet rescalings are taken into account as soon as bits are appended to the tag.

To properly end the coding process, once x_k has been encoded and all rescaling have been performed,

- if $\ell'_k < 2^{p-2}$, then 0 and $f+1$ 1s are appended to c;
- else 1 and $f+1$ 0s are appended to c.

This ensures that f is flushed at the end of the encoding process and the tag generated belongs to the coding interval. Figure 3.7 shows some examples of the encoding of a bit when $p=3$ and $f=2$.

Finite-precision arithmetic codes can be represented as a state machine, the state being defined as the concatenation of the quantities that allows the machine to run during the next event and, for example, to compute the next output when an input arises. Obviously, here, this set of quantities is composed of the lower and upper value of the current interval, and the current value of the variable f, since this knowledge allows the next step of the algorithm to perform. Such descriptions are used in subsequent chapters to evaluate the redundancy and to perform efficient decoding.

Therefore, the state of the FP-AC is defined as $\left(\ell'_{\kappa}, h'_{\kappa}, f_{\kappa}\right)$, where f_{κ} represents the value of f after encoding the κth source symbol.

The decoding is done in a way similar to the IP-AC. It starts with the p most significant bits of the tag. Again, the selections have to be determined to estimate the source outcomes. Each time a scaling of the coding interval is performed, the

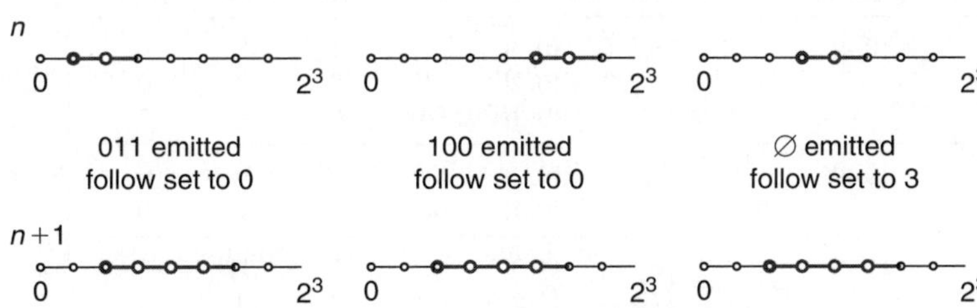

FIGURE 3.7 Finite-precision arithmetic encoding.

most significant bit of the tag is dropped, and the p remaining bits are considered for further decoding.

Quasi-Arithmetic Codes

In FP-ACs, the coding interval $[0, 2^p[$ contains a finite number of integers, ℓ'_K and h'_K, which may only take a finite number of values, more precisely, $3 \times 2^p/16$ values. Nevertheless, this is not the case for f_K, which may take any positive integer value. To be able to describe finite-precision arithmetic coder/decoders with finite look-up tables or finite state machines (FSMs), Howard and Vitter (1992) introduced *quasi-arithmetic codes*. In such codes, f does not belong to the state but is a variable resulting from state transitions (output variable). Thus, f is not considered in the computation of the coder states, and the state can take only a finite number of values, which characterizes an FSM.

Example 3.23: Table 3.2 gives the states, outputs, and all possible transitions of a quasi-arithmetic coder precomputed for a binary source with $2^p = 4$. The value of the variable f is not considered in this state model. Only the action of incrementing f when a rescaling takes place is signaled by the letter f in the table, also referred to as *follow-up* in a study by Howard and Vitter (1992). The encoder has three states corresponding to integer subdivisions of the interval [0, 4[. The subdivisions that can possibly take place next are function of the source probability distribution. They are chosen in such a way that the corresponding distribution approximation minimizes the excess rate (Howard and Vitter, 1992) due to the finite precision of the coding interval. Assume, for example, that the FSM is in state $S_K = 1$, (defined by the interval $[\ell'_K, h'_K[= [0, 3[$). Depending on the probability of the input binary symbol 0, the interval [0, 3[is further subdivided into corresponding to an approximated probability 2/3, if $p_0 > 1/2$, or into [0, 1[, corresponding

TABLE 3.2 Quasi-Arithmetic Code for $2^p = 4$, Taken from Guionnet and Guillemot (2004)

State		p_0 and Interval		$x_K = 0$		$x_K = 1$	
S_K	$[\ell'_K, h'_K[$	Subdivision		Out	S_{K+1}	Out	S_{K+1}
0	[0, 4[	$2/3 \leqslant p_0$	[0, 3[	–	1	11	0
		$1/3 \leqslant p_0 < 2/3$	[0, 2[	0	0	1	0
		$p_0 \leqslant 1/3$	[0, 1[	00	0	–	2
1	[0, 3[	$1/2 \leqslant p_0$	[0, 2[	0	0	10	0
		$p_0 < 1/2$	[0, 1[	00	0	f	0
2	[1, 4[	$1/2 \leqslant p_0$	[1, 3[	f	0	11	0
		$p_0 < 1/2$	[1, 2[	01	0	1	0

to an approximated probability of 1/3, if $p_0 \leqslant 1/2$. Both subdivisions result, after appropriate bit emission and scaling, into the state $S_{\kappa+1} = 0$.

Alternatively, to limit f, an emission of bits may be forced as soon as $f \geqslant f_{\max}$, by temporarily modifying p_0, as done, e.g., in a study by Ben-Jamaa et al. (2008), see also Example 3.24. The state vector is then $\left(\ell'_\kappa, h'_\kappa, f_\kappa\right)$ and the number of states is bounded by $3f_{\max} \times 2^p/16$. Example 4.14, page 98, provides a FSM description of such quasi-arithmetic coder.

Example 3.24: Consider a quasi-arithmetic code with $q = 4$, $p_0 = 0.2$, and $f_{\max} = 1$. As long as $f < f_{\max}$, the division of the coding interval is done according to p_0 with rounding toward the nearest integer. As soon as $f \geqslant f_{\max}$, the division is done with respect to 2^{q-1}, the midpoint of the coding interval, to force an emission of bits. Table 3.3 lists the states $([\ell_\kappa, h_\kappa[, f_\kappa)$, transitions, and outputs.

Adaptive Arithmetic Codes

Adaptive versions of the arithmetic coder are easily derived from the conventional arithmetic coder by adapting the bisection of the current coding interval to the current probability of the source outcome. In the preceding sections, these probabilities were assumed constant. In the adaptive version, they are allowed to vary, but the way the probabilities are estimated is the same at the encoder and the decoder to ensure decodability.

TABLE 3.3 States, Transitions, and Outputs of a Quasi-Arithmetic Code with $q = 4$, $p_0 = 0.2$, and $f_{\max} = 1$

		Input: 0		Input: 1	
Index	State	Next State	Output	Next State	Output
0	$([0, 16[, 0)$	$([0, 12[, 0)$	00	$([3, 16[, 0)$	–
1	$([0, 12[, 0)$	$([0, 16[, 0)$	000	$([2, 12[, 0)$	–
2	$([3, 16[, 0)$	$([4, 16[, 1)$	0	$([6, 16[, 0)$	–
3	$([2, 12[, 0)$	$([0, 16[, 0)$	001	$([0, 16[, 1)$	–
4	$([4, 16[, 1)$	$([0, 16[, 0)$	011	$([0, 16[, 0)$	10
5	$([6, 16[, 0)$	$([0, 16[, 0)$	011	$([0, 16[, 0)$	1
6	$([0, 16[, 1)$	$([0, 16[, 0)$	01	$([0, 16[, 0)$	10

Usually, the probability vector of the source is indeed estimated and updated on-the-fly at the encoder and at the decoder, see Marpe et al. (2003), Osorio and Bruguera (2004), and Rissanen (1986) for more details. Such a strategy allows very easily a context-based encoding and decoding, since several sets of probabilities can be computed, depending on predetermined conditions (the context).

3.2.7 Universal Codes

These codes are mostly used for compressing texts (including *html* files used for transmitting Web pages) or programs, and are routinely used for that purpose in everyday's life. Even if JSCD techniques are seldom applied to this context, they can improve the transmission performance of such signals over wireless channels (Jaoua et al., 2008a,b). The methods used in this context are of the same nature as for video files.

An adaptive or universal source code is a code that needs no a priori assumptions about the statistical dependencies of the data to be encoded. There are many types of universal codes. For example, the dynamic Huffman code (see, e.g., Gallager, 1978) is an adaptive version of the Huffman code. There are several ways to use arithmetic coding in an adaptive way (see Marpe et al., 2003; Osorio and Bruguera, 2004; Rissanen, 1986). Nevertheless, among the existing universal codes, the encoding techniques motivated by the 1977 Lempel–Ziv algorithm (Ziv and Lempel, 1977) and the 1978 Lempel–Ziv algorithm (Ziv and Lempel, 1978) are virtually standard in practical lossless data compression because they empirically achieve good compression, and they are computationally efficient. This section, largely inspired by Savari (1997a), briefly describes these two universal codes.

The Lempel–Ziv-based coding techniques use a dictionary, see Figure 3.8. This dictionary may be explicit, as in the Lempel–Ziv–Welch (LZW) algorithm or implicit, i.e., formed by the previously encoded symbols.

Consider a source X with values in an alphabet $\mathcal{A}$ of J symbols. The Lempel–Ziv incremental parsing rule starts with a dictionary consisting of the J source symbols. At any parsing point, the next parsed phrase σ is the longest

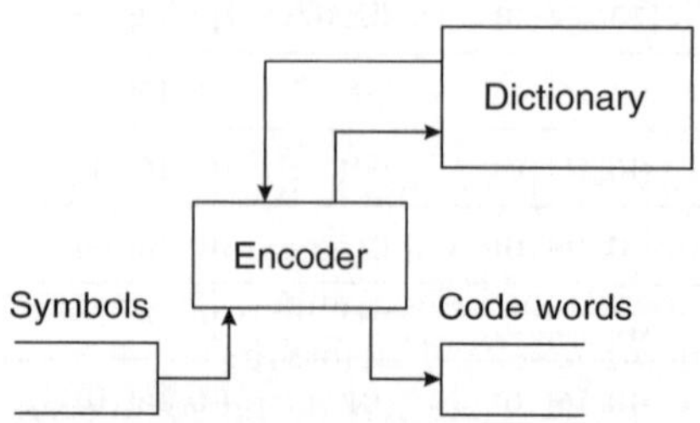

FIGURE 3.8 Dictionary-based encoder.

dictionary entry, which is a prefix of the unparsed source output. Once this phrase has been selected, the dictionary for the Lempel–Ziv incremental parsing rule is enlarged by replacing σ with its J single-letter extensions. For all three of the encoding procedures that will be considered, if the dictionary contains M entries, then $\lceil \log_2 M \rceil$ bits are used to encode the next parsed phrase.

Example 3.25 (LZ78): Assume that $\mathcal{A}=\{0,1,2\}$ and that source output is the string 0 0 0 0 2... Initially, the dictionary is $\{0,1,2\}$. With LZ78, the first parsed string is 0, and the dictionary is updated to $\{\mathbf{00},\mathbf{01},\mathbf{02},1,2\}$. At this point, the unparsed source output is 0 0 0 2... The second parsed string is 00, and the revised dictionary is $\{\mathbf{000},\mathbf{001},\mathbf{002},01,02,1,2\}$. Now, the unparsed source sequence is 0 2... The third parsed string is 02, resulting in the dictionary $\{000,001,002,01,\mathbf{020},\mathbf{021},\mathbf{022},1,2\}$.

Practical implementations of the Lempel–Ziv incremental parsing rule often differ somewhat from the original 1978 Lempel–Ziv algorithm. In the LZW algorithm (Welch, 1984), initially, the dictionary entries are the J source symbols. At any parsing point, the next parsed phrase is the longest dictionary entry, which is a prefix of the unparsed source output. Thus far, the parsing rule is identical to the one used by 1978 Lempel–Ziv algorithm. Nevertheless, instead of replacing the last parsed phrase by its J single-letter extensions, for LZW algorithm, the dictionary is enlarged by adding the last parsed phrase concatenated with the first symbol of the unparsed source output. According to Welch (1984), LZW algorithm achieves very similar compression to 1978 Lempel–Ziv algorithm but is easier to implement. Note that any string can appear as a parsed phrase at most once for the original Lempel–Ziv incremental parsing rule, whereas it can occur as a parsed phrase up to J times for LZW algorithm.

Example 3.26 (LZW): Consider now the LZW parser and the same source output sequence 0 0 0 0 2... Initially, the dictionary is $\{0,1,2\}$. The first parsed string is 0, and the remaining source output is 0 0 0 2.... The dictionary is enlarged to $\{0,\mathbf{00},1,2\}$. The next parsed string is 00, and the unparsed source sequence is now 0 2.... The dictionary is expanded to $\{0,00,\mathbf{000},1,2\}$. The third parsed string is 0, and the rest of the source output 2.... The new dictionary is $\{0,00,\mathbf{02},000,1,2\}$, and the fourth parsed phrase is 2.

For the 1978 Lempel–Ziv algorithm, it is clear that the decoder can use the sequence of code symbols to simulate the evolution of the parsers' dictionary and subsequently reconstruct the source output. This is less obvious for the LZW decoder. For any string σ and letter j, define the string $\sigma \circ j$ as the string formed by appending j to the string σ. The LZW decoder can easily determine the first source-output symbol u_1. The new dictionary entry is of the form $u_1 \circ j$ for some source symbol j. To find j, the decoder looks at the code letters corresponding to the second phrase. If these code letters indicate that the second parsed phrase is u_1 or $u_1 \circ j$, then j and u_1 are the same symbol. Otherwise, the second parsed phrase is some u_2 which is distinct from u_1, and therefore, j is the same as u_2.This argument can be extended to show that it is possible to accurately decode any source string from its corresponding string of code letters.

3.3 QUANTIZATION

In Section 3.2, the source outcomes were assumed to belong to some finite alphabet $\mathcal{A}$. This is however not the case for natural signals, which have to be analog-to-digital converted (sampling in time) and quantized (sampling in amplitude) so that they can be processed digitally. Moreover, once this is performed, if the source encoder has a first prediction step, or transform/subband processing step, the resulting signal can take an almost continuous set of values. Therefore, some kind of quantization has to take place to tune the quality/bitrate tradeoff. This also allows some entropy coding to take place after quantization.

The aim of this section is to present the principle and performance of *scalar quantizers*, where each source sample is quantized independently from the other samples and to provide a very short introduction to multidimensional *vector quantization*, where several samples are quantized simultaneously. Scalar quantizers are most widely used in current coding standards. For more details, see Berger (1971), Gray and Neuhoff (1998), and Sayood (2005).

3.3.1 Introduction

Consider a memoryless source X with values in a set $\mathcal{X}$, e.g., the set of reals $\mathbb{R}$, and described by the probability density function (pdf) $p(x)$. A J-level (scalar) quantizer Q performs a partition of $\mathcal{X}$ into J quantization cells $\mathcal{B}_j$, $j=1,\ldots,J$ and associates to any $x \in \mathcal{X}$ the *quantization index* j if $x \in \mathcal{B}_j$, i.e.,

$$Q(x) = j \Longleftrightarrow x \in \mathcal{B}_j. \tag{3.27}$$

When $\mathcal{X} = \mathbb{R}$, the partition usually consists in nonoverlapping quantization intervals $\left]b_{j-1}, b_j\right]$, with $b_0 = -\infty$ and $b_J = +\infty$. In this case,

$$Q(x) = j \Longleftrightarrow b_{j-1} < x \leqslant b_j.$$

The b_js are the decision boundaries of the quantizer. The inverse quantizer Q^{-1} aims at providing an estimate (reconstruction level) $a_j \in \mathcal{X}$ for the source outcome x from the knowledge of an index j and possibly of $p(x)$

$$Q^{-1}(j) = a_j. \tag{3.28}$$

Note that Q^{-1} is only a formal notation and does not correspond to the reciprocal function of Q since quantization is noninvertible.

The quantization function q is the composition of the quantizer and the inverse quantizer

$$q(x) = Q^{-1}(Q(x)), \tag{3.29}$$

To evaluate the error introduced by a quantizer, one may use a distortion measure, which is a function of the source sample and of its image by the quantization function. In what follows, quadratic distortion measures are considered, i.e., for any $x \in \mathcal{X}$,

$$d(x, q(x)) = (x - q(x))^2, \tag{3.30}$$

where $x - q(x)$ is the *quantization error.*

Example 3.27: Figure 3.9 shows the quantization function for a six-level quantizer. Except for the first and last intervals of the partition of the real line, the reconstruction values have been taken at the middle of the decision intervals. Figure 3.10 displays the quantization error. This error may grow unbounded as the absolute value of x increases.

The *distortion D* is the expected value of the distortion measure

$$D = E[d(X - q(X))]. \tag{3.31}$$

It is the first figure of merit to evaluate the performance of a quantization scheme. The second is the minimum amount of bits required to represent a quantization index, i.e., the entropy of the quantized source $Q(X)$ or $q(X)$.

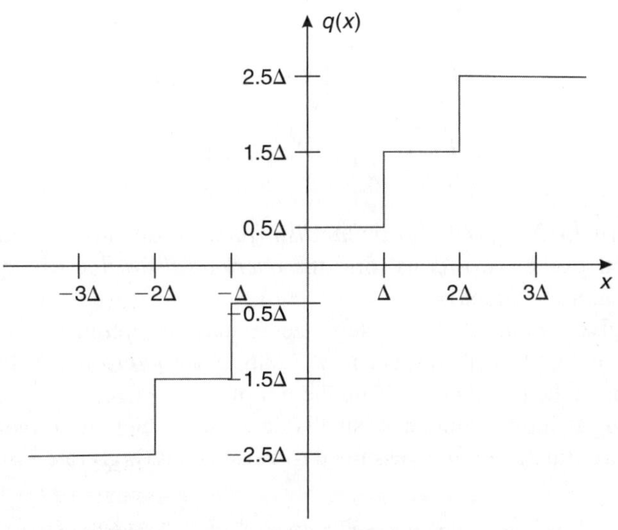

FIGURE 3.9 Quantization function with six possible quantization levels.

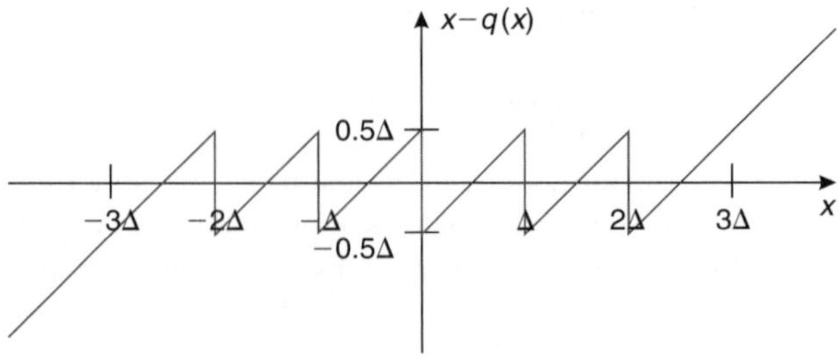

FIGURE 3.10 Quantization error for the six-level quantizer of Figure 3.9.

3.3.2 Uniform Quantizer

Uniform quantizers are defined by quantization intervals, which have all the same step size Δ, except for the first, last, and maybe the central quantization interval containing zero, which may be larger (leading to quantizers with a *dead zone*). The location of the reconstruction levels may be optimized, as explained in Section 3.3.3, but usually they are taken at the middle of the quantization interval, to keep the description of the uniform quantizer as simple as possible, see Figure 3.9.

Assuming that $\mathcal{X}=\mathbb{R}$, the distortion (3.31) introduced by a J-level uniform quantizer, with J even, centered around the origin and without dead zone, is expressed as

$$\begin{aligned} D_{\mathrm{u}}(J,\Delta)=&\sum_{j=2}^{J-1}\int_{(j-J/2-1)\Delta}^{(j-J/2)\Delta}\left(x-\left(j-\frac{J+1}{2}\right)\Delta\right)^2 p(x)\,\mathrm{d}x\\ &+\int_{-\infty}^{(-J/2+1)\Delta}\left(x+\frac{J-1}{2}\Delta\right)^2 p(x)\,\mathrm{d}x\\ &+\int_{(J/2-1)\Delta}^{+\infty}\left(x-\frac{J-1}{2}\Delta\right)^2 p(x)\,\mathrm{d}x. \end{aligned} \tag{3.32}$$

The first term in (3.32) is the *granular distortion* (the quantization error remains bounded), the other two terms form the *overload distortion* (the quantization error is no more bounded).

For a given value of J, the step size Δ may be optimized by taking the derivative of (3.32) with respect to Δ, setting it to zero and solving for Δ. Unfortunately, the result depends on the pdf of x.

Allowing an infinite number of small quantization intervals, on each of which $p(x)$ is approximated by $\widetilde{p}(x)$ assumed to be constant (high-rate assumption),

$$\forall x\in[(j-1)\Delta,j\Delta[,\tilde{p}(x)=\frac{1}{\Delta}\int_{(j-1)\Delta}^{j\Delta}p(e)\,\mathrm{d}e, \tag{3.33}$$

one can perform a high-rate analysis of the distortion of the uniform scalar quantizer. Under these assumptions, and further assuming that the resulting indexes are entropy-coded, one may show that

$$D_{\mathrm{u}}(H)=\frac{1}{12}2^{-2(H-h(X))}, \tag{3.34}$$

where

$$h(X)=-\int_{-\infty}^{\infty} p(x)\log_2 p(x)\,\mathrm{d}x \tag{3.35}$$

is the *differential entropy* of the source X. The fact that quantizer outputs are assumed to be entropy-coded is underlined in (3.34) by denoting the rate by H instead of R.

Example 3.28: Assume that X is zero-mean Gaussian, with variance σ^2, then

$$h(X)=\frac{1}{2}\log_2 2\pi e\sigma^2$$

and (3.34) becomes

$$D_{\mathrm{u}}(H)=\frac{\pi e}{6}\sigma^2\cdot 2^{-2H}\approx 1.42\sigma^2 2^{-2H}. \tag{3.36}$$

The performance is very close (about 1.5 dB) to the rate-distortion function

$$D_{\mathrm{g}}^{*}(H)=\sigma^2 2^{-2H}, \tag{3.37}$$

in the Gaussian case, see Cover and Thomas (1991).

Scalar quantizers, provided that their output is entropy-coded, thus provide good high-rate performance.

3.3.3 Distortion-Minimizing Quantizers

Assume again that $\mathcal{X}=\mathbb{R}$. The aim now is to find a quantizer with variable step size that minimizes the distortion. One may express (3.31) as a function of the decision boundaries and reconstruction levels as

$$D_{\mathrm{nu}}(J,a_1,\ldots,a_J,b_0,\ldots,b_J)=\sum_{j=1}^{J}\int_{b_{j-1}}^{b_j}\left(x-a_j\right)^2 p(x)\,\mathrm{d}x. \tag{3.38}$$

The distortion is minimized when both derivatives with respect to both unknowns are zero. Since this is difficult to obtain in a single step, let us characterize separately both conditions.

For a given set of boundaries, one has to take the derivative of (3.38) with respect to a_j, set it equal to zero, and solve for a_j to get

$$a_j = \frac{\int_{b_{j-1}}^{b_j} xp(x)\,\mathrm{d}x}{\int_{b_{j-1}}^{b_j} p(x)\,\mathrm{d}x}, \quad j = 1 \dots J. \tag{3.39}$$

This is the *centroid update* equation.

Similarly for a given set of reconstruction levels (centroids), one has to take the derivative of (3.38) with respect to b_j, thus obtaining

$$b_j = \frac{a_{j+1} + a_j}{2}, \quad j = 1 \dots J-1. \tag{3.40}$$

Except for $b_0 = -\infty$ and $b_J = +\infty$, the decision boundaries are at the middle of the reconstruction levels. This is the so-called nearest-neighbor condition. Any value of the signal should be reconstructed at the closest centroid.

One thus gets a set of necessary conditions (3.39) and (3.40), which have to be satisfied by a quantization function minimizing the distortion. The Lloyd–Max algorithm (Lloyd, 1957, 1982; Max, 1960) is able to obtain a locally optimal quantization function. It iteratively updates the reconstruction levels using (3.39) and the decision boundaries using (3.40), starting, e.g., from a uniform quantizer.

Again, high-rate assumptions simplify the performance analysis of the quantization scheme. If, again, one assumes the indexes resulting from the optimization are entropy encoded, one may show[1] that the asymptotic rate-distortion function is

$$D_{\mathrm{nu}}(H) = \frac{1}{12}\left(\int_{-\infty}^{+\infty} (p(x))^{1/3}\,\mathrm{d}x\right) 2^{-2\left(H - \frac{2}{3}h(X)\right)}. \tag{3.41}$$

Example 3.29: Consider again the case of a zero-mean Gaussian source with variance σ^2. Then

$$\int_{-\infty}^{+\infty} (p(x))^{1/3}\,\mathrm{d}x = \left(2\pi\sigma^2\right)^{1/3}\sqrt{3}$$

and (3.41) becomes

$$D_{\mathrm{nu}}(H) = \frac{\pi e^{2/3}\sqrt{3}}{6}\sigma^2 2^{-2H}. \tag{3.42}$$

One gets a behavior similar to (3.36). Moreover, since $\pi e^{2/3}\sqrt{3}/6 > \pi e/6$, one may deduce that the uniform quantizer provides better performance than the nonuniform quantizer.

As illustrated by Examples 3.28 and 3.29 on a Gaussian source, entropy-coded nonuniform scalar quantizers do not perform as good as entropy-coded uniform quantizers. Moreover, the former are more complex to describe. This is why

1. More materials (Appendices X, Y and Z) can be found at http://www.elsevierdirect.com/companions/9780123744494.

most of the quantization schemes used in source-coding standards are uniform quantizers.

This situation may seem amazing since in this case, performing a dedicated distortion minimization is less efficient than a plain uniform quantizer. This is due to the fact that this distortion minimization was performed without taking the rate into account, and that a coding problem is always a tradeoff between these two conflicting goals: better quality and smaller rate.

3.3.4 Entropy-Constrained Quantizers

The question now would naturally be: what is the best quantizer when the rate is measured through entropy? The answer to such a question is obtained by minimizing the distortion under a *rate constraint*. One has to search for a quantizer q_{nurc} such that (3.38) is minimized under the constraint

$$H(q_{\mathrm{nurc}}(X)) \leqslant H_0 \tag{3.43}$$

for some rate H_0. The solution to this problem involves a set of $J-1$ nonlinear equations

$$\ln \frac{p_{j+1}}{p_j} = \lambda \left(a_{j+1} - a_j\right)\left(a_{j+1} + a_j - 2b_j\right), \quad j = 1, \ldots, M-1, \tag{3.44}$$

where λ is adjusted to obtain the desired rate, and where

$$p_j = \int_{b_{j-1}}^{b_j} p(x)\,\mathrm{d}x, \tag{3.45}$$

see Berger (1972) and Noll and Zelinski (1978). The best reconstruction levels are still obtained from (3.39). Even if the analysis is quite complex, it may be shown that again, the uniform quantizer is the optimal one, at high and even at low rates (Farvardin and Modestino, 1984).

3.3.5 Vector Quantizers

In vector quantization, several source samples are quantized simultaneously and a single index is associated to a vector of source samples. Vector quantization allows to account for the correlation between source samples directly at the quantizer, which improves its efficiency. The set of available rates is also much larger than with scalar quantizers, for example, four samples may be quantized with a single bit, leading to a rate of 0.25 bit/sample).

Many notions introduced for scalar quantization extend in a straightforward way to vector quantization. Consider a source $\mathbf{X}$ generating vectors $\mathbf{x} \in \mathbb{R}^k$ with a pdf $p(\mathbf{x})$. For a quantizer q with quantization cells $\mathcal{B}_1, \mathcal{B}_2 \ldots$ and reconstruction points $\mathbf{a}_1, \mathbf{a}_2 \ldots$, one may introduce the quadratic distortion measure

$$\mathrm{d}(\mathbf{x}, q(\mathbf{x})) = \frac{1}{k} (\mathbf{x} - q(\mathbf{x}))^{\mathrm{T}} (\mathbf{x} - q(\mathbf{x})) \tag{3.46}$$

$$= \frac{1}{k} |\mathbf{x} - q(\mathbf{x})|^2. \tag{3.47}$$

With this average per component distortion measure, the distortion becomes

$$D = \sum_{j=1}^{J} \int_{\mathcal{B}_j} \mathrm{d}(\mathbf{x}, q(\mathbf{x})) p(\mathbf{x}) \mathrm{d}\mathbf{x} \tag{3.48}$$

$$= \frac{1}{k} \sum_{j=1}^{J} \int_{\mathcal{B}_j} \left|\mathbf{x} - \mathbf{a}_j\right|^2 p(\mathbf{x}) \mathrm{d}\mathbf{x}. \tag{3.49}$$

The optimal (distortion-minimizing) vector quantizer has to satisfy conditions that are very close to (3.39) and (3.40) in the scalar case. Assuming that the quantization cells have been adjusted, the reconstruction points have to satisfy the centroid condition

$$\mathbf{a}_j = \frac{\int_{\mathcal{B}_j} \mathbf{x} p(\mathbf{x})\, \mathrm{d}\mathbf{x}}{\int_{\mathcal{B}_j} p(\mathbf{x})\, \mathrm{d}\mathbf{x}}. \tag{3.50}$$

Assuming that the reconstruction points have been found, one may get optimal quantization cells defined as follows:

$$q(\mathbf{x}) = \mathbf{a}_j \text{ or } \mathbf{x} \in \mathcal{B}_j,$$

implicitly defined by the nearest neighbor condition

$$d(\mathbf{x}, \mathbf{a}_j) \leqslant d(\mathbf{x}, \mathbf{a}_\ell), \text{ for all } \ell \neq j, 1 \leqslant \ell \leqslant L. \tag{3.51}$$

An extension of the Lloyd–Max algorithm may be considered with (3.50) and (3.51). As for the scalar case, only locally optimal vector quantizers may be achieved.

The Lindé–Buzo–Gray (LBG) algorithm (Lindé et al., 1980) allows to get a locally optimal quantizer when only source outcomes of the source to quantify are available. Typical quantization cells and reconstruction level obtained with the LBG algorithm are represented in Figure 3.11.

3.3.6 Example: Quantization in H.264/AVC

H.264/AVC (ITU-T and ISO/IEC JTC 1, 2003) involves only uniform scalar quantizers. The size of the quantization steps is adjusted by the parameter Q_{P}, for *quantization parameter*, which may take 52 different values. The size of the quantization step doubles when Q_{P} is increased by 6. Basically, if x is the input coefficient, the quantized coefficient is

$$Q(x) = \mathrm{round}\left(\frac{x}{Q_{\mathrm{step}}}\right), \tag{3.52}$$

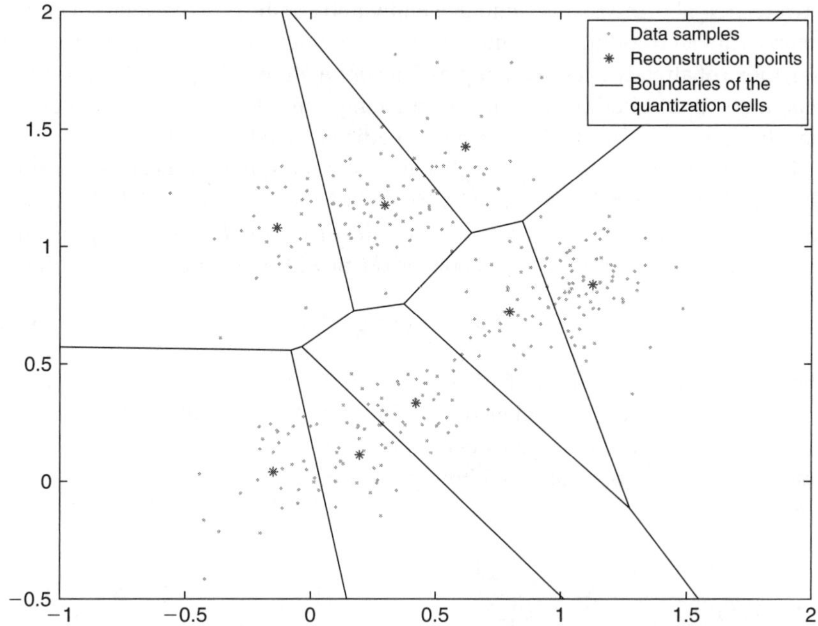

FIGURE 3.11 Vector quantization (2D), decision regions, and reconstruction points.

where Q_{step} is expressed as

$$Q_{\text{step}} = 2^{\frac{Q_P - 4}{6}}, \tag{3.53}$$

see Ma et al. (2005). In fact, the quantization process involves scaling to adjust the quantization step of the coefficient to quantize.

3.3.7 Summary

Under the high-resolution assumption, uniform quantization followed by entropy coding is the most efficient in terms of rate-distortion tradeoff compared to distortion-optimized quantization, even followed by entropy coding.

Moreover, under the same assumption, whatever the type of quantizer, the shape of the rate-distortion function for a source X is proportional to the variance of X. Therefore, any method allowing to decrease the variance of the signal to be quantized, without increasing the distortion, will improve the rate-distortion tradeoff. A first example is given below.

3.4 DIFFERENTIAL CODING

Differential coding plays an important role in audio, still image, and video coding. It is a way to exploit the temporal correlation between samples (in audio coding), between neighboring pixels (in image and video coding), or between successive pictures (in video coding). The main idea is to predict the value taken by a sample from those of already encoded samples. The *innovation* is the *prediction residual*, sometimes also called *texture* in image coding, and it contains all elements or contributions of elements that were not predicted, see Figure 3.12b.

Usually, when the source samples are correlated, the variance of the innovation may be much less than the variance of the original signal. As seen in Section 3.3, for a source X of variance σ_X^2, the distortion D may be expressed as a function of the rate R used to encode the quantized source outcomes as follows

$$D(R) = \varepsilon_X \sigma_X^2 2^{-2R}, \tag{3.54}$$

where ε_X depends on the pdf of the source. This result has been shown for X Gaussian but extends to other pdfs. Differential coding allows to reduce the variance of the samples to be encoded. Thus, it may allow the distortion to reduce compared to a coding scheme without prediction.

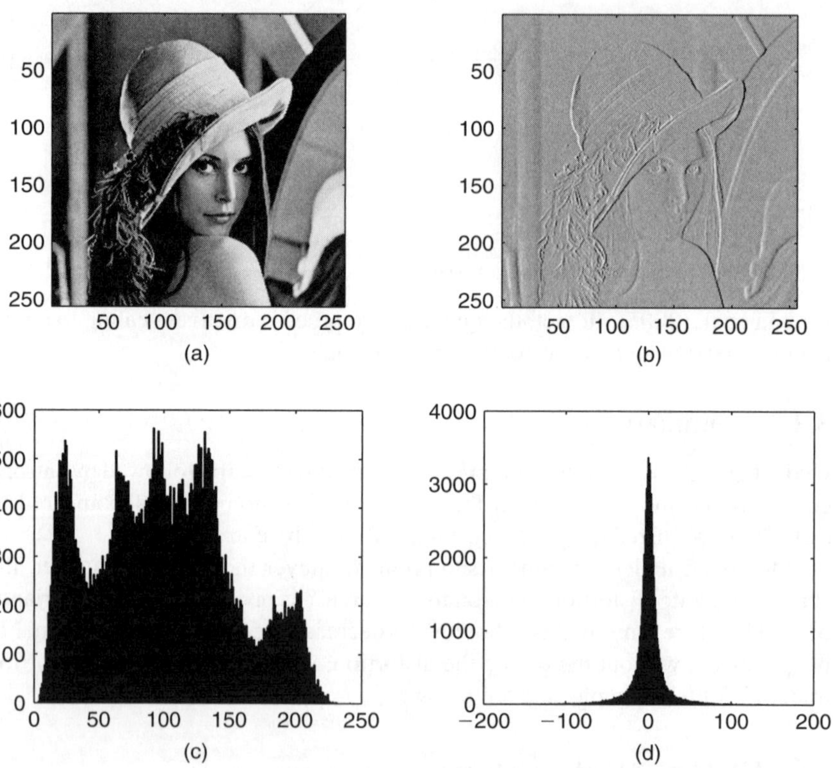

FIGURE 3.12 Original picture (a), prediction residual when column i is predicted from column $i-1$ (b), histogram of the original picture, $\sigma_X^2 = 2800$ (c), and histogram of the prediction residual, $\sigma_E = 466$ (d).

In what follows, the signal to be encoded is assumed to be a realization of a sequence of random variables X_κ, assumed to be identically distributed, stationary, with autocorrelation function

$$\gamma_X(\ell) = E\left[X_\kappa X_{\kappa+\ell}\right].$$

The structure of the differential coder has to account for the fact that the global scheme is lossless or lossy.

3.4.1 Simple Differential Coding Scheme

The basic components of a differential coding scheme are described in Figure 3.13. The memory stores previously considered source samples, which are used by the predictor to build an estimate of X_κ. To allow decoding, the predictor has to be causal, i.e., prediction of the κth sample has to involve only previously processed samples, which have to be also available at decoder side when performing the reconstruction of the κth sample.

Assuming that the prediction function f_κ involves p past samples, the prediction residual becomes

$$E_\kappa = X_\kappa - f_\kappa\left(X_{\kappa-1}, \ldots, X_{\kappa-p}\right), \tag{3.55}$$

Considering only lossless operations after evaluating E_κ, the latter is available at receiver side. At decoder side, assuming that $X_{\kappa-1}, \ldots, X_{\kappa-p}$ have been previously decoded, one may evaluate

$$\widetilde{X}_\kappa = E_\kappa + f_\kappa\left(X_{\kappa-1}, \ldots, X_{\kappa-p}\right) = X_\kappa, \tag{3.56}$$

see Figure 3.14. It is thus possible to reconstruct perfectly X_κ at decoder side.

Assume now that E_κ is quantized with a uniform scalar quantizer Q of step size Δ. Only a noisy version $\widetilde{E}_\kappa$ of E_κ is then available at decoder side. There is

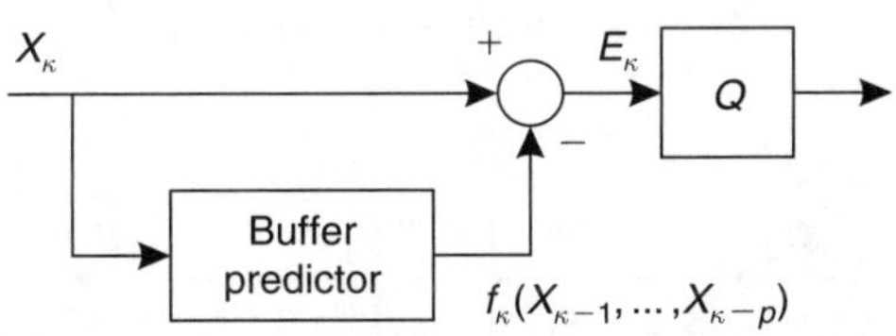

FIGURE 3.13 Differential coder involving a forward prediction for X_κ.

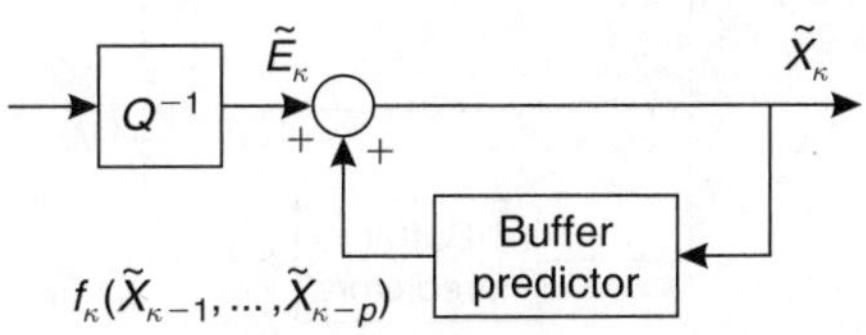

FIGURE 3.14 Differential decoder.

no way to recover E_κ, and thus, the availability at decoder side of the samples used to perform prediction at encoder is no more satisfied. Nevertheless, one may assume that noisy versions $\widetilde{X}_{\kappa-1},\dots,\widetilde{X}_{\kappa-p}$ of $X_{\kappa-1},\dots,X_{\kappa-p}$ are available at decoder side. The reconstruction of the κth sample may be done as follows:

$$\begin{aligned}\widetilde{X}_\kappa &= \widetilde{E}_\kappa + f_\kappa(\widetilde{X}_{\kappa-1},\dots,\widetilde{X}_{\kappa-p})\\ &= Q^{-1}(Q(X_\kappa - f_\kappa(X_{\kappa-1},\dots,X_{\kappa-p}))) + f_\kappa(\widetilde{X}_{\kappa-1},\dots,\widetilde{X}_{\kappa-p})\\ &= X_\kappa + f_\kappa(\widetilde{X}_{\kappa-1},\dots,\widetilde{X}_{\kappa-p}) - f_\kappa(X_{\kappa-1},\dots,X_{\kappa-p}) + Z_\kappa,\end{aligned} \tag{3.57}$$

where Z_κ is some random variable with support $[-\Delta/2, \Delta/2]$ representing the quantization error (only granular distortion is considered).

From (3.57), one may deduce that the reconstruction error for X_κ depends on the quantization error Z_κ of E_κ and on the reconstruction error of previous samples. Reconstruction errors may grow unbounded or may compensate, depending on the structure of the prediction function. This drift effect, i.e., the fact that encoder and decoder do not perform prediction with the same information, is the main reason for not using the scheme of Figure 3.13 when some lossy coding scheme follows the prediction loop.

3.4.2 Differential Coding Scheme with Decoder at the Encoder

To fight against the drift effect, the coder has to perform prediction with samples that are available at decoder side. Supplementing the scheme of Figure 3.13 with the decoder of Figure 3.14 allows to satisfy this constraint, see Figure 3.15. The prediction residual (3.55) then becomes

$$E_\kappa = X_\kappa - f_\kappa(\widetilde{X}_{\kappa-1},\dots,\widetilde{X}_{\kappa-p}), \tag{3.58}$$

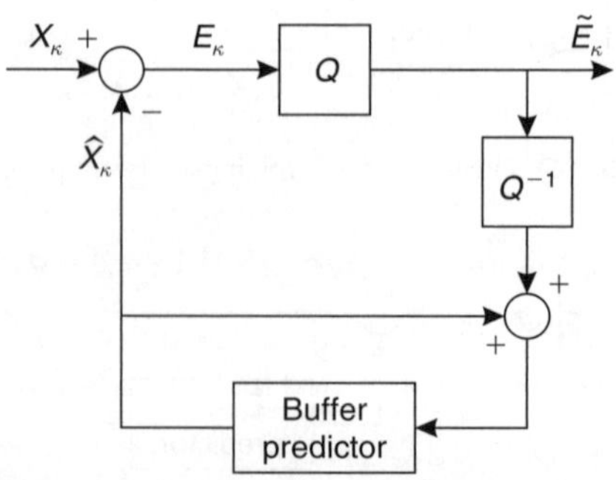

FIGURE 3.15 Differential coder with local decoder.

and at decoder side,

$$\begin{aligned}\widetilde{X}_\kappa &= \widetilde{E}_\kappa + f_\kappa(\widetilde{X}_{\kappa-1},\ldots,\widetilde{X}_{\kappa-p})\\ &= Q^{-1}(Q(X_\kappa - f_\kappa(\widetilde{X}_{\kappa-1},\ldots,\widetilde{X}_{\kappa-p}))) + f_\kappa(\widetilde{X}_{\kappa-1},\ldots,\widetilde{X}_{\kappa-p})\\ &= X_\kappa + f_\kappa(\widetilde{X}_{\kappa-1},\ldots,\widetilde{X}_{\kappa-p}) - f_\kappa(\widetilde{X}_{\kappa-1},\ldots,\widetilde{X}_{\kappa-p}) + Z'_\kappa\\ &= X_\kappa + Z'_\kappa, \end{aligned} \tag{3.59}$$

where Z'_κ represents the quantization error bounded in $[-\Delta/2, \Delta/2]$ (again only granular distortion is considered).

From (3.59), one observes that the reconstruction error of the κth sample depends now only on the quantization error of the κth prediction residual. This is a very interesting property; the quantization errors do not propagate with time, contrary to the scheme of Section 3.4.1.

3.4.3 Prediction Function

The efficiency of a differential-coding scheme depends on the choice of the prediction function. The simplest ones involve only the previously encoded sample

$$f_\kappa(\widetilde{X}_{\kappa-1},\ldots,\widetilde{X}_{\kappa-p}) = \widetilde{X}_{\kappa-1}, \tag{3.60}$$

one of the previously encoded samples (not necessarily the last one)

$$f_\kappa(\widetilde{X}_{\kappa-1},\ldots,\widetilde{X}_{\kappa-p}) = \widetilde{X}_{\kappa'}, \tag{3.61}$$

with $\kappa' \in \{\kappa-p,\ldots\kappa-1\}$, or a linear combination of samples

$$f_\kappa(\widetilde{X}_{\kappa-1},\ldots,\widetilde{X}_{\kappa-p}) = a_1^\kappa \widetilde{X}_{\kappa-1} + \cdots + a_p^\kappa \widetilde{X}_{\kappa-p}. \tag{3.62}$$

The prediction functions (3.60) and (3.61) are special cases of (3.62). The parameters of (3.61) and (3.62) have to be optimized to minimize the variance of the prediction residual. The parameters of the prediction function have then to be transmitted to the decoder, except if the optimization is done using only quantities that are also available to the decoder. There is thus some tradeoff to be found between prediction accuracy, which is better when the prediction function becomes more complex, computational complexity, and amount of side information to be transmitted so that encoder and decoder behave similarly. See Sayood (2005) for more details.

3.4.4 Example: Motion-Compensated Prediction

In the case of H.264/AVC in INTER mode (ITU-T and ISO/IEC JTC 1, 2003), the macroblocks of a picture are predicted from parts of already encoded pictures and involve a motion estimation and compensation step. Contrary to techniques working at a pixel level, such as *optical flow* estimators, in a study by ITU-T and ISO/IEC JTC 1 (2003), the prediction is done at a block level. A tradeoff between efficiency and complexity is obtained by adaptively dividing each macroblock into blocks of 16×16, 16×8, 8×16, or 8×8 pixels. Blocks of 8×8 pixels may be additionally divided into subblocks of 8×4, 4×8, or 4×4 pixels. The possible partitions of macroblocks and blocks are illustrated in Figure 3.16.

In the first generation of standards like H.263 or MPEG-4, only blocks of 8×8 and 16×16 pixels were considered during inter coding. For each block (or subblock), the area that is most similar to this block in previously encoded reference pictures is searched for. A motion vector indicating the translation required between the found reference area and the current block has to be encoded. The index of the picture used as reference for the considered block has also to be transmitted. In fact, for the same picture, *multiple references* are allowed, the picture used as reference may depend on the considered block in the picture to be encoded. The prediction function is thus described by (3.61). The prediction for a block may also linearly combine several areas in previously encoded pictures, in which case, the prediction function is closer to (3.62). Figure 3.17 illustrates these concepts that also extends to B pictures, i.e., picture for which the

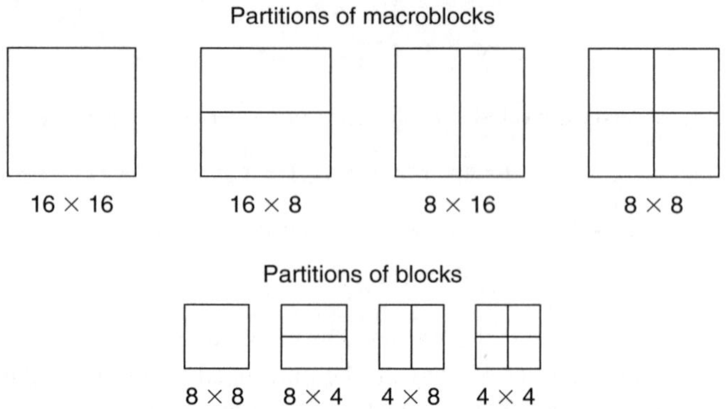

FIGURE 3.16 Partitions of a macroblock or block into subblocks used for motion-compensated prediction.

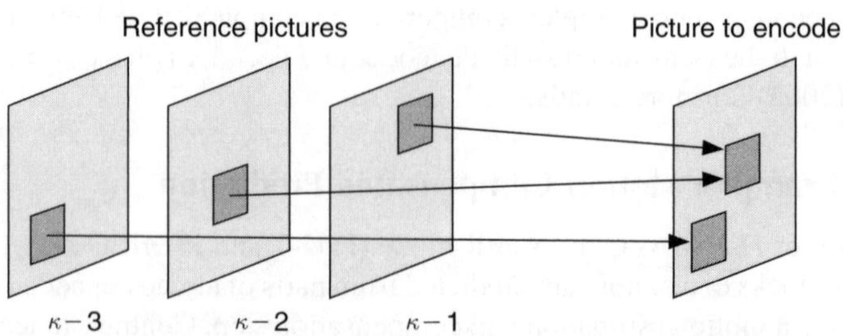

FIGURE 3.17 Motion-compensated prediction with multiple reference pictures; the motion vectors and the index of the references have to be transmitted for each block.

motion compensation may involve future pictures, but the coding has to remain causal.

The accuracy of motion vectors may be up the quarter a pixel by interpolating the reference picture. To reduce the amount of information to be stored for the motion information, the spatial correlation between motion vectors is exploited by differential encoding.

3.4.5 Example: Intra Prediction

Intra prediction allows a differential coding of intracoded pictures. The pixels of a macroblock are predicted from pixels belonging to macroblocks surrounding the current macroblock and already encoded. H.264/AVC provides two intracoding modes, namely INTRA4 $\times$ 4 and INTRA16 $\times$ 16.

With the INTRA4 $\times$ 4 mode, a macroblock is divided into 16 subblocks of 4 $\times$ 4 pixels, and prediction is done for each of these subblocks independently. The prediction functions are described by (3.62). Nevertheless, to simplify the optimization, only nine prediction modes are available, which determine nine sets of parameters for the prediction function. In the DC mode, all pixels of a subblock are predicted from the average of the values of the pixels, which are above and on the left of the current subblock, see Figure 3.18. Two other directional prediction modes among the eight available modes are also provided in Figure 3.18.

With the INTRA16 $\times$ 16 mode, a single-prediction function is applied for the whole macroblock and only four sets of parameters are available, corresponding to vertical, horizontal, DC, and plane, see ITU-T and ISO/IEC JTC 1 (2003) for more details.

Macroblocks of the chrominance planes U and V are predicted in the same way as the INTRA16 $\times$ 16 mode for the luminance. Since chrominance varies quite slowly in most cases, the four prediction modes of INTRA16 $\times$ 16 are efficient enough.

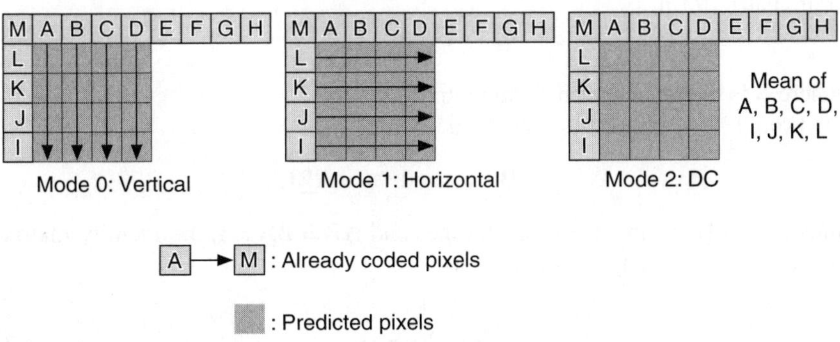

FIGURE 3.18 Three of the nine prediction modes for INTRA4 $\times$ 4 prediction.

3.5 TRANSFORM CODING

Transform coding is another way to exploit the temporal correlation between samples, or the spatial correlation between pixels of an image. Taking a vector of 256 sound samples or a matrix of 16×16 pixels of an image, there is no sample or pixel that is significantly more important than one of its neighbors. The energy of the samples or of the pixels is spread across the 256 vectors or matrices of the canonical basis used to represent the vector of sound samples or the matrix of pixels. The main idea of transform coding is to express vectors or matrices in basis, which are better suited for compression, i.e., in basis in which the most important features of a signal may be captured by a reduced number of basis vectors or matrices. One tries to find a basis in which the signal is as sparse as possible.

3.5.1 Introduction

Vector Transforms

Consider a vector $\mathbf{x} = (x_1, \ldots, x_k)^{\mathrm{T}} \in \mathbb{R}^k$ of samples extracted, for example, from some sound record. The vector $\mathbf{x}$ may be linearly transformed into a vector $\mathbf{t} = (t_1, \ldots, t_k)^{\mathrm{T}}$ as follows:

$$t_\kappa = \sum_{i=1}^{k} a_{\kappa,i} x_i, \tag{3.63}$$

or in matrix form

$$\mathbf{t} = \underline{\mathbf{A}}\mathbf{x}, \tag{3.64}$$

where $\underline{\mathbf{A}}$ is the forward transform matrix (it should be of full rank). The vector $\mathbf{x}$ may be recovered from $\mathbf{t}$ using the inverse transform

$$x_\kappa = \sum_{i=1}^{k} b_{\kappa,i} t_i, \tag{3.65}$$

or in matrix form

$$\mathbf{x} = \underline{\mathbf{B}}\mathbf{t}, \tag{3.66}$$

where $\underline{\mathbf{B}}$ is the inverse transform matrix.

From (3.64) and (3.66), one deduces that

$$\mathbf{x} = \underline{\mathbf{BA}}\mathbf{x} \quad \text{and} \quad \mathbf{t} = \underline{\mathbf{AB}}\mathbf{t},$$

thus $\underline{\mathbf{A}}$ and $\underline{\mathbf{B}}$ are inverse of each other and $\underline{\mathbf{AB}} = \underline{\mathbf{BA}} = \underline{\mathbf{I}}$, the identity matrix. Moreover, (3.66) may be rewritten as

$$\mathbf{x} = \sum_{i=1}^{k} \mathbf{b}_i t_i, \tag{3.67}$$

showing that $\mathbf{x}$ is a linear combination of the columns $\mathbf{b}_i$ of $\underline{\mathbf{B}}$, which forms a basis of $\mathbb{R}^k$ usually different from the canonical basis. The t_is are the coordinates of $\mathbf{x}$ in this basis.

Matrix Transforms, Separable Transforms

Consider now a matrix $\underline{\mathbf{X}} \in \mathbb{R}^{k \times k}$ representing the pixels of a block of size $k \times k$ of an image. A linear transform of this matrix may be written as

$$t_{\kappa,\nu} = \sum_{i=1}^{k} \sum_{j=1}^{k} a_{i,j,\kappa,\nu} x_{i,j}, \tag{3.68}$$

where $x_{i,j}$ and $t_{\kappa,\nu}$ are, respectively, the pixel at position (i,j) in $\underline{\mathbf{X}}$ and the transformed coefficient at position (κ, ν) in the transformed matrix $\underline{\mathbf{T}}$.

A matrix transform is separable if it may be evaluated by first transforming the rows of $\underline{\mathbf{X}}$ and then the columns of the resulting matrix. In the case of a separable transform, (3.68) may be written as

$$t_{\kappa,\nu} = \sum_{i=1}^{k} \sum_{j=1}^{k} a_{\kappa,i} a_{\nu,j} x_{i,j}, \tag{3.69}$$

or in matrix form

$$\underline{\mathbf{T}}(:) = \left(\underline{\mathbf{A}} \otimes \underline{\mathbf{A}}\right) \underline{\mathbf{X}}(:), \tag{3.70}$$

where $\otimes$ denotes the Kronecker product of two matrices, $\underline{\mathbf{A}}$ is a matrix of size $k \times k$, and $\underline{\mathbf{X}}(:)$ is the vector obtained by stacking the columns of $\underline{\mathbf{X}}$ (Jain, 1989). The inverse transform may be written as

$$\underline{\mathbf{X}}(:) = \left(\underline{\mathbf{B}} \otimes \underline{\mathbf{B}}\right) \underline{\mathbf{T}}(:), \tag{3.71}$$

with $\underline{\mathbf{B}} = \underline{\mathbf{A}}^{-1}$.

Separable transforms are very interesting in terms of computational complexity. Evaluating a transform using (3.68) involves $O\left(k^4\right)$ operations, whereas with (3.70), the complexity is only $O\left(k^3\right)$. Fast algorithms may still further reduce these quantities by very large factors.

Orthonormal Transforms

A transform is orthonormal when all rows of the forward transform matrix $\underline{\mathbf{A}}$ form an orthonormal set of vectors. Then $\underline{\mathbf{A}}^{\mathrm{T}}\underline{\mathbf{A}} = \underline{\mathbf{A}}\underline{\mathbf{A}}^{\mathrm{T}} = \underline{\mathbf{I}}$ and $\underline{\mathbf{B}} = \underline{\mathbf{A}}^{-1}$. A very interesting feature of orthonormal transforms is that they are energy-preserving.

Let $\underline{\mathbf{A}}$ be an orthonormal transform matrix, $\mathbf{x} \in \mathbb{R}^k$, and $\mathbf{t} = \underline{\mathbf{A}}\mathbf{x}$. Then the energy of the transformed vector

$$\begin{aligned} \mathbf{t}^{\mathrm{T}}\mathbf{t} &= \left(\underline{\mathbf{A}}^{\mathrm{T}}\mathbf{x}\right)^{\mathrm{T}} \left(\underline{\mathbf{A}}^{\mathrm{T}}\mathbf{x}\right) \\ &= \mathbf{x}^{\mathrm{T}}\underline{\mathbf{A}}\underline{\mathbf{A}}^{\mathrm{T}}\mathbf{x} = \mathbf{x}^{\mathrm{T}}\mathbf{x}, \end{aligned} \tag{3.71'}$$

is equal to the energy of the initial vector.

Usually, transform coefficients are quantized. This operation introduces some noise in the transform domain. When performing inverse transform, the noise will be of same energy in the initial domain.

Choosing a Good Transform

As already mentioned, the aim of transform coding is to find a transform corresponding to a basis in which the signal may be represented with as few components as possible. The *energy compaction* ability of a transform is usually measured by its *coding gain* (Jain, 1989)

$$G_{\mathrm{TC}} = \frac{\frac{1}{k}\sum_{\kappa=1}^{k}\sigma_\kappa^2}{\left(\prod_{\kappa=1}^{k}\sigma_\kappa^2\right)^{1/k}}, \tag{3.72}$$

where σ_κ^2 is the variance of the κth transform coefficient. One has $G_{\mathrm{TC}} = 1$ when all σ_κ^2s are equal (this is the case, for example, when the initial vector consists of realization of a white noise), else one has $G_{\mathrm{TC}} > 1$.

A second parameter to be optimized is the size k of the transform. Increasing k allows to get better coding gains, the price to be paid is a higher computational complexity. Moreover, when k is too large, the characteristics of the vector or of the matrix to be transformed may vary, which degrades the performance of the transform. A tradeoff has thus to be found between complexity and efficiency.

3.5.2 Karhunen–Loève Transform

The discrete *Karhunen–Loève transform* (KLT) (Hotelling, 1933; Karhunen, 1947; Loève, 1948) is the optimal transform in terms of energy compaction, as it maximizes the coding gain (3.72) among any transform-coding method (Jain, 1989).

Consider a zero-mean random vector $\mathbf{X}$ with correlation matrix

$$\underline{\mathbf{R}}_{\mathbf{x}} = E\left[\mathbf{X}\mathbf{X}^{\mathrm{T}}\right]. \tag{3.73}$$

The forward KLT matrix

$$\Phi^{\mathrm{T}} = [\varphi_1, \ldots, \varphi_k] \tag{3.74}$$

is formed by the k eigenvectors of $\underline{\mathbf{R}}_{\mathbf{x}}$ defined as

$$\underline{\mathbf{R}}_{\mathbf{x}}\varphi_\kappa = \lambda_\kappa\varphi_\kappa,\ \kappa = 1, \ldots, k. \tag{3.75}$$

These eigenvectors form an orthonormal basis, and the inverse KLT matrix is Φ.

Example 3.30: The Windows XP sound `tada.wav` is a stereo signal sampled at 22.05 kHz and consisting of 42752 pairs of data points. Figure 3.19 represents the mean of the left and right stereo channels and the

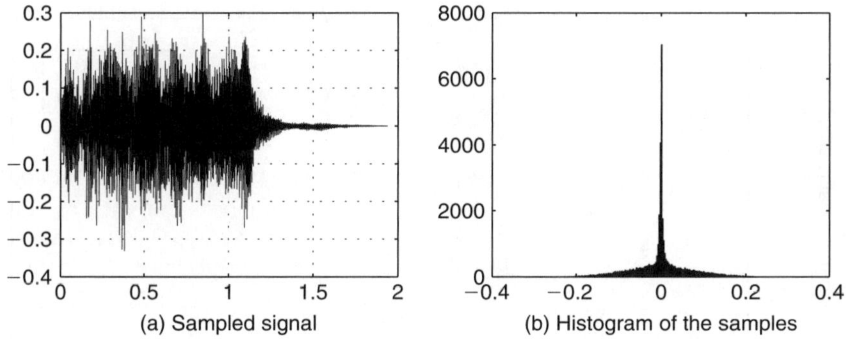

FIGURE 3.19 Tada.wav sound (mean between the left and right channels).

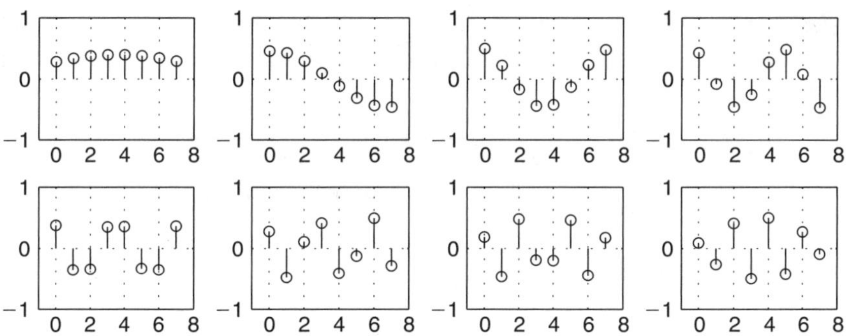

FIGURE 3.20 Vectors forming the basis of the KLT of length 8 for the sound `tada.wav`.

corresponding histogram. The mean signal is divided into vectors of eight samples, from which an estimate $\widehat{\underline{\mathbf{R}}}_{\mathbf{x}}$ of the correlation matrix of the random vector generating these samples is built. The corresponding eigenvectors, which form the KLT matrix, are represented in Figure 3.20. The estimated variance of each component of the vectors of transformed coefficients is represented in Figure 3.21, illustrating the energy compaction property of the KLT. The coding gain is $G_{\mathrm{TC}} = 1.99$.

The optimality in terms of energy compaction of the KLT is a useful property for coding applications. Nevertheless, its basis vectors depend on the signal to be coded, and thus, either the basis or the correlation matrix have to be transmitted to the decoder. The resulting overhead makes the KLT suboptimal in terms of signal-to-noise ratio for a given bitrate and even less efficient than signal-independent suboptimal transforms. Moreover, the KLT is not separable, which makes it less interesting for image coding applications.

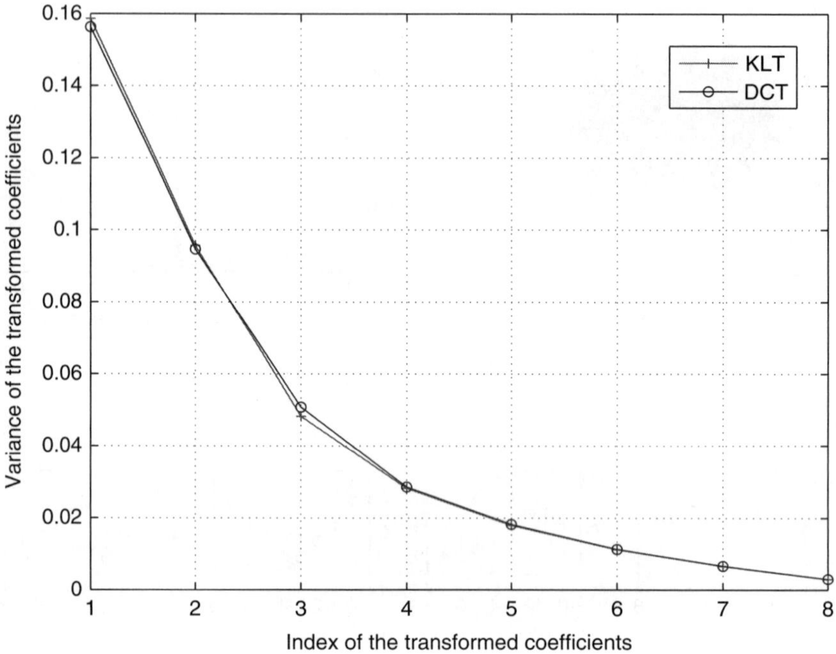

FIGURE 3.21 Partition of the energy among the transform coefficients for a KLT and a DCT of length 8 for the sound `tada.wav`.

3.5.3 Discrete Cosine Transform

The components of the DCT (Ahmed et al., 1974) matrix are defined as follows

$$a_{i,\kappa} = \alpha_i \cos \frac{\pi(2\kappa - 1)(i-1)}{2k}, \quad i, \kappa = 1, \ldots, k,$$

with

$$\alpha_1 = \sqrt{\frac{1}{k}} \quad \text{and} \quad \alpha_i = \sqrt{\frac{2}{k}} \text{ for } i > 1.$$

The DCT is closely related to the discrete Fourier transform (DFT) from which it may be deduced (Narasimha and Peterson, 1978). The basis vectors of the DCT for $k = 8$ are presented in Figure 3.22. Comparing Figures 3.20 and 3.22, one sees that except for the first vector, the other vectors of the DCT are very similar to those of the KLT of Example 3.30.

Example 3.31: Consider the same signal as in Example 3.30. Figure 3.21 shows that the performance in terms of energy compaction of the DCT is very close to that of the KLT.

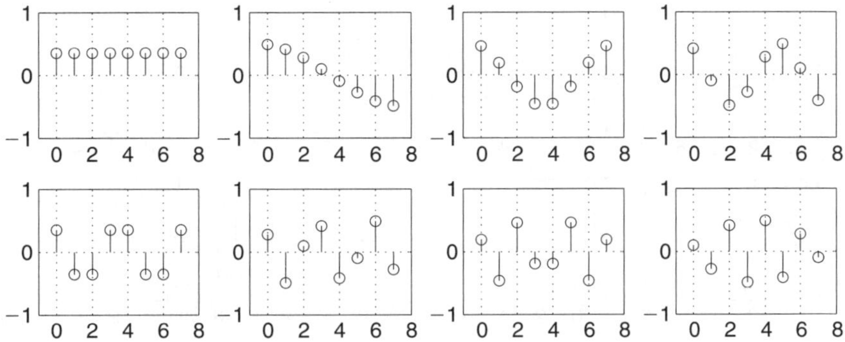

FIGURE 3.22 Vectors forming the basis of the DCT of length 8.

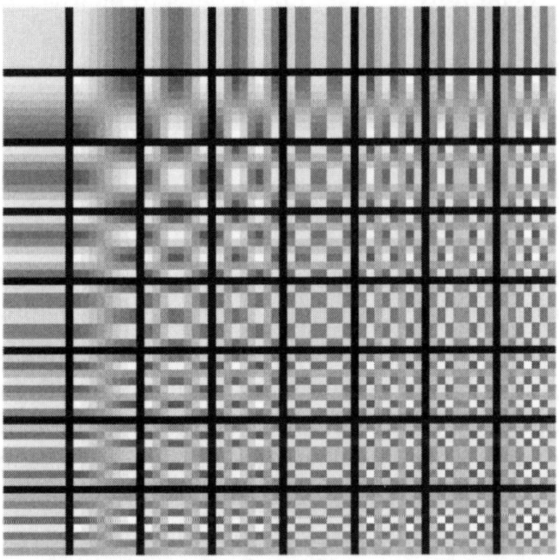

FIGURE 3.23 Two-dimensional DCT of size 8.

The DCT is a separable transform. Its basis matrices for the case of transforms of size 8×8 are represented in Figure 3.23. The top-left basis matrix captures the average (DC) value of the matrix. The first row corresponds to vertical contours, whereas the first column represents horizontal contours.

3.5.4 Discrete Walsh–Hadamard Transform

When computational complexity is the main concern, the DCT may be inappropriate. The Discrete Walsh–Hadamard transform (DWHT) may be implemented using only additions and subtractions. It decomposes an arbitrary input vector into a sum of Walsh functions (Walsh, 1923). The DWHT matrix $\underline{\mathbf{H}}_k$ of size

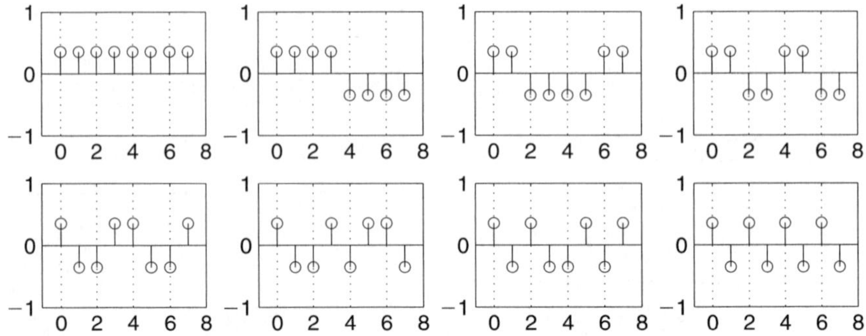

FIGURE 3.24 Vectors forming the basis of the DWHT of length 8.

$k \times k$, with $k = 2^m$ is built by rearranging Hadamard matrices $\underline{\mathbf{H}}'_k$, which may be defined recursively as follows:

$$\underline{\mathbf{H}}'_{2k} = \frac{1}{\sqrt{2}} \begin{pmatrix} \underline{\mathbf{H}}'_k & \underline{\mathbf{H}}'_k \\ \underline{\mathbf{H}}'_k & -\underline{\mathbf{H}}'_k \end{pmatrix} \tag{3.76}$$

with

$$\underline{\mathbf{H}}'_1 = 1. \tag{3.77}$$

The DWHT matrix $\underline{\mathbf{H}}_k$ is then obtained by reordering the rows in increasing *sequency* order, where the sequency of a vector is half the number of sign changes of that vector. The basis vectors of the DWHT $\underline{\mathbf{H}}_8$ are represented in Figure 3.24. Two other examples of DWHT matrices are provided in the next section.

The amount of energy compaction of this transform is significantly less than with the two previously considered transforms, see also Example 3.32.

Example 3.32: Table 3.4 illustrates the evolution of the coding gain G_{TC} as a function of the size k of the transform for the KLT, the DCT, and the DWHT applied on the signal of Example 3.30. On this example, for small values of k, the coding gains of the KLT and DCT are quite similar. The DWHT is significantly worse. When k is larger than 32, the DCT is less efficient than the KLT but does not require the transmission of any side information, contrary to the KLT.

3.5.5 Example: H.264 Transform Coding

Transform coding is involved in all video-coding standards to allow the prediction residuals to be efficiently encoded. Former video coders such as MPEG-1 or H.263 use a 2D DCT of size 8×8 on each block of the picture.

In H.264, integer transforms are involved (ITU-T and ISO/IEC JTC 1, 2003). The size of these transforms is mainly 4×4 pixels to fit the size of the smallest

TABLE 3.4 Coding Gain as a Function of the Size of the Transform

k	8	16	32	64
KLT	1.99	2.22	2.79	3.57
DCT	1.97	2.21	2.49	2.86
DWHT	1.38	1.39	1.40	1.43

subblocks, see Section 3.4.3. Three types of transforms are available. The first is applied to the prediction residuals of the luminance Y and of the chrominance U and V, whatever the prediction mode (inter or intra). The transform matrix is

$$\mathbf{A} = \begin{bmatrix} 1 & 1 & 1 & 1 \\ 2 & 1 & -1 & -2 \\ 1 & -1 & -1 & 1 \\ 1 & -2 & 2 & -1 \end{bmatrix}. \tag{3.78}$$

If the macroblock is encoded in INTRA16 $\times$ 16, a second transform is applied after the first one. The 16 DC coefficients obtained from the first transform of a luminance macroblock are transformed using the DWHT of size 4×4

$$\mathbf{H}_4 = \begin{bmatrix} 1 & 1 & 1 & 1 \\ 1 & 1 & -1 & -1 \\ 1 & -1 & -1 & 1 \\ 1 & -1 & 1 & -1 \end{bmatrix}. \tag{3.79}$$

The four DC coefficients obtained from the first transform of the chrominance macroblock are transformed using a DWHT of size 2×2

$$\mathbf{H}_2 = \begin{bmatrix} 1 & 1 \\ 1 & -1 \end{bmatrix}. \tag{3.80}$$

The way the DC coefficients are grouped before being transformed and the encoding order are represented in Figure 3.25. If a macroblock is predicted with the INTRA16 $\times$ 16 mode, the block indexed -1 is encoded first. This block contains all DC coefficients of the luminance. The blocks indexed from 0 to 25 are encoded next.

Compared to the DCT, the transform matrix (3.78) used by H.264/AVC consists only of integers ranging from -2 to 2. At the price of a slight suboptimality compared to the floating-point DCT, this allows to evaluate the transform and its inverse with only binary shifts, additions, and subtractions. In the case of the DWHT (3.79) and (3.80), only additions and subtractions are required. Another advantage of these transforms is their perfect invertibility. Using a floating-point

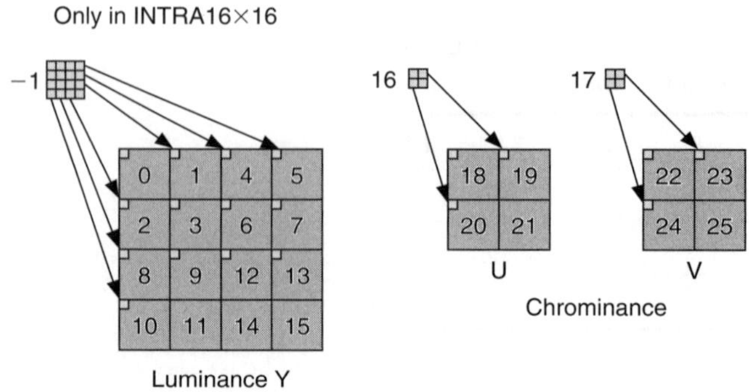

FIGURE 3.25 Transforms used in H.264/AVC and encoding order.

implementation of a DCT results always in the introduction of some (small) numerical noise due to the finite precision of floating-point numbers.

3.6 WAVELET-BASED CODING

This section, largely inspired from the review paper by Usevitch (2001), introduces some concepts of wavelet-based coding. Such techniques are mainly used in still-image (ISO/IEC JTC 1, 2000) and video coding. The main idea is again to perform an energy compaction of the signal to facilitate its coding. For more details, see Daubechies (1992); Vetterli and Kovačević (1995). This section is here for the sake of completeness, since wavelets are not currently used in video standards, but it is likely that they will be used in the near future. Moreover, the tools explained in this book also apply when such transforms are used.

3.6.1 Wavelet Transform

Figure 3.26 illustrates the basic structure of a wavelet decomposition scheme. The input signal is processed by through a low-pass filter h and a high-pass filter g. The two filtered signals are then subsampled of a factor 2. This operation is a one-level transform. A transformation with several levels (or *scales*) is obtained by iterating the one-level transform on the low-frequency signal. This procedure may be repeated L times to get the wavelet coefficients $d_\kappa^{\ell 1}$, $\ell \in \{1, \ldots, L\}$ and d_κ^{L0}. Since a wavelet transform may also be interpreted as some filter bank processing (Daubechies, 1992; Vetterli and Kovačević, 1995), the coefficients $\left\{d_\kappa^{\ell 1}\right\}$ and $\left\{d_\kappa^{L0}\right\}$ are often called subbands.

This 1D transform may be extended to the 2D case using separable wavelet filters (Jain, 1989). In this case, the 2D filtering is obtained by processing first the rows and second the columns of pictures. Figure 3.27a illustrates such 2D decomposition on $L=1$ level. The name of each subband is shown in Figure 3.27b. Most of the information is contained in the low-frequency subband. Figure 3.28a and b represent a decomposition on $L=3$ levels.

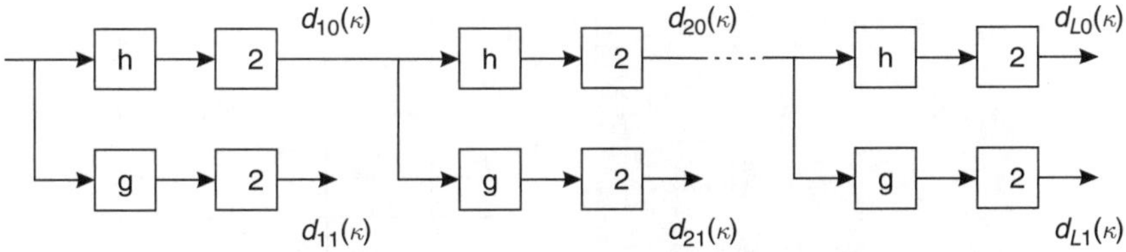

FIGURE 3.26 1D wavelet decomposition on L levels; the coefficient $d_\kappa^{\ell j}$ corresponds to the ℓth subband (with $j=0$ for the low frequencies and $j=1$ for the high frequencies).

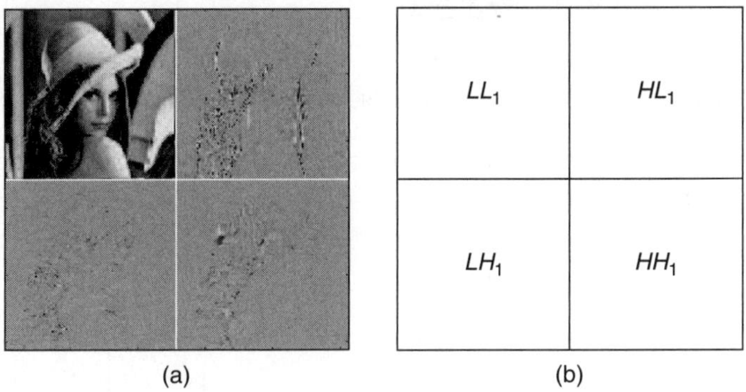

FIGURE 3.27 Two-dimensional separable wavelet decomposition on first level.

Wavelet decomposition schemes first made use of orthogonal filters, i.e., filters satisfying

$$\begin{cases} \sum_\kappa h_{\kappa-2i} h_{\kappa-2j} = \delta(i-j), \\ \sum_\kappa g_{\kappa-2i} g_{\kappa-2j} = \delta(i-j), \\ \sum_\kappa g_{\kappa-2i} h_{\kappa-2j} = 0. \end{cases} \tag{3.81}$$

The main feature of orthogonal filters is that they are energy-preserving. The distortion introduced during quantization of the subbands $\{d_\kappa^{\ell 1}\}$, $\ell \in \{1, \ldots, L\}$, and $\{d_\kappa^{L0}\}$ allows to predict very easily the distortion on the reconstructed picture.

When implementing directly the filtering operation, the amount of coefficients to be transmitted may increase. Consider, for example, a signal of k samples passing through a filter with impulse response of length n. One obtains $k+n-1$ coefficients before each decimator and thus $k+n$ coefficients at the output of each level. To solve this problem, many authors chose to perform a *circular convolution*: the signal to be filtered is periodized, and the output of the filter is thus also periodic. The output of a filter fed with a signal of period k would also be periodic with period k samples. The drawback of a simple periodization is that some abrupt changes may occur at the beginning and end of the periodic pattern, introducing artificially high-frequency contents, which may reduce

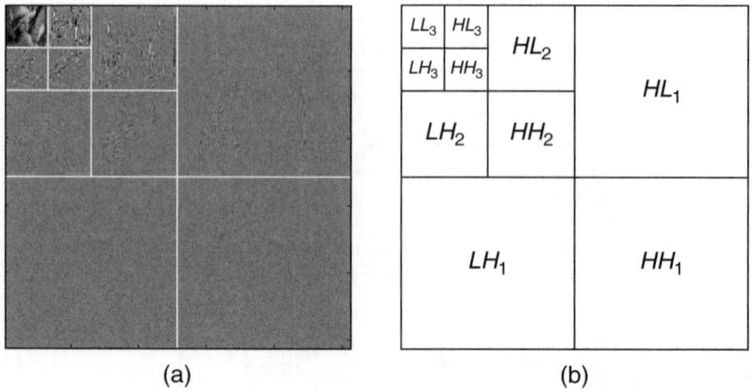

FIGURE 3.28 Two-dimensional separable wavelet decomposition on three levels.

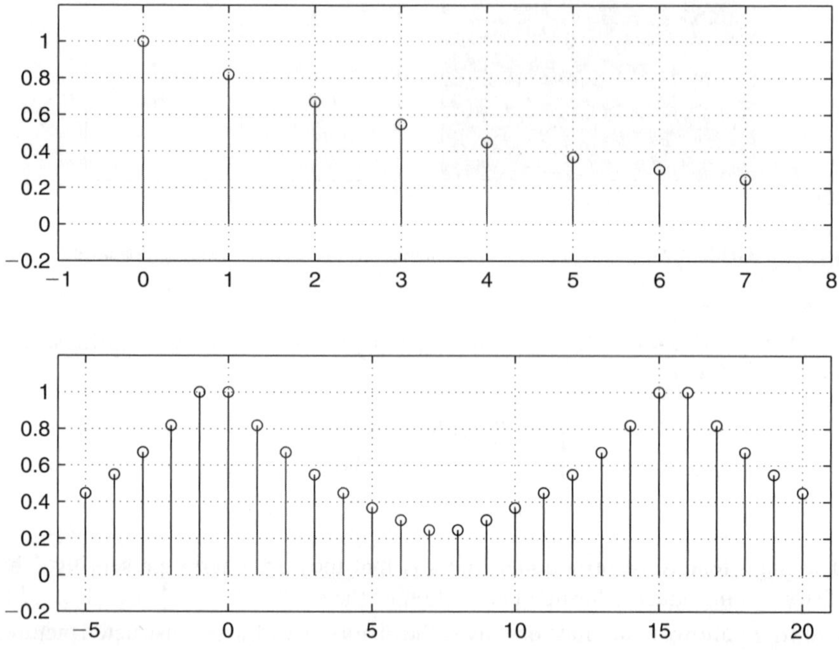

FIGURE 3.29 Initial signal (top) and signal after symmetrization and periodization (bottom).

the compression efficiency. Thus, the signals are often first symmetrized before being periodized, as illustrated by Figure 3.29.

This operation allows to get a period of length $2k$ when the original signal is of length k. The periodized signal is symmetric, but this property is not necessarily preserved after filtering. To avoid having to transmit $2k$ samples, the impulse response of the filters have also to be *symmetric* or *antisymmetric* (such filters are of *linear phase*). The output signal will then have the same property.

TABLE 3.5 Two Pairs of Biorthogonal Wavelet Filters

9/7 Filters		5/3 Filters		
h	g	h	g	Index
0.852699	0.788486	1.060660	0.707107	0
0.37402	0.418092	0.353553	0.353553	−1, 1
−0.110624	−0.040689	−0.176777		−2, 2
−0.023849	−0.064539			−3, 3
0.037828				−4, 4

Nevertheless, the only filters which are orthogonal, have linear phase and finite-impulse response, are the *Haar* filters $h=(1,1)$ and $g=(1,-1)$ (Daubechies, 1988). Thus, the orthogonality constraints (3.81) have been relaxed and replaced by a *biorthogonality* (Vetterli and Herley, 1992). Biorthogonal filters are no more energy preserving. In the family of biorthogonal wavelets, some are very close to orthogonality, such as those provided in Table 3.5. This justifies the use of the Daubechies 9/7 filters, for example, in the JPEG-2000 standard.

Note that the filtering operations can be efficiently implemented using *lifting* schemes, which allow to reduce the amount of operations that have to be done on the various subbands (Daubechies and Sweldens, 1998).

3.6.2 Subband Quantization and Entropy Coding

Once the subbands have been obtained, they have to be quantized and entropy-coded. First version of wavelet-based image coders perform (scalar or vector) quantization and entropy coding of each subband independently of the other ones (Antonini et al., 1992). The performance of the resulting schemes were in the order of magnitude of JPEG (ITU, 1992).

Significant performance improvements have been obtained thanks to bitplane and tree coders, such as embedded zerotree coding (EZW) (Shapiro, 1993), set partitioning in hierarchical trees (SPIHT) (Said and Pearlman, 1996), embedded block coding with optimized truncation (EBCOT) (Taubman, 2000), or embedded zeroblocks of wavelet coding based on context modeling (EZBC) (Hsiang and Woods, 2000). These techniques model the distribution of amplitudes of subband coefficients using generalized Gaussian distributions

$$p(w)=a\exp\left(-\left(\frac{|w|}{\sigma^2}\right)^{\beta}\right), \tag{3.82}$$

where $\beta < 2$. They also take into account the residual correlation within a subband and the correlation between subbands. In fact, since all wavelet coefficients are obtained by high-pass filtering of the initial image, when a quantized coefficient is null in a lower frequency subband, it is very likely that the quantized coefficients in the higher frequency subbands corresponding to the same location in the original image will also be null; contours will likely result in high values along the frequency domain, whereas smooth regions will likely provide zero (after quantization) values along the frequency domain. A coding using trees of zeros may then be very efficient to represent this large amount of zeros in various subbands. See Sayood (2005) for more details.

3.7 PACKETIZATION OF COMPRESSED DATA

To perform an efficient transmission of compressed multimedia contents, the compressed data have to be made compliant with the requirements of the communication standard used to transport data. Adaptation of the compressed data to the network layers has been previously considered not only in the MPEG standards (ISO/IEC JTC 1/SC 29, 1996) but also in ITU standards like H.320 or H.324. However, only a close interaction between the network adaptation and the source coder allows to get the best performances in terms of compression and robustness.

This is why the H.264/AVC video coder consists of two layers represented in Figure 3.30. The VCL (*video coding layer*) is responsible for the compression of data and, the NAL (*network abstraction layer*) performs the adaptation of output

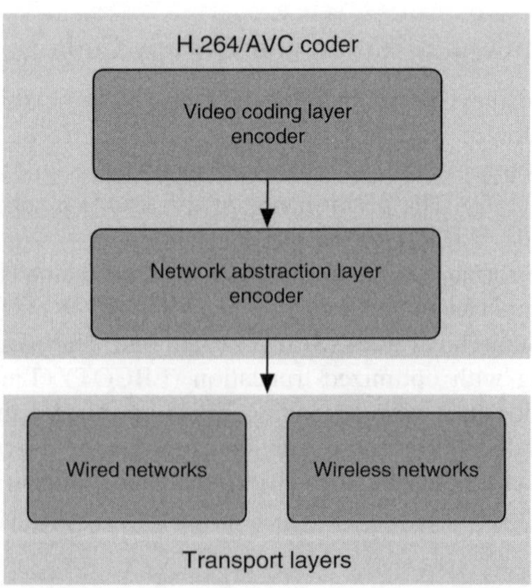

FIGURE 3.30 Interaction between the H.264/AVC coder and a network.

of the VCL to the target communication system. For commuted networks (H.320, H.324, or MPEG-2), the NAL delivers an ordered stream of bytes containing synchronization codes so that the transport layers and the video decoder can easily identify the structure of the stream. For datagram network (UDP/IP or TCP/IP), the NAL produces packets of data, without inserting any synchronization code.

More details on the protocol layers used to transport multimedia contents are provided in Chapter 7.

3.8 CONCLUSION

This chapter described in a concise way the main tools used in actual image coding standards, and insisted on the mechanisms that made them useful. Obviously, a given standard makes use of carefully chosen combination of tools. For example, a transform image-coding algorithm cannot be made from a transform directly followed by a VLC code, whose entries are the transform coefficients. The reason is that without resorting to high-order extensions (which would not be reasonable), this process would result in at least 1 bit per pixel, which is much higher than the expected performance. In this case, some kind of run-length code has to be used (see Section 4.3.2 for an example). The situation is different when the transform is followed by an arithmetic coder since several samples may be encoded in the same code word. However, high-order ACs are quite complex, and one prefers the use of binary ACs. Therefore, some translation to binary values has to be performed, this is known as *binarization*.

Chapter 4

Identifying Residual Redundancy

Joint source-channel decoding (JSCD) is possible, thanks to the presence of residual redundancy in the noisy bitstream available at receiver side. This redundancy facilitates the recovery at receiver side of the message generated by the source. Obviously, in a practical system, redundancy is intentionally introduced through channel codes, but many other sources of redundancy can be identified. A large amount of redundancy comes from the network layers (see Chapter 9) and more unexpectedly from the source bitstream itself (its content and its structure). This chapter describes these various redundancies, analyzes them, and provides means for evaluating them.

For people familiar with source coding, redundancy is understood as some suboptimality of the compression process. This inefficiency may be due to the source coder itself, but also to the model of the source. In fact, the adequation of the model significantly impacts the compression performance. The more accurate the model, the shorter the compressed bitstream.

From a channel coder point of view, redundancy is usually defined independently of the source model (all code words are assumed equally likely) and is rather understood as a structural property. Consider, e.g., a block code mapping source sequences of k bits to code words of n bits. Among the 2^n sequences of n bits, only 2^k corresponds to code words of n bits. Thus, $2^n - 2^k$ sequences of n bits are not code words. As soon as $n-k>2$, there are many more sequences of n bits that are not code words than those that are. The redundancy of the code is evidenced by the fact that all sequences are not code words. Usually, it is measured as the difference of the number of bits used to represent the code words (here n) and the minimum amount of bits that would be required to represent the code words, e.g., with a variable-length code, which corresponds to the base-2 logarithm of the size of the set of code words (here k), assuming that all code words are equally likely. The redundancy is thus estimated as $n-k$ bits. In fact, this estimate is an upper bound of the actual redundancy, since it does not take into account the actual probability mass function (pmf) of the source.

A similar approach is considered in the second part of the chapter. When considering vectors of n bits, some could have been generated by a given source coder, some not, revealing a residual redundancy similar to that introduced by channel codes. Such redundancy can be used without prior knowledge on the source model, even if its actual evaluation should involve the a priori probabilities of each code word.

Section 4.1 introduces residual redundancy from a source-coding point of view. Section 4.2 focuses on residual redundancy seen from a channel-coding point of view. The various sources of redundancy are then illustrated in Section 4.3 on examples involving practical sound and video coders.

4.1 STOCHASTIC REDUNDANCY

This type of redundancy is closely linked to classical probabilistic source models.

Consider a source described by a random vector $\mathbf{X}_{1:k}$ generating message $\mathbf{x}_{1:k}$ belonging to $\mathcal{A}^k$, the set of all sequences of length k over the alphabet $\mathcal{A}$. Given a probability model P for the source, $p(\mathbf{X}_{1:k}=\mathbf{x}_{1:k})$ (denoted by $p(\mathbf{x}_{1:k})$ when there is no ambiguity) is the probability that $\mathbf{X}_{1:k}$ emits $\mathbf{x}_{1:k} \in \mathcal{A}^k$.

4.1.1 Definitions and Examples

As seen in Chapter 3, the entropy of the source

$$H(\mathbf{X}_{1:k}) = -\sum_{\mathbf{x}_{1:k}} p(\mathbf{x}_{1:k}) \log_2(p(\mathbf{x}_{1:k})) \tag{4.1}$$

is the absolute lower bound on the expected code length when performing lossless encoding of $\mathbf{x}_{1:k}$ with a source code (Cover and Thomas, 1991; Shannon, 1948; Yeung, 2002). In (4.1), the self-information $-\log_2(p(\mathbf{x}_{1:k}))$ associated to the message $\mathbf{x}_{1:k}$ is sometimes also referred to as the *optimal length* of a code associated to $\mathbf{x}_{1:k}$.

Definition 4.1: *A* block source code *is defined by a mapping C from $\mathcal{A}^k$ to the set $\{0,1\}^*$ of finite-length binary sequences (C is the* encoding function*).* ∎

The set of code words is then the set $\mathcal{C}_k = C(\mathcal{A}^k)$ of all encoded sequences of $\mathcal{A}^k$. For a particular sequence $\mathbf{x}_{1:k}$, $\ell(C(\mathbf{x}_{1:k}))$ denotes the length of the code associated to $\mathbf{x}_{1:k}$. The *stochastic redundancy* of a source code measures the inefficiency of the source code with respect to the entropy of the source. This type of redundancy is thus closely linked (i) to the model of the source, (ii) to the model used to design the source code, and (iii) to the efficiency of source code.

Definition 4.2: *The* pointwise stochastic redundancy, *defined as*

$$R_k(C,P;\mathbf{x}_{1:k}) = \ell(C(\mathbf{x}_{1:k})) + \log_2 p(\mathbf{x}_{1:k}), \tag{4.2}$$

determines the redundancy of the code for a particular value $\mathbf{x}_{1:k}$ generated by the source, given the probability model P. ■

$R_k(C,P;\mathbf{x}_{1:k})$ compares $\ell(C(\mathbf{x}_{1:k}))$ to the optimal length $-\log_2(p(\mathbf{x}_{1:k}))$. For some source codes and some $\mathbf{x}_{1:k}$, (4.2) may be negative, but in average, $R_k(C,P;\mathbf{x}_{1:k})$ is positive or null.

Definition 4.3: *The* average stochastic redundancy

$$\begin{aligned}\overline{R}_k(C,P) &= E_P[R_k(C,P;\mathbf{x}_{1:k})] \\ &= E_P[\ell(C(\mathbf{x}_{1:k}))] - H_k(\mathbf{X}_{1:k})\end{aligned} \tag{4.3}$$

is the expectation of the pointwise stochastic redundancy (4.2) *with respect to P.* ■

$\overline{R}_k(C,P)$ evidences, as expected, the role of the entropy of the source.

Example 4.4 (Fixed-length encoding of a memoryless source): Assume that the source X is memoryless, stationary, and characterized by the pmf $p_i = p(X = x_i)$, $x_i \in \mathcal{A}$, $i = 1\ldots|\mathcal{A}|$. A fixed-length encoding function C_F is used to represent each source outcome with $L = \lceil \log_2|\mathcal{A}| \rceil$ bits. The entropy of the source is

$$H(X) = -\sum_i p_i \log_2 p_i, \tag{4.4}$$

and thus, the average stochastic redundancy (4.3) becomes

$$\overline{R}_k(C_F,P) = \lceil \log_2|\mathcal{A}| \rceil + \sum_i p_i \log_2 p_i. \tag{4.5}$$

The redundancy (4.5) comes from the suboptimality of the fixed-length code. It may be very small when the source symbols are equally likely, but it may become large when the pmf of the source is skewed.

Variable-length codes may also leave some redundancy in the bitstream, as illustrated by the following example.

Example 4.5 (Redundancy of a Huffman code): Consider a source X with alphabet $\mathcal{A} = \{1,2,3\}$ such that $p_1 = p(X=1) = 0.6$, $p_2 = p(X=2) = 0.25$, and

$p_3 = p(X=3) = 0.15$. The entropy of X is

$$H(X) = -\sum_{i=1}^{3} p_i \log_2 p_i = 1.357 \text{ bits/symbol}.$$

A Huffman code $\mathcal{C}_H$ is designed for the source X such that $C_H(1)=0$, $C_H(2)=10$, and $C_H(3)=11$. The pointwise stochastic redundancies for this code are

$$\begin{aligned} R_1(C_H, (p_1, p_2, p_3); 1) &= 1 + \log_2(0.6) = 0.263 \text{ bits}, \\ R_1(C_H, (p_1, p_2, p_3); 2) &= 2 + \log_2(0.25) = 0 \text{ bit}, \\ R_1(C_H, (p_1, p_2, p_3); 3) &= 2 + \log_2(0.15) = -0.7370 \text{ bits}. \end{aligned}$$

The redundancy for symbol 2 is null, since the optimal length $-\log_2(p_2)$ is integer and can be matched by a single-letter Huffman code. The pointwise stochastic redundancy for symbol 3 is negative but is compensated by the positive redundancy for symbol 1. The average stochastic redundancy is

$$\begin{aligned} \overline{R}_1(C_H, (p_1, p_2, p_3)) &= \sum_{i=1}^{3} p_i \big(\ell(C_H(i)) + \log_2 p_i\big) \\ &= 0.0473 \text{ bits/symbol}. \end{aligned}$$

The redundancy here is clearly due to the fact that the Huffman code has been designed for single-letter model of the source. It may be reduced by using a code designed for groups of two or more source outcomes.

Definition 4.6: *The redundancy-rate problem consists in determining the growth rate with k of*

$$\overline{R}_k(\mathcal{P}) = \min_{C_k} \max_{P \in \mathcal{P}} \overline{R}_k(C_k, P) \tag{4.6}$$

or of

$$R^*(\mathcal{P}) = \min_{C_k} \max_{P \in \mathcal{P}} R_k^*(C_k, P), \tag{4.7}$$

for a class $\mathcal{P}$ of probability models for a source X. ■

This measure of redundancy is useful in practice, when the probability model of the source is not perfectly known or when its characteristics are estimated and only known with some uncertainty. In the following example, only the probability of the most likely symbol is assumed to be known.

Example 4.7 (Source with partial knowledge of the pmf): For a source encoded with a Huffman code, it is well known (Gallager, 1968) that $\overline{R}_k(C_H, P)$ is less than 1 bit for any k and whatever the distribution P. Consider the class $\mathcal{P}(p_1)$ of pmfs such that p_1 is the probability of the most likely

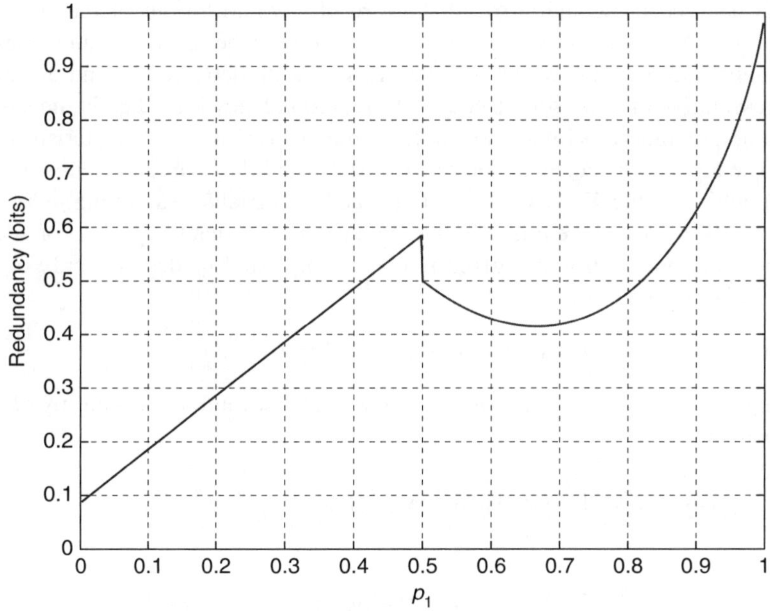

FIGURE 4.1 Upper bound of the redundancy of a Huffman code when the probability of the most likely sequence is p_1.

sequence in $\mathcal{A}^k$, where $\mathcal{A}$ is a finite-length alphabet. For $\mathcal{P}(p_1)$, Gallager (1978) provides a more accurate upper bound for $\overline{R}_k(\mathcal{P}(p_1))$ given by

$$\begin{aligned}\overline{R}_k(\mathcal{P}(p_1)) &\leqslant p_1 + 1 - \log_2 e + \log_2 \log_2 e \text{ if } p_1 < 0.5 \\ &\leqslant 2 - H_{\mathrm{b}}(p_1) - p_1 \text{ if } p_1 \geqslant 0.5,\end{aligned}$$

with $H_b(x) = -x\log_2 x - (1-x)\log_2(1-x)$. Figure 4.1 represents $\overline{R}_k(\mathcal{P}(p_1))$ as a function of p_1. The 1 bit upper bound is reached when p_1 gets close to 1.

4.1.2 Example of a Residual Redundancy Evaluation

The residual redundancy has been evaluated for many source coders. Redundancy of Huffman codes has been addressed by Capocelli and de Santis (1989), Gallager (1978) and Manstetten (1992). Szpankowski (2000) provides, a generalization to other fixed-to-variable length codes. Variable-to-fixed length codes, such as Lempel–Ziv codes, are considered by Plotnik et al. (1992), Savari (1997b), Stubley (1994), and Yang et al. (2005). Redundancy of arithmetic codes has been analyzed by Drmota et al. (2002). Finally, for various universal codes, i.e., codes that adapt to the possibly nonstationary characteristics of the source, an interesting study of the interplay between delay and redundancy is studied, e.g., by Krichevsky and Trofimov (1981).

A detailed presentation of all these results would be out of the scope of this book. Nevertheless, it is interesting to consider the residual redundancy of Shannon–Fano codes, which are the basis of arithmetic codes, and are quite representative of other source coders. Here, redundancy is studied in the simple case of a memoryless binary source (Bernoulli) with known characteristics.

The alphabet is taken as binary, $\mathcal{A}=\{0,1\}$, and the source is assumed to be Bernoulli, i.e., the $X_1 \ldots X_k \ldots$ are i.i.d. random variables satisfying $p(X_\kappa=1)=p$ for all $\kappa>0$, where p is a known parameter such that $0<p<1$. For such a source, the probability of a given realization $\mathbf{x}_{1:k}$ of $X_{1:k}$ depends only on the number of ones in $\mathbf{x}_{1:k}$

$$p(X_{1:k}=\mathbf{x}_{1:k})=p^{\sum_\kappa x_\kappa}(1-p)^{k-\sum_\kappa x_\kappa}. \tag{4.8}$$

Using the fact that $\sum_{\kappa=1}^{k} X_\kappa$ follows a binomial distribution, the entropy of $X_{1:k}$ is evaluated as

$$\begin{aligned} H_k(p) &= -\sum_{\kappa=0}^{k}\binom{k}{\kappa}p^\kappa(1-p)^{k-\kappa}\log_2\left(p^\kappa(1-p)^{k-\kappa}\right) \\ &= -\sum_{\kappa=0}^{k}\binom{k}{\kappa}p^\kappa(1-p)^{k-\kappa}\left(\kappa\log_2 p+(k-\kappa)\log_2(1-p)\right). \end{aligned} \tag{4.9}$$

Since the source is memoryless, (4.9) may also be evaluated as

$$H_k(p)=k\left(-p\log_2 p-(1-p)\log_2(1-p)\right). \tag{4.10}$$

By definition, the Shannon–Fano encoding function C_{SF} (Cover and Thomas, 1991) assigns a code word to $\mathbf{x}_{1:k}$ of length

$$\ell(C_{\mathrm{SF}}(\mathbf{x}_{1:k})) = \left\lceil -\log_2 p(X_{1:k}=\mathbf{x}_{1:k})\right\rceil \tag{4.11}$$

$$= \left\lceil -\log_2 p^{\sum_\kappa x_\kappa}(1-p)^{k-\sum_\kappa x_\kappa}\right\rceil \tag{4.12}$$

$$= \left\lceil -\left(\sum_\kappa x_\kappa\right)\log_2 p - \left(k-\sum_\kappa x_\kappa\right)\log_2(1-p)\right\rceil. \tag{4.13}$$

One may then deduce the average redundancy from (4.9) and (4.13)

$$\begin{aligned} \overline{R}_k(C_{\mathrm{SF}},p) &= \sum_{\kappa=0}^{k}\binom{k}{\kappa}p^\kappa(1-p)^{k-\kappa}\big(\left\lceil-\kappa\log_2 p-(k-\kappa)\log_2(1-p)\right\rceil \\ &\quad -\left(-\kappa\log_2 p-(k-\kappa)\log_2(1-p)\right)\big) \\ &= 1-\sum_{\kappa=0}^{k}\binom{k}{\kappa}p^\kappa(1-p)^{k-\kappa}\left\langle-\kappa\log_2 p-(k-\kappa)\log_2(1-p)\right\rangle, \end{aligned} \tag{4.14}$$

where $\langle x\rangle=(x-\lfloor x\rfloor)$ stands for the fractional part of x.

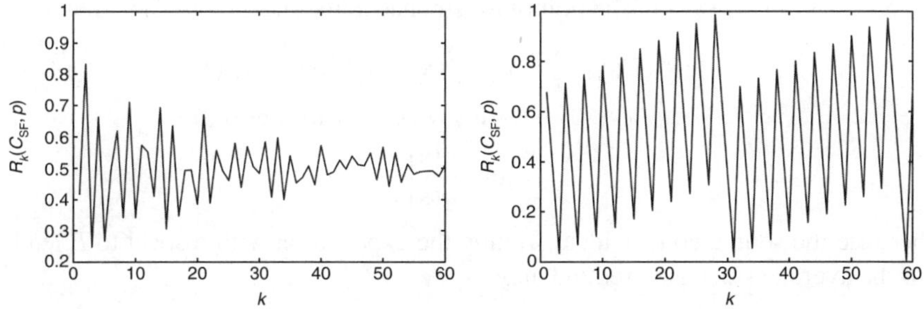

FIGURE 4.2 Redundancy of the Shannon–Fano code for $p = 1/\pi$ (left) and $p = 0.2$ (right).

Figure 4.2 illustrates the evolution of (4.14) as a function of the length k of the sequence to encode for two values of p, namely $p = 1/\pi$ and $p = 0.2$. A convergent ($k = 1/\pi$) and an oscillatory ($k = 0.2$) behavior may be identified. They have been studied in detail by Szpankowski (2000), which establishes that

$$\overline{R}_k(C_{SF}, p) = 1/2 + o(1)$$

if $\alpha = \log_2(p/(1-p))$ is irrational and

$$\overline{R}_k(C_{SF}, p) = 1/2 - (\langle Mk\beta \rangle - 1/2)/M + O(\rho^k)$$

if $\alpha = N/M$ is rational, where gcd $(N, M) = 1$, $\beta = \log_2(1/(1-p))$, and $0 < \rho < 1$.

Both types of behavior have been evidenced for other block codes by Szpankowski (2000).

Usually, the residual redundancy due to the inefficiency of the source coder is not larger than 1 or 2 bits when the characteristics of the source are well known by the source coder.

4.1.3 Redundancy Due to Mismatched Source Models

Many widely used source-coding applications, such as JPEG (ITU, 1992), MPEG2 AAC (ISO/IEC, 1997), or H.263 (ITU-T, 2005), embed entropy coders that are not designed for a particular source but optimized in such a way as to provide satisfying performance on typical data they have to encode. The entropy code is thus not matched to the source, since it does not perform an estimation of the source characteristics. This leads to some redundancy in the compressed bitstream.

Consider again a source emitting realizations $\mathbf{x}_{1:k} \in \mathcal{A}^k$ of a random vector $\mathbf{X}_{1:k}$ with probability model P. Assume that a wrong probability model Q for the source has been used to design an optimal (entropy-achieving) source code $\mathcal{C}_k^Q$ with encoding function C_k^Q. The code has, thus, been designed assuming that

$p(\mathbf{X}_{1:k}=\mathbf{x}_{1:k})=q(\mathbf{x}_{1:k})$. The pointwise stochastic redundancy (4.2) is then

$$\begin{aligned} R_k\left(C_k^Q, P; \mathbf{x}_{1:k}\right) &= \ell\left(C_k^Q(\mathbf{x}_{1:k})\right)+\log_2 p(\mathbf{x}_{1:k}) \\ &= -\log_2 q(\mathbf{x}_{1:k})+\log_2 p(\mathbf{x}_{1:k}) \\ &= \log_2 \frac{p(\mathbf{x}_{1:k})}{q(\mathbf{x}_{1:k})}, \end{aligned} \tag{4.15}$$

because the source code is ideal. Taking the expectation with respect to P leads to the average stochastic redundancy

$$\overline{R}_k\left(C_k^Q, P\right)=E_P\left[\log_2 \frac{p(\mathbf{X}_{1:k})}{q(\mathbf{X}_{1:k})}\right]=D(P\|Q), \tag{4.16}$$

where

$$D(P\|Q)=\sum_{\mathbf{x}_{1:k}} p(\mathbf{x}_{1:k}) \log_2 \frac{p(\mathbf{x}_{1:k})}{q(\mathbf{x}_{1:k})}$$

is the *Kullback–Leibler divergence* or relative entropy between the probability models P and Q (Cover and Thomas, 1991). In (4.16), it is necessary to indicate the probability model that will serve to evaluate the expectation, since several probability models are considered.

Assume now that the source-encoding function C_k^Q is suboptimal for the wrong probability model Q so that it introduces the pointwise stochastic redundancy $R_k(C_k^Q, Q; \mathbf{x}_{1:k})$ for a given source sequence $\mathbf{x}_{1:k}$. Thus, (4.15) becomes

$$R_k\left(C_k^Q, P, Q; \mathbf{x}_{1:k}\right)=\ell\left(C_k^Q(\mathbf{x}_{1:k})\right)+\log_2 p(\mathbf{x}_{1:k}) \tag{4.17}$$

$$=R_k(C_k, Q; \mathbf{x}_{1:k})+\log_2 \frac{p(\mathbf{x}_{1:k})}{q(\mathbf{x}_{1:k})}. \tag{4.18}$$

Using (4.17), one gets an average stochastic redundancy expressed as

$$\begin{aligned} \overline{R}_k\left(C_k^Q, P, Q\right) &= E_P\left[\ell\left(C_k^Q(\mathbf{X}_{1:k})\right)+\log_2 p(\mathbf{X}_{1:k})\right] \\ &= E_P\left[\ell\left(C_k^Q(\mathbf{X}_{1:k})\right)\right]-H(\mathbf{X}_{1:k}). \end{aligned} \tag{4.19}$$

Using (4.18), one gets

$$\begin{aligned} \overline{R}_k\left(C_k^Q, P, Q\right) &= E_P\left[R_k\left(C_k^Q, Q; \mathbf{X}_{1:k}\right)+\log_2 \frac{p(\mathbf{X}_{1:k})}{q(\mathbf{X}_{1:k})}\right] \\ &= D(P\|Q)+E_P\left[R_k\left(C_k^Q, Q; \mathbf{X}_{1:k}\right)\right]. \end{aligned} \tag{4.20}$$

In (4.20), because $R_k(C_k^Q, Q; \mathbf{x}_{1:k})$ may be positive or negative, depending on $\mathbf{x}_{1:k}$, and because the expectation is taken over an other pmf than that used to design the code, nothing can be said about $E_P[R_k(C_k^Q, Q; \mathbf{X}_{1:k})]$, except that

$$D(P\|Q)+E_P\left[R_k\left(C_k^Q, Q; \mathbf{X}_{1:k}\right)\right] \geqslant 0.$$

Example 4.8: Consider again the source of Example 4.5 and assume that the Huffman code has been designed for the probability model Q, with $q_1 = p(X=a) = 0.5$, $q_2 = p(X=b) = 0.25$, and $q_3 = p(X=c) = 0.25$. For this source model, an optimal Huffman code may be generated, $C^Q = \{0, 10, 11\}$. The average stochastic redundancy for this code is

$$\overline{R}_k\left(C_k^Q, P\right) = D(P||Q) = 0.0473 \text{ bits/symbol.}$$

This value is the same as that obtained in Example 4.5.

Albeit Examples 4.5 and 4.8 provide the same value of the redundancy; one may give two interpretations to the results. In Example 4.5, the redundancy is due to the suboptimality of the source code. In Example 4.8, the source code is optimal, but not for the correct source model. A way to determine the redundancy of a source code may then be to determine a probability model Q for which the source code is optimal and then to determine the Kullback–Leibler distance between the actual source models P and Q.

4.2 DETERMINISTIC REDUNDANCY

This interpretation of redundancy is closely linked to the structure of the source coder and no more to a too coarse statistical model of the source considered by the source coder. The redundancy mainly results from constraints imposed on the generated bitstream by the source coder itself and by its environment.

To illustrate this type of redundancy, one may examine the analogy between a packet of compressed data and the way a book chapter is organized. The bitstream has to be produced using code words satisfying some *syntax* constraints, see Section 4.2.3. Any succession of bits may not necessarily correspond to a code word, like any succession of letters is not necessarily a word in English. Since compressed data are usually sent over packet-switched networks, the source coder has to satisfy some *packetization* constraints (Section 4.2.4), which are very similar to the typesetting rules a writer has to comply with (period at the end of a sentence, starting new paragraphs on a new line...). The author usually employs words that put together have a meaning. Similarly, the bitstream generated by the source coder has also to satisfy some *semantic* constraints: compressed data represent, *e.g.*, blocks of a given number of pixels, of a given number of sound samples, and of some text. This constraint prevents from using some meaningless successions of code words, see Section 4.2.5.

Before examining these types of redundancy, a new measure of redundancy has to be introduced in Section 4.2.1.

4.2.1 Definition and Examples

Consider a linear (n, k) binary channel code $\mathcal{C}$ (Blahut, 1983; McEliece, 2002; Richardson and Urbanke, 2008). This code represents a k-dimensional subspace of the vector space of binary vectors of n bits. Using a generator matrix $\underline{\mathbf{G}}$, *information* vectors of k bits are transformed into *code* vectors of n bits. $\mathcal{C}^{\perp}$, the subspace orthogonal to $\mathcal{C}$ is of dimension $n-k$ and contains 2^{n-k} elements. The redundancy of $\mathcal{C}$ is the number of bits added to information vectors to get code vectors

$$R(\mathcal{C}(n,k)) = n-k \text{ bits.} \tag{4.21}$$

It corresponds to the dimension of $\mathcal{C}^{\perp}$. The redundancy may also be interpreted as the difference between the number of bits used to represent the code words and the number of bits that would be required to represent each code word, e.g., with a variable-length code, assuming that all are equally likely and that sequences which are not code words have a null probability. Under these assumptions, one would get

$$R(\mathcal{C}) = n - \log_2 |\mathcal{C}| = n-k \text{ bits.} \tag{4.22}$$

Example 4.9: Consider the systematic binary Hamming code $\mathcal{H}(7,4)$. A generator matrix for this code is

$$\underline{\mathbf{G}} = \begin{pmatrix} 1 & 0 & 0 & 0 & 1 & 1 & 0 \\ 0 & 1 & 0 & 0 & 1 & 0 & 1 \\ 0 & 0 & 1 & 0 & 0 & 1 & 1 \\ 0 & 0 & 0 & 1 & 1 & 1 & 1 \end{pmatrix}.$$

Using $\underline{\mathbf{G}}$, the information vector of 4 bits $\mathbf{u} = (0101)$ is transformed into a code vector $\mathbf{x} = (0101010)$ of 7 bits. The redundancy introduced is $7-4=3$ bits. The vector

$$\mathbf{x} = \mathbf{u}\underline{\mathbf{G}} = (0101010)$$

belongs, thus, to $\mathcal{H}(7,4)$, whereas $\mathbf{x}' = (0001010)$ does not belong to $\mathcal{H}^{\perp}(7,4)$, since there is no vector of 4 bits with which $\mathbf{x}'$ may be generated using $\underline{\mathbf{G}}$.

This form of redundancy has its counterpart for source codes. Consider a source code $\mathcal{C}$ with encoding function C transforming sequences from $\mathcal{A}^*$, the set of all finite sequences over the alphabet $\mathcal{A}$ into $\{0,1\}^*$. Consider the set $\{0,1\}^n$ of binary sequences of n bits. Among the 2^n sequences of this set, some are contained in $\mathcal{C}$ and some not. Thus, as for channel codes, it is, thus, possible to define, independently of the source, the subset Ω_n of $\{0,1\}^n$ of all sequences of exactly n bits, which are contained in $\mathcal{C}$

$$\Omega_n = \left\{ \mathbf{b} \in \{0,1\}^n \,|\, \exists \mathbf{x} \in \mathcal{A}^*, C(\mathbf{x}) = \mathbf{b} \right\}. \tag{4.23}$$

Ω_n is thus the intersection of the code $\mathcal{C}$ with $\{0,1\}^n$ or its restriction to sequences of n bits. The subset Φ_n of $\mathcal{A}^*$ corresponding to Ω_n is defined as

$$\Phi_n = \left\{ \mathbf{s} \in \mathcal{A}^* | C(\mathbf{s}) \in \Omega_n \right\}; \tag{4.24}$$

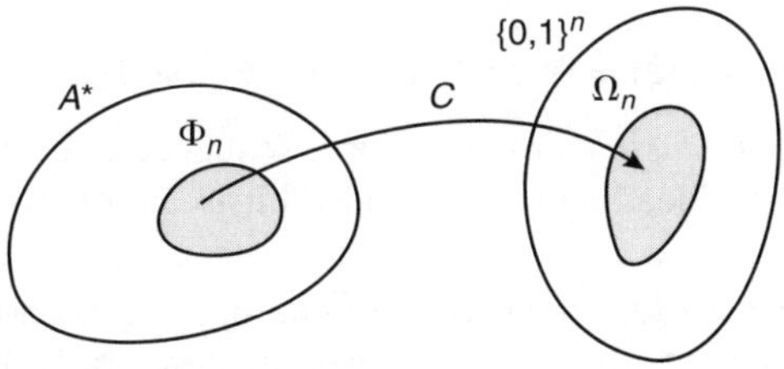

FIGURE 4.3 Subset Ω_n of sequences of n bits belonging to the code C and set Φ_n of associated source sequences.

thus, one has

$$\Omega_n = C(\Phi_n). \tag{4.25}$$

Ω_n is the set of all sequences of n bits that are consistent with the code C and Φ_n is the set of all corresponding code sequences, see Figure 4.3.

Example 4.10: Consider a variable-length code C_1 over the alphabet $\mathcal{A} = \{a, b, c\}$ such that $C_1(a) = 0$, $C_1(b) = 100$, and $C_1(c) = 111$. Among the sequences of 4 bits, 0111 is a possible outcome of the code corresponding to the sequence of source symbols *ac*, but 0110 could not be generated by the code and does not belong to Ω_4. One may easily construct Ω_4 and Φ_4 by testing the 2^4 combinations of 4 bits to get

$$\Omega_4 = \{0000, 0100, 1000, 0111, 1110\}$$

and

$$\Phi_4 = \{aaaa, ab, ba, ac, ca\}.$$

Thus, among the 16 possible encoded sequences of 4 bits, only 5 correspond to possible outcomes of the source coder. Using (4.22), one gets $R(\Omega_4) = 4 - \log_2 |\Omega_4| = 1.68$ bits of redundancy. This redundancy comes from the *syntax* of the code words and from the fact that the code has been restricted to sequences of 4 bits.

As pointed out in the introduction of Section 4.2, a source may have to satisfy some constraints, which are often difficult to take entirely into account within the source coder. Some of these constraints may be embedded in the probability model P, as described in Section 4.1, and some are more easily described by tests that have to be satisfied.

Consider a constraint and the associated test function $\xi : \mathcal{A}^* \to \{0, 1\}$. To determine whether a sequence $\mathbf{x} \in \mathcal{A}^*$ complies with the constraint imposed on the source, one has to simply verify whether $\xi(\mathbf{x}) = 1$. Taking ξ into account leads to the definition of two new sets,

$$\Omega_n(\xi) = \{\mathbf{b} \in \{0, 1\}^n \mid \exists \mathbf{x} \in \mathcal{A}^*, \xi(\mathbf{x}) = 1, C(\mathbf{x}) = \mathbf{b}\} \tag{4.26}$$

and

$$\Phi_n(\xi) = \{\mathbf{x} \in \mathcal{A}^* \mid \xi(\mathbf{x}) = 1, C(\mathbf{x}) \in \Omega_n\}. \tag{4.27}$$

$\Phi_n(\xi)$ is the set of all *feasible* or *valid* source sequences, i.e., compliant with the source constraints and leading to an encoded bitstream of n bits.

Example 4.11: Consider the code $\mathcal{C}_1$ of Example 4.10 and assume that the source is known not to output two consecutive symbols that are identical. This property is easily implemented into a test function ξ_1 and one gets, for example, $\xi_1(aabca) = 0$ and $\xi_1(abcac) = 1$. Accounting for this constraint leads to the following sets

$$\Omega_4(\xi_1) = \{0100, 1000, 0111, 1110\} \subset \Omega_4$$

and

$$\Phi_4(\xi_1) = \{ab, ba, ac, ca\} \subset \Phi_4,$$

reducing the size of the set of valid source sequences. Note that part of the reduction comes from the fact that two identical consecutive symbols are forbidden and part comes from the knowledge of the number of bits of the sequence.

From (4.22), for a source code $\mathcal{C}$, depending on whether a constraint ξ is taken into account, one may introduce the *unconstrained deterministic redundancy*

$$R_n(\mathcal{C}) = n - \log_2 |\Phi_n| = n - \log_2 |\Omega_n| \tag{4.28}$$

and the *constrained deterministic redundancy*

$$R_n(\mathcal{C}, \xi) = n - \log_2 |\Phi_n(\xi)| = n - \log_2 |\Omega_n(\xi)|, \tag{4.29}$$

both defined for encoded bitstreams of length n. In (4.28) and (4.29), $|\cdot|$ stands for the cardinal number of a set.

From (4.23) and (4.26), one deduces that

$$\Omega_n(\xi) \subset \Omega_n, \tag{4.30}$$

and as a consequence,

$$R_n(\mathcal{C}, \xi) \geqslant R_n(\mathcal{C}). \tag{4.31}$$

Example 4.12: The unconstrained deterministic redundancy of the source code $\mathcal{C}_1$ defined in Example 4.10 is $R_4(\mathcal{C}_1) = 4 - \log_2 |\Omega_4| = 1.68$ bits, whereas the constrained one is $R_4(\mathcal{C}_1, \xi_1) = 4 - \log_2 |\Omega_4(\xi)| = 2$ bits.

The difficulty, compared to channel codes, is that usually, neither the subsets Ω_n and $\Omega_n(\xi)$ nor consequently the subsets Φ_n and $\Phi_n(\xi)$ have the nice property of being vector subspaces. Their structure is, thus, much more difficult to identify.

The main difference between (4.3) and (4.29) is that the definition of stochastic redundancy implies the consideration of source vectors of k symbols, whereas deterministic redundancy involves sequences of n bits, as available at the input of a transmission channel. Deterministic redundancy is, thus, much closer to the needs in JSCD, since one tries to estimate an element of Ω_n or $\Omega_n(\xi)$ from measurements of a sequence of n outcomes of a source coder. As will be seen in Chapter 5, the organization of the *code words* in Ω_n or $\Omega_n(\xi)$ will impact significantly on the performance of JSCD techniques. Here, the focus is on the cardinality of Ω_n or $\Omega_n(\xi)$. As for channel codes, more precise properties such as minimum distances or the distance spectrum (Anderson, 2005) of a *source code* can be derived.

4.2.2 Evaluation of the Deterministic Redundancy

Because of the lack of an evident structure in the sets involved in the definitions of $R_n(\mathcal{C})$ and $R_n(\mathcal{C}, \xi)$, their evaluation often requires brute-force enumeration algorithms. One may either test the 2^n sequences of $\{0, 1\}^n$, isolating those belonging to Ω_n, or isolate in $\mathcal{A}^*$ the sequences which belong to Φ_n. As soon as n becomes larger than some tens, this task becomes cumbersome. For example, with $n=40$, more than 10^{12} sequences of 40 bits have to be tested.

In many situations, efficient recursive algorithms allow to evaluate $R_n(\mathcal{C})$ and $R_n(\mathcal{C}, \xi)$. This section considers source codes that can be described by *finite-state machines* (FSMs), and for which such recursive algorithms can be constructed. Huffman-like and finite-precision arithmetic codes, see Chapter 3 and the studies by Bi et al. (2006), and Gormish and Allen (1993) are two examples of such source codes.

Finite-State Machines

An FSM or finite-state automation is a model of a system composed of a finite number of *states*, *transitions* between those states, and *actions*. A state keeps the information about the past necessary to determine, maybe with some input, the evolution of the FSM. In other words, the state reflects the impact of past inputs on the system from the system start to the present moment, i.e., the memory of the system. A transition indicates a state change and is usually described by a condition that needs to be fulfilled to enable the transition. An action is a description of an activity that is performed at a given moment, e.g., during a transition.

Among the class of FSM, *transducers* are best suited to represent the behavior of many types of source coders, since they generate an output based on a given input and state using actions. Formally, a transducer FSM is a seven-tuple $(\mathcal{I}, \mathcal{O}, \Sigma, \sigma_0, t, \omega, \mathcal{F})$, where $\mathcal{I}$ is the input alphabet, $\mathcal{O}$ is the output alphabet, Σ is the finite, nonempty set of states, and σ_0 is the initial state (an element of Σ). The state transition function $t: \Sigma \times \mathcal{I} \rightarrow \Sigma$ determines the evolution with time of the states when a given input is applied on the FSM and ω is the output function.

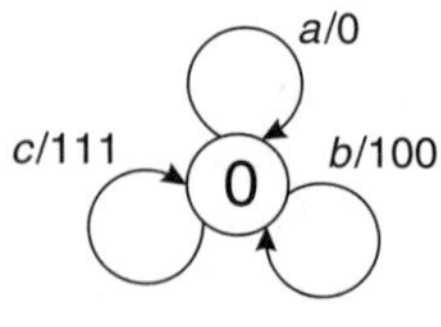

FIGURE 4.4 Finite-state machine for the code C_1 (encoder).

The *Mealy* transducer (Mealy, 1955), in which $\omega: \Sigma \times \mathcal{I} \to \mathcal{O}$, i.e., the output depends on the state and on the input, is considered here. Finally, $\mathcal{F}$ is the set of admissible final states (also a subset of Σ). Note that when at a given time instant, the state is in $\mathcal{F}$, the FSM does not necessarily stop, but when it stops, its state has to belong to $\mathcal{F}$.

Mealy transducers can be used to represent source encoders, as illustrated by Examples 4.13 and 4.14, where two *symbol-clock* transducers are considered, i.e., a transducer for which each transition is driven by a single input symbol.

Example 4.13: Consider the code $C_1 = \{0, 100, 111\}$ of Example 4.10. A Mealy transducer for the coder can be built with $\mathcal{I} = \{a, b, c\}$ and $\mathcal{O} = \{0, 1\}^*$, see Figure 4.4. There is only a single state 0, since the output of the coder at a given time instant κ does not depend on the previous outputs. Transitions are represented by arrows on which is attached a label indicating the input symbol and the corresponding output sequence of bits. For example, $t(0, c) = 0$ and $\omega(0, c) = 111$ are represented in Figure 4.4 by $c/111$ on a arrow linking the state 0 to itself.

Example 4.14: Consider now a quasi-arithmetic code, such as that described in Example 3.24, page 53, with $q = 4$, $p_0 = 0.2$, and $F_{\max} = 1$. The Mealy transducer of Figure 4.5 with $\mathcal{I} = \{0, 1\}$, $\mathcal{O} = \{0, 1\}^*$, and initial state 0 describes the way the arithmetic coder is working. Note that some transitions do not produce any output bit (*mute transitions*), which is necessary to allow some compression. For example, $\omega(0, 1) = \omega(3, 1) = \emptyset$.

Constraints on the source can be taken into account; the price to be paid is a modification of the structure of the transducer, as described in Example 4.15. This is again a symbol-clock transducer.

Example 4.15: Consider now the code C_1 and the constraint ξ_1 of Example 4.11 (ξ_1 indicates that the emission of two consecutive identical symbols is not possible). The Mealy transducer of Figure 4.6 takes ξ_1 into account. The states 1–3 of the FSM keep in memory the last emitted symbol by the source (the state 1 is associated to a, 2 to b, and 3 to c). Only two transitions

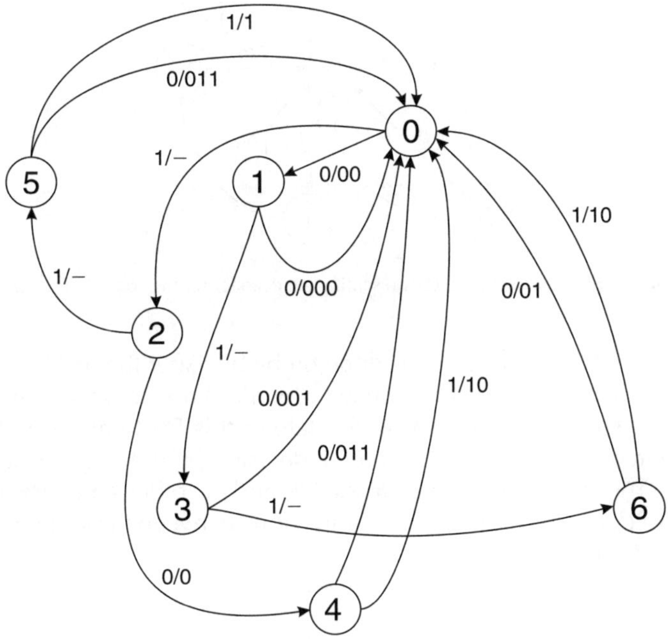

FIGURE 4.5 Finite-state machine representing an integer arithmetic coder with $q=4$, $p_0=0.2$, and $F_{\max}=1$.

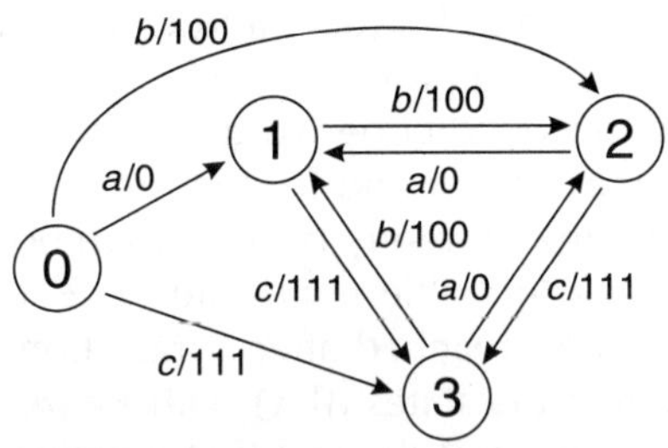

FIGURE 4.6 Finite-state machine for the code C_1 (encoder) accounting for the constraint ξ_1.

are emerging from states 1–3 (for example, from state 1, there is no transition corresponding to the input symbol a).

Transducers can also represent the source decoder, as illustrated by Examples 4.16 and 4.17, where a bit-clock transducer is considered, i.e., a transducer for which each transition is driven by a single input bit.

Example 4.16: Consider now the code $\mathcal{C}=\{00, 01, 10, 110, 111\}$ associated to the source alphabet $\mathcal{A}=\{a, b, c, d, e\}$. The transducer associated to the decoder for $\mathcal{C}$ has $\mathcal{I}=\{0, 1\}$ as input bits and $\mathcal{O}=\{a, b, c, d, e\}^*$ as

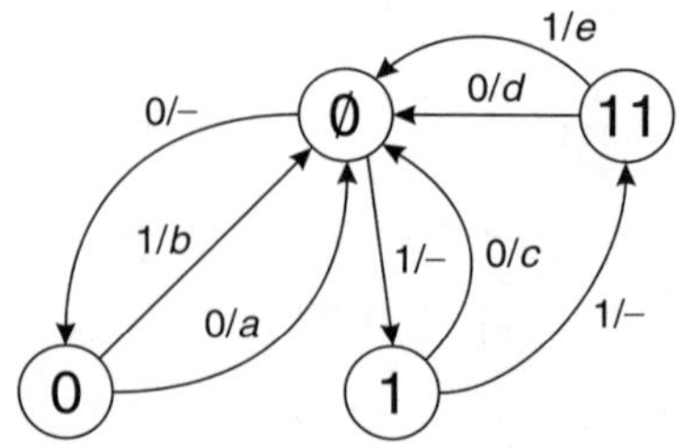

FIGURE 4.7 Finite-state machine representing the decoder for the code C of Example 4.16.

output symbols. The decoding is done bit by bit. After the first bit has been processed, it is only possible to have a partial knowledge about the corresponding source symbol. Thus, the state of this FSM corresponds to the accumulated decoder input bits until a decision on the output symbol can be made. The initial state is thus labeled $\emptyset$, and the other states are the set of prefixes of $\mathcal{C}$, *i.e.*, 0, 1, and 11. The transitions between these states depend on the input bits, see Figure 4.7.

Example 4.17: As in Example 4.15, consider $\mathcal{C}_1$ and ξ_1, but now focus on the source decoder. Input and output alphabets are now exchanged. The corresponding transducer has $\mathcal{I}=\{0,1\}$ and $\mathcal{O}=\{a,b,c\}^*$, see Figure 4.8. Again, some transitions do not produce any output symbol, since they correspond to a partially decoded code word. The state corresponds now to the last decoded source symbol and for the processed bits which do not allow yet to determine the currently decoded symbol. The initial state σ_0 is $(\emptyset,\emptyset)$, and the set of admissible final states is $\mathcal{F}=\{(a,\emptyset),(b,\emptyset),(c,\emptyset)\}$, corresponding to the states reached after a transition with emission of an output symbol. Note that the states $(\emptyset,x)$ with $x\in\{a,b,c\}$ are only a *transient* states, taking into account the fact that no symbol has already been decoded, and that all symbols are thus allowed.

If the FSM does not describe with full accuracy the structure of the source, it may happen that from given states, some inputs are not allowed in the FSM, see Example 4.15. The set of transitions emerging from such states may be supplemented with transitions to an additional *error* state ε (such transitions are not represented in what follows). Whenever ε is reached, the FSM would remain stuck in that state, indicating that the constraints on the input sequence are not satisfied. Such FSMs obviously allow error detection.

Evaluation of the Deterministic Redundancy Using FSM

One aims at determining the deterministic redundancy of a source code $\mathcal{C}$ (possibly with constraints represented by ξ) for a given value of n and the length

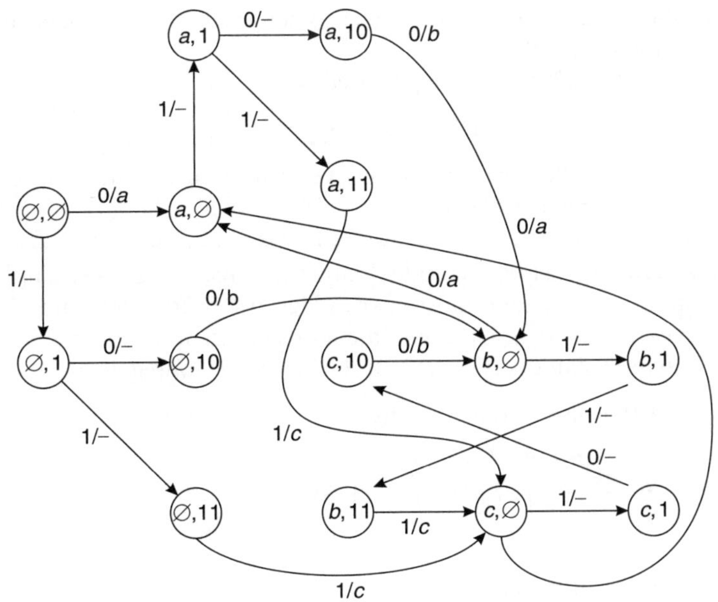

FIGURE 4.8 Finite-state machine for the code C_1 (decoder), accounting for the constraint ξ_1.

of the bitstream generated by the source coder. For that purpose, according to (4.28) and (4.29), one has to evaluate $|\Omega_n|$ or $|\Phi_n|$ (or $|\Omega_n(\xi)|$ or $|\Phi_n(\xi)|$ in the constrained case).

Consider a symbol-clock FSM $(\mathcal{I}, \mathcal{O}, \Sigma, \sigma_0, t, \omega, \mathcal{F})$ representing the coder associated to C (possibly accounting for ξ). A sequence $\mathbf{x}$ belongs to Ω_n (or $\Omega_n(\xi)$) if it leads to a *path*, i.e., a succession of states, starting from σ_0 and ending in $\mathcal{F}$, for which the cumulated length of each transition output is n. $|\Omega_n|$ (or $|\Omega_n(\xi)|$) is thus equal to the number $\eta(n, \sigma_0)$ of paths starting from σ_0 and ending in $\mathcal{F}$, with n bits as output. A recursive evaluation of $\eta(n, \sigma_0)$ is possible by considering all transitions that are stemming from σ_0

$$\eta(n, \sigma_0) = \sum_{x \in \mathcal{I}} \eta(n - \ell(\omega(\sigma_0, x)), t(\sigma_0, x)) \tag{4.32}$$

with the following boundary conditions.

- $\eta(\nu, \varepsilon) = 0$ for any $\nu \geqslant 0$
- $\eta(\nu, \sigma) = 0$ for any $\nu < 0$ (the considered path has a too long output)
- $\eta(0, \sigma) = 0$ if $\sigma \notin \mathcal{F}$ (the considered path does not end in an admissible final state)
- $\eta(0, \sigma) = 1$ if $\sigma \in \mathcal{F}$

In (4.32), $\ell(\omega(\sigma_0, x))$ corresponds to the length of the output bit sequence when a transition from state σ_0 to state $t(\sigma_0, x)$ occurs. When the input symbol

x is not valid, $t(\sigma_0, x) = \varepsilon$ and the path needs not be taken into account, and thus $\eta(\nu, \varepsilon) = 0$ for any $\nu \geqslant 0$.

Using (4.32), (4.29), for example, becomes

$$R_n(\mathcal{C}, \xi) = n - \log_2 \eta(n, \sigma_0)\,. \tag{4.33}$$

Example 4.18 illustrates this recursive evaluation for the code $\mathcal{C}_1$ with constraint ξ_1 taken into account.

Example 4.18: Consider the code $\mathcal{C}_1$ and the constraint ξ_1 of Example 4.11 represented by the Mealy transducer of Figure 4.6. To determine $|\Omega_4(\xi_1)|$, one has to evaluate $\eta(4,0)$. For the considered transducer, the set of admissible final states is $\mathcal{F} = \{1,2,3\}$. Using (4.32), one gets

$$\begin{aligned}\eta(4,0) &= \eta(4-\ell(\omega(0,a)), t(0,a)) + \eta(4-\ell(\omega(0,b)), t(0,b)) \\ &\quad + \eta(4-\ell(\omega(0,c)), t(0,c)) \\ &= \eta(3,1) + \eta(1,2) + \eta(1,3).\end{aligned}$$

Now

$$\begin{aligned}\eta(3,1) &= \eta(3-\ell(\omega(1,a)), t(1,a)) + \eta(3-\ell(\omega(1,b)), t(1,b)) \\ &\quad + \eta(3-\ell(\omega(1,c)), t(1,c)) \\ &= \eta(3,\varepsilon) + \eta(0,2) + \eta(0,3) \\ &= 0+1+1 = 2.\end{aligned}$$

Similarly, one may obtain $\eta(1,2) = 1$ and $\eta(1,3) = 1$. Finally, one gets $\eta(4,0) = 4$, and thus $|\Omega_4(\xi_1)| = 4$, as detailed in Example 4.11.

Even if recursive, a straightforward implementation of (4.32) would lead to a significant computational complexity, as soon as n becomes large. Storing and using intermediate results, i.e., values of $\eta(\nu, \sigma)$ with $\nu < n$ and $\sigma \in \Sigma$, may provide an important complexity reduction, since it is very likely that terms like $\eta(\nu, \sigma)$ for a given value of (ν, σ) appear several times in the recursion (4.32). The price to be paid being a memory of $(n-1)|\Sigma|$ cells to keep the already evaluated $\eta(\nu, \sigma)$s.

In the next sections, a more accurate characterization of two main sources of redundancy actually met in recent standards, namely redundancy due to the syntax of the code words and to their packetization is discussed.

4.2.3 Syntax of the Code Words

This specific type of redundancy is present when some sequences of bits cannot be interpreted as successions of code words of the considered code, like any succession of letters cannot be interpreted as a succession of English words.

Prefix codes, such as Huffman or arithmetic codes, are very often used to perform data compression. For such codes, decoding is *instantaneous*, i.e., it is not necessary to receive the whole encoded bitstream to start decoding. This property results in some constraint on the length of the code words forming a prefix code. Consider a binary prefix code $\mathcal{C}$ with word lengths $\{\ell_1 \dots \ell_m\}$. Since $\mathcal{C}$ is prefix, the *Kraft number* $K(\mathcal{C})$ of $\mathcal{C}$ has to satisfy

$$K(\mathcal{C}) \leqslant 1, \tag{4.34}$$

see Chapter 3, page 42.

When (4.34) is an equality, the code $\mathcal{C}$ is said to be *complete*. In such cases, any sequence of $\{0, 1\}^*$ can be interpreted as a succession of code words, possibly terminated by the prefix of a code word. When (4.34) is a strict inequality, the code is not complete or *redundant*. For redundant codes, Proposition 4.19 is satisfied.

Proposition 4.19: *Consider a binary prefix code $\mathcal{C}$ with word length $\{\ell_1, \dots, \ell_m\}$ such that $K(\mathcal{C}) < 1$, then there exist arbitrarily long sequences in $\{0, 1\}^*$ which cannot be decoded into sequences of code words.* ■

Proof. Let $\ell_{\max} = \max\{\ell_1, \dots, \ell_m\}$. There exist $2^{\ell_{\max}}$ sequences of $\ell_{\max}$ bits, and among these sequences, $2^{\ell_{\max}-\ell_\kappa}$ sequences start with the κth code word. Because of the prefix condition, no sequence can start with two distinct code words. Summing over all code words and using the fact that $K(\mathcal{C}) < 1$, the number of sequences starting with one of the code words of $\mathcal{C}$ is

$$\sum_{\kappa=1}^{m} 2^{\ell_{\max}-\ell_\kappa} = 2^{\ell_{\max}} \sum_{\kappa=1}^{m} 2^{-\ell_\kappa} < 2^{\ell_{\max}}. \tag{4.35}$$

As a consequence, some sequences start with a succession of bits that cannot be interpreted as a code word. ■

Example 4.20 illustrates Proposition 4.19 on a simple variable-length code.

Example 4.20: Consider again the code $\mathcal{C}_1 = \{0, 100, 111\}$. The Kraft number of this code is $K(\mathcal{C}_1) = 2^{-1} + 2^{-3} + 2^{-3} = 0.75 < 1$. Thus, there exists finite sequences, for example 0110, which cannot be decoded using $\mathcal{C}_1$.

For a code $\mathcal{C}$ with encoding function C, a first idea to evaluate the redundancy due to the syntax of the code words in sequences of n bits would be to determine $|\Omega_n|$, as defined by (4.23). Nevertheless, considering finite sequences of n bits, which have to be mapped to an integer number of code words, as required by (4.23), introduces implicitly a termination constraint which is not directly linked to syntax, but more to the packetization of data, as detailed in Section 4.2.4.

To evaluate only the redundancy due to the syntax of the code words, consider the set of all sequences of n bits, which are prefixes of sequences of code words

$$\Omega_n^{\mathrm{s}} = \left\{ \mathbf{b} \in \{0, 1\}^n \mid \exists \mathbf{x} \in \mathcal{A}^*, \mathbf{b} \sqsubseteq C(\mathbf{x}) \right\}, \tag{4.36}$$

where $\mathbf{a} \sqsubseteq \mathbf{b}$ indicates that $\mathbf{a}$ is a prefix of $\mathbf{b}$ (see Hopcroft and Ullman (1979), Chapter 3). The deterministic redundancy due to the syntax of the code words directly follows from (4.28)

$$R_n^{\mathrm{s}}(\mathcal{C}) = n - \log_2 |\Omega_n^{\mathrm{s}}|. \tag{4.37}$$

Its evaluation can be obtained through an enumeration of all code words of n bits or more realistically with the help of a bit-clock FSM describing the decoder associated to C. Considering a bit-clock transducer is important here, since sequences of bits not necessarily corresponding to an integer number of symbols have to be considered. This is not possible with symbol-clock FSM. The recursions (4.32) and the termination constraints are easily described with a bit-clock FSM, as illustrated by Example 4.21.

Example 4.21: Consider the code $\mathcal{C}_1$ of Example 4.10. Focusing on the transducer of the source decoder, input and output alphabets are $\mathcal{I} = \{0, 1\}$ and $\mathcal{O} = \{a, b, c\}^*$, see Figure 4.9. The states still correspond to the processed bits that do not allow yet to decode a symbol. They have been labeled with integers instead of binary numbers. The initial state σ_0 is 0. For the evaluation of the redundancy due to syntax, all states are allowed as final states, *i.e.*, $\mathcal{F} = \{0, 1, 2, 3\}$. Now, $|\Omega_n^{\mathrm{s}}| = \eta(n, \sigma_0)$, where $\eta(n, \sigma_0)$ may be recursively evaluated as

$$\eta(n, \sigma_0) = \sum_{b \in \mathcal{I}} \eta(n-1, t(\sigma_0, b)) \tag{4.38}$$

with the same limit conditions as in Section 4.2.2. From each state of the considered FSM, at most two branches may stem. Each branch has a single input bit; thus in (4.38), $\eta(n, \sigma_0)$ is the sum of the number of path of $n-1$ bits starting from $t(\sigma_0, b)$, with $b \in \mathcal{I}$. Applying (4.38) to $\mathcal{C}_1$, one gets the results of Table 4.1, which shows that for sequences of $n = 32$ bits, more than 7 bits of redundancy are due to the syntax of the code words. The relative (per bit) redundancy $R_n^{\mathrm{s}}(\mathcal{C}_1)/n$ is about 0.235.

Bit-clock FSM, such as that described in Example 4.21, may also be considered to evaluate the deterministic redundancies $R_n(\mathcal{C})$ or $R_n(\mathcal{C}, \xi)$. The main

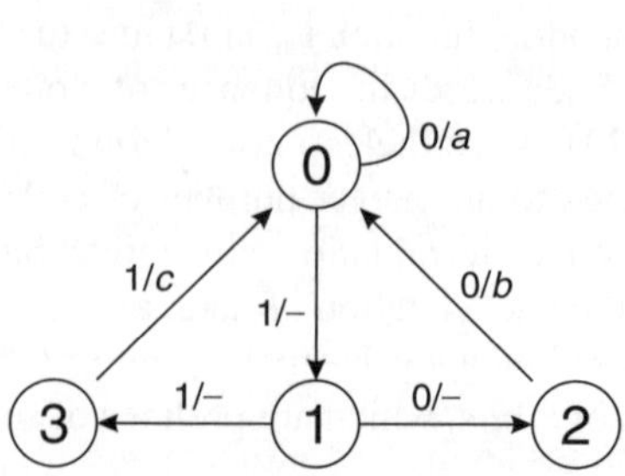

FIGURE 4.9 Finite-state machine for the code $\mathcal{C}_1$ (decoder), no constraint being taken into account.

TABLE 4.1 Redundancy due to the syntax of the code words for the code $\mathcal{C}_1$

n	8	12	16	20	32
$\lvert\Omega_n^{\mathrm{s}}\rvert$	73	601	4969	41081	23.2×10^6
$R_n^{\mathrm{s}}(\mathcal{C}_1)$	1.81	2.77	3.72	4.67	7.53
$R_n^{\mathrm{s}}(\mathcal{C}_1)/n$	0.226	0.231	0.233	0.234	0.235

difference between the evaluation of R_n^{s} and R_n lies in the set of final states, which has to correspond to end of code words when considering R_n. Thus, with the FSM of Example 4.21, one would take $\mathcal{F} = \{0\}$.

Usually, for well-designed codes, the deterministic redundancy due to syntax is much smaller than that shown in Example 4.21. Notable exceptions are error-detecting or correcting variants of Huffman codes (Huffman, 1952), such as Reversible variable-length codes (RVLC) (Takishima et al., 1995) or Universal variable-length codes (UVLC) (Itoh and Cheung, 2000). These codes have been designed to introduce some amount of redundancy, which can be used at decoder side to perform error detection or correction, see Section 10.2.2.

4.2.4 Packetization of the Coded Bitstream

Compressed data are usually put into packets to allow transmission over packet-switched networks, see Chapter 9 for more details. The number of bits n of a packet is usually known at receiver side, since it is contained in the header of the packet, or it may be deduced after segmentation of the received data into packets.

Successive packets arc usually requested to contain an integer number of encoded symbols to allow independent decoding of data contained in each of them. In some cases, even the number of symbols k within a packet is known. Packetization of data thus imposes constraints on the bitstream generated by the source coder; the resulting deterministic redundancy due to *packetization* is evaluated in this section. This type of redundancy is independent of the completeness of the source code. As will be seen, even for complete Huffman codes designed for sources with dyadic probability distribution, knowing k and n may lead to significant redundancy in the bitstream.

Two cases are considered, depending on whether only the length of the packet is known or the length and the number of symbols it contains are known.

Known Length of the Packet

In this case, the number of symbols of a packet is not known. One may only assume that a packet contains an integer number of symbols. When performing the evaluation of $\eta(n, \sigma_0)$ using (4.32) to evaluate $R_n(\mathcal{C})$, both redundancy due to the syntax of the code words and the packetization are taken into account

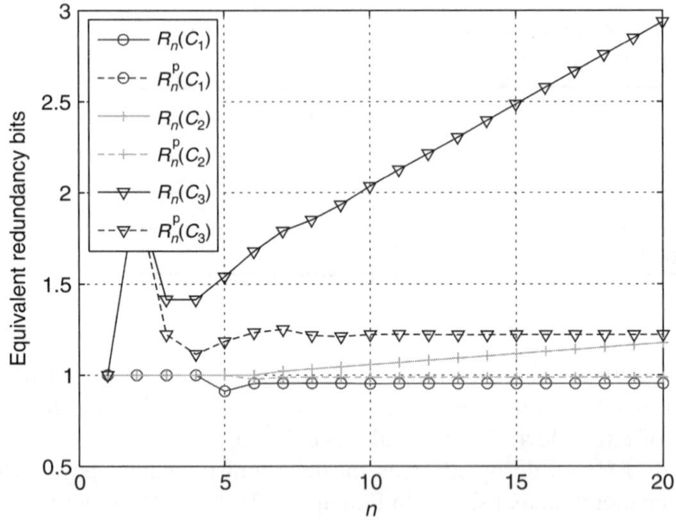

FIGURE 4.10 Deterministic redundancy (syntax and packetization) for a complete ($\mathcal{C}_1$) and two incomplete ($\mathcal{C}_2, \mathcal{C}_3$) variable-length codes as a function of the length of the packet.

simultaneously. To isolate the part due to packetization, one has to evaluate $R_n^{\mathrm{s}}(\mathcal{C})$, as described in Section 4.2.3. Then, the redundancy due to packetization $R_n^{\mathrm{p}}(\mathcal{C})$ may be obtained as follows

$$R_n^{\mathrm{p}}(\mathcal{C}) = R_n(\mathcal{C}) - R_n^{\mathrm{s}}(\mathcal{C}). \tag{4.39}$$

A bit-clock FSM representing the decoder allows to evaluate $R_n(\mathcal{C})$ and $R_n^{\mathrm{s}}(\mathcal{C})$.

Example 4.22: Consider three variable-length codes $\mathcal{C}_1$, $\mathcal{C}_2$, and $\mathcal{C}_3$ with code-word lengths $\{1,2,3,4,5,5\}$ for $\mathcal{C}_1$, $\{1,2,3,4,5,6\}$ for $\mathcal{C}_2$, and $\{1,3,3,4,5,5\}$ for $\mathcal{C}_3$. One may easily verify that $\mathcal{C}_1$ is complete, whereas $\mathcal{C}_2$ and $\mathcal{C}_3$ are not complete. $\mathcal{C}_2$ has less redundancy than $\mathcal{C}_3$. Figure 4.10 illustrates the evaluation of $R_n(\mathcal{C}_i)$ and $R_n^{\mathrm{p}}(\mathcal{C}_i)$, $i=1,\ldots,3$ as a function of n. For the complete code $\mathcal{C}_1$, $R_n(\mathcal{C}_1)$ remains constant around 0.95 bit when n increases. On the contrary, $R_n(\mathcal{C}_2)$ and $R_n(\mathcal{C}_3)$ increase linearly, whereas $R_n^{\mathrm{p}}(\mathcal{C}_2)$ and $R_n^{\mathrm{p}}(\mathcal{C}_3)$ remain also constant around 1.2 bit.

Example 4.22 shows that even for a complete Huffman code, there is some (small) redundancy introduced by the knowledge of the length of the encoded bitstream. This redundancy remains constant as soon as n is sufficiently large.

Known Length of the Packet and Number of Symbols It Contains

When both the number of bits n and the number of symbols k in a packet are known, the results of Section 4.2.2 again apply, provided that some adaptations are performed on the function η used to evaluate the cardinal number of Ω_n.

Consider a code $\mathcal{C}$ and its associated bit-clock FSM $(\mathcal{I}, \mathcal{O}, \Sigma, \sigma_0, t, \omega, \mathcal{F})$ representing the decoder. This FSM accounts for some constraint ξ but not for the number of encoded symbols. Let $\eta(n, k, \sigma_0)$ be the number of paths in this FSM starting from σ_0 and ending in $\mathcal{F}$, with n bits as input and k symbols as output. Each transition from σ_0 is driven by a single bit. Any path considered in $\eta(n, k, \sigma_0)$ starts with a transition from σ_0. Thus, $\eta(n, k, \sigma_0)$ is recursively evaluated as

$$\eta(n, k, \sigma_0) = \sum_{b \in \mathcal{I}} \eta(n-1, k - \ell(\omega(\sigma_0, b)), t(\sigma_0, b)) \tag{4.40}$$

with the following limit conditions:

- $\eta(n, k, \varepsilon) = 0$ for any n and k
- $\eta(n, k, \sigma) = 0$ for any $n < 0$ or $k < 0$ (the considered path has too long input or output)
- $\eta(0, 0, \sigma) = 0$ if $\sigma \notin \mathcal{F}$ (the considered path does not end in an admissible final state)
- $\eta(0, 0, \sigma) = 1$ if $\sigma \in \mathcal{F}$

The constraint on the number of symbols is taken into account by the second argument of $\eta(n, k, \sigma_0)$. When evaluating $|\Omega_n(\xi)| = \eta_b(n, k, \sigma_0)$, redundancy due to the syntax of $\mathcal{C}$, ξ, and due to the packetization are taken into account. Example 4.23 illustrates the redundancy due to the packetization of data when both n and k are known.

With a symbol-clock FSM representing the encoder, $\eta(n, k, \sigma_0)$ would be evaluated using the following recursion

$$\eta(n, k, \sigma_0) = \sum_{x \in \mathcal{I}} \eta(n - \ell(\omega(\sigma_0, x)), k-1, t(\sigma_0, x)); \tag{4.41}$$

obviously, the functions ω and t in (4.41) are not the same ones as in (4.40).

Example 4.23: Consider again the variable-length code $\mathcal{C}_2$ with code-word lengths $\{1, 2, 3, 4, 5, 6\}$ of Example 4.22. The associated FSM is similar to that of Figure 4.4. Figure 4.11 illustrates the evaluation of $R_n(\mathcal{C}_2) = R_n^{\mathrm{s}}(\mathcal{C}_2) + R_n^{\mathrm{p}}(\mathcal{C}_2)$ and $R_n^{\mathrm{p}}(\mathcal{C}_2)$ as a function of n for various values of k. $R_n^{\mathrm{s}}(\mathcal{C}_2)$ is deduced from the results of Example 4.22. The redundancy due to syntax is negligible when compared to that induced by the knowledge of n and k. For a given value of k, there exists some n for which $R_n^{\mathrm{p}}(\mathcal{C}_2)$ is minimized. This value corresponds more or less to $k\overline{\ell}$, where $\overline{\ell}$ is the average length of the code $\mathcal{C}_2$, assuming a probability for the ith symbol equal to $2^{-\ell_i}$, with ℓ_i the length of the code associated to the ith symbol. For $\mathcal{C}_2$, $\overline{\ell} = 1.875$ bits.

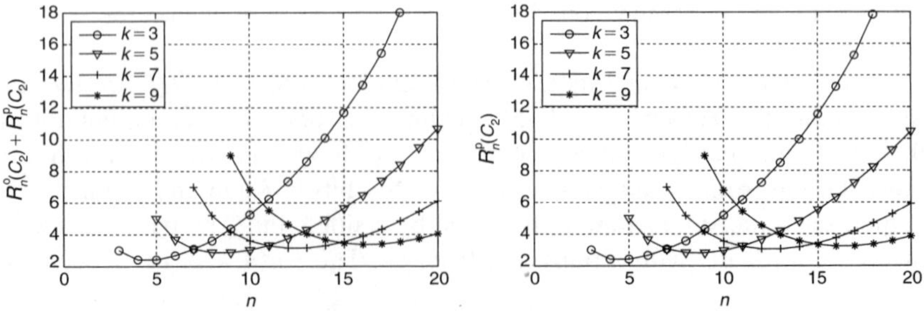

FIGURE 4.11 Deterministic redundancy due to syntax and packetization (left) and to packetization only (right) when n and k are known for the variable-length code C_2.

The knowledge of n and k results in a significant amount of redundancy. Nevertheless, if assuming n available is quite realistic in many situations, since it is stored in packet headers, considering k as known is more questionable. It could also be provided in the packet header, at a price of $\lceil \log_2 k_{\max} \rceil$ bits, if one knows that $k \leqslant k_{\max}$. The availability of n, as well as the corresponding redundancy, comes thus usually for free, but this is no more the case in general for k.

4.2.5 Semantics of the Source

This type of redundancy comes from the fact that the succession of data to be compressed has usually a meaning. Decoding some random bitstream may not necessarily produce a consistent succession of source symbols, just like succession of randomly chosen English words does not necessarily result in a meaningful sentence.

A first way to account for the semantic of a source is by considering a Markov model of it (Guivarch et al., 2000; Jeanne et al., 2005). With such model, successions of source symbols that are not consistent with some very basic syntax rules or which have no valid semantic interpretation could be avoided. Nevertheless, even with complex high-order Markov models, it is hardly possible to get a global description of the semantic of the source. For example, the fact that a compressed bitstream should represent

- less than a given number of transformed coefficient from a block of pixels of an image, see Section 4.3.2,
- samples of a sound record, with long-term correlation,
- an HTML file, with its specific syntax rules,
- a natural image, with smooth regions and with contours

is very difficult to describe only with a Markov model or require very sophisticated mixtures of models, see, e.g., Winkler (2005) for images.

A second possibility is to consider an FSM designed to verify some basic or more sophisticated syntax rules. Such FSMs are quite easy to design to check some syntax rules, e.g., in HTML files, where each HTML tag should start with

< and end with >, the main part of the HTML text should start with the `<html>` tag and end with the `</html>` tag.

Again, when an FSM is available to account for the constraints imposed by the semantic of the source, tools such as those described in Section 4.2.2 can be used to quantify the redundancy due to the semantic of the source. More details are discussed in Section 4.3.

4.2.6 Other Sources of Redundancy

Redundancy due to the packetization of the bitstream has been studied in Section 4.2.4. This type of redundancy is more linked to requirements imposed by the transmission scheme than to constraints on the source or the inefficiency of the source coder. In fact, the layered architecture of many communication systems involve packet encapsulation at several levels (Kurose and Ross, 2005). Many headers are appended by the lower protocol layers to packets generated by the source coder at the upper application layer. These headers are usually strongly correlated from one packet to the next. Moreover, since they contain error-prone information, they are protected by cyclic redundancy checks (CRCs) or checksums. As a result, a large amount of redundancy is introduced by the various encapsulations. This redundancy can again be exploited to allow JSCD. Chapter 9 is dedicated to the exploitation of the redundancy introduced by the network to perform JSCD of packets at various protocol layers.

4.3 COMPARING VARIOUS SOURCES OF REDUNDANCY

Two examples are considered in this section. First, we consider the residual redundancy in the coding of an important part of the bitstream in the MPEG 4 AAC audio coder (ISO/IEC, 2005). Then, we examine the redundancy in the encoding of blocks of pixels in the H.263+ video coder (ITU-T, 2005).

4.3.1 Scale Factors in MPEG 4-AAC

High-quality audio codecs, such as the MPEG-AAC (ISO/IEC, 1997), are used to get high compression rates without perceptual distortion. Originally designed for data transmission (or data storage) over reliable channels, these codecs are highly vulnerable to transmission errors, largely because the bitstream they generate consists of variable-length codes. Our aim here is to evaluate the residual redundancy in some parts of the bitstream generated by an MPEG-AAC.

System Overview

The MPEG-AAC standard specifies a bitstream format for encoded audio data. The bitstream is segmented in *frames* of fixed length in the case of fixed bitrate encoding or variable length in the case of variable bitrate encoding. We consider here the most simple AAC bitstream: a monophonic audio signal encoded with the low complexity profile. A frame is made up of several parts:

- A *fixed header*: It contains mainly the following information: number of audio channels, sampling frequency, and encoder profile. The same header is repeated at the beginning of each frame.
- A *variable header*: It contains mainly the length of the current frame in bytes. This header is of fixed length, but its content may vary from one frame to the other.
- An *individual channel stream* (ICS) field: It contains mainly the shape and length of the transform analysis-window and the first scale factor (called *global gain*).
- A *section* field: It describes the grouping of frequency subbands in so-called *sections*. For each section, the section length and the variable-length code used for coding quantized transform coefficients are specified.

The length of the ICS and section fields is not fixed a priori. Among the other fields, one may find the following:

- A *scale-factor* field: Scale factors determine the quantization step for each subband. It uses a differential code followed by a VLC.
- A *spectral data* field: It corresponds to the quantized transform coefficients. Eleven VLCs can be used, one per section. Code words can be interleaved with escape sequences for coding high magnitudes.
- *Data stream element* (DSE): It is an optional field that allows the inclusion of additional raw binary data in the bitstream. These data are ignored by a standard AAC decoder but can be used by more elaborate decoding devices, which are thus still standard compliant.

Three categories of data can be identified in AAC frames, according to their sensitivity to errors (Korhonen and Wang, 2003). *Critical data* consists of headers, ICS, and section fields. Without these data, decoding is almost impossible. Thus, errors on critical data are usually considered equivalent to a frame loss. *Intermediate data* are the scale factors. If missing, the audio output is highly distorted. The remaining part of the bitstream (spectral data) is much less sensitive to errors.

Redundancy in the Scale Factors

In a packet of AAC-encoded data, the scale factors are usually not that easily isolated in a noisy bitstream, since they are preceded by a header which size is not fixed. Assuming that the fixed and variable headers, the ICS and the section fields have been received correctly, the location of the beginning of the scale-factor field can be identified without ambiguity.

The scale factors are encoded with a VLC $\mathcal{C}_{\text{AAC}}$ consisting of 121 code words. The Kraft number of this code is $K(\mathcal{C}_{\text{AAC}}) = 1$, indicating that the code is complete. Using the results of Section 4.2.3, one may easily verify that the redundancy

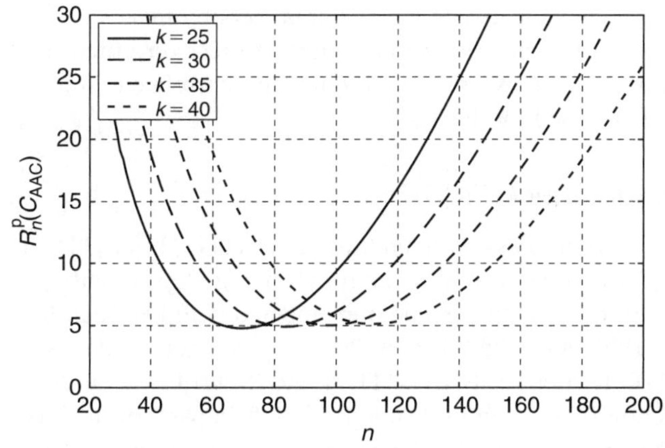

FIGURE 4.12 Redundancy due to packetization in MPEG AAC scale-factor fields, as a function of the size n of the field and of the number k of scale factors it contains.

due to syntax is null, which is consistent with the value of the Kraft number. Now, assuming that the length of the scale-factor field is known, one may evaluate the redundancy due to packetization, according to Section 4.2.4. One gets about 1.5 bits of redundancy, whatever the length of the scale-factor field. The size of the scale-factor field can easily be stored in the optional DSE field. However, the price to pay for getting these 1.5 bits of redundancy is usually too high.

The number of scale factors in each scale-factor field may be inferred from the header information (ICS and section fields). As shown in Section 4.2.4, this information provides a significant amount of redundancy. It can be evaluated considering a simple FSM not accounting for the number of symbols and a recursive evaluation of η using (4.40). Figure 4.12 presents the evolution of the redundancy due to packetization as a function of the size n of the scale-factor field and of the number of coefficients k it contains. The minimum residual redundancy is about 5 bits for k between 25 and 40. Nevertheless, for a fixed value of k, the redundancy increases significantly as soon as n becomes different from the value $\widetilde{n}(k)$, which minimizes the residual redundancy for a given k. In such situations, the price to be paid for the transmission of n as side information may be worthwhile.

Discussion

Here, only the redundancy left in the scale factors has been analyzed. Similar evaluations could be done for the spectral data field, using exactly the same set of tools. The main difference comes from the fact that several sets of variable-length code words have to be considered, depending on the value of each previously decoded scale factor. This complicates significantly the analysis.

Other sources of redundancy may also be considered for this application. For example, Alajaji et al. (1996) as well as Fingscheidt and Vary (2001) model the statistical dependency between coded parameters by means of Markov chains. A similar approach could also be considered to recover efficiently the headers of each frame, since they are also significantly correlated. Tools developed in Chapter 9 can be used for this task.

4.3.2 Texture Blocks in H.263

Image and video compression standards, such as ITU (1992), ITU-T (1990), and ITU-T (2005), involve a lossless entropy coding step, usually after a quantization step. Entropy coding can be implemented with Huffman-like codes as in JPEG (ITU, 1992) and H263 (ITU-T, 2005) or arithmetic codes as in JPEG-2000 (ISO/IEC JTC 1, 2000) or H.264 (ITU-T and ISO/IEC JTC 1, 2003).

This section focuses on the redundancy left in parts of the bitstream by the H.263 video encoder (ITU-T, 2005). In many image and video compression systems, each image is divided into blocks which are encoded independently, see Chapter 3. For images that are intra encoded, the blocks are processed directly. For inter-encoded pictures, a motion estimation/compensation step is done first to get a *residual image*, containing the motion compensation residuals.

The pixels of each block of an intra encoded image or a residual image are processed by a 2D discrete cosine transform (DCT) (ITU-T, 2005). The number of transformed coefficients is equal to the number of pixels of the image block. These DCT coefficients are then quantized and zig-zag scanned to get vectors that are likely to end with many zeros (corresponding to high-frequency coefficients that are coarsely quantized). These vectors are then entropy coded to get coded *texture data*. More precisely, in the H.263 standard (ITU-T, 2005), each VLC code word corresponds to a triple (*run*, *level*, and *last*): *run* represents the number of zeros till the next nonzero vectorized DCT coefficient, *level* represents the value of this nonzero DCT coefficient, and *last* indicates whether this DCT coefficient is the last nonzero value of the vector. The last zeros of a vector are thus no more coded. As a consequence, each VLC code word corresponds to (run+1) DCT coefficients (*run* zero DCT coefficients and one nonzero coefficient).

The encoder produces packets associated to one or several blocks of pixels, where header information, motion vectors, and texture data are encoded consecutively for each block. When *data-partitioning* is activated (ITU-T, 2005), these data are grouped within three partitions (headers, motion vectors, and texture data), each of which is separated by a synchronization marker. This marker facilitates the separation of each partition into three virtual packets. Considering the packet containing the texture data, several constraints have to be satisfied.

Constraint 4.24: *For each image block, the sum of the number of DCT coefficients represented by its corresponding VLC code words should be less or equal* k_P *(the number of pixels or DCT coefficients of an image block) (ITU-T, 2005).*

So, the VLC code words of a compressed image block should be such that

$$\sum_{\kappa=1}^{k_V}(r_\kappa+1)\leqslant k_P,$$

where r_κ is the run *value associated to the κth VLC code word and k_V is the number of VLC code words used to represent the image block.* ■

Constraint 4.25: *Only the last VLC code word corresponding to a compressed image block should represent a triple* (run,level, *and* last) *with* last *equal to one (ITU-T, 2005).* ■

These two constraints are clearly related to the semantic of the source coder, see Section 4.2.5. Combined with the syntax of the code words used to represent the triples, they result in a redundancy that is evaluated in what follows.

Single Image Block

In a first time, the number of bits n of the encoded texture part of a given image block is assumed to be known. We also consider that these n bits correspond to $k_D \geqslant 1$ DCT coefficients (the $k_P - k_D$ last coefficients being null).

As in Section 4.2.4, one may adapt the definition η to account for the various constraints imposed by the coding of H.263 texture blocks. To facilitate computations, the VLC code words used to encode the transformed and quantized blocks of pixels are grouped into two codebooks $\mathcal{C}_0$ and $\mathcal{C}_1$ corresponding to code words associated with triples, with *last* equal to 0 and to *last* equal to 1. The length in bits of a code word $\mathbf{c}$ is denoted by $\ell(\mathbf{c})$ and the number of DCT coefficients it represents by $\kappa(\mathbf{c})$.

The number $\eta_{\text{TB}}(n)$ of sequences of n bits consisting of VLC code words encoding a *texture block* and satisfying Constraints 4.24 and 4.25 and the constraints due to the syntax and packetization is then

$$\eta_{\text{TB}}(n)=\sum_{\kappa=1}^{k_P}\eta_{\text{TB}}^1(\kappa,n), \tag{4.42}$$

where $\eta_{\text{TB}}^1(\kappa,n)$ is the number of sequences of n bits that VLC code words represent *exactly* κ DCT coefficients. Evaluating $\eta_{\text{TB}}^1(\kappa,n)$ requires to account for Constraint 4.25, which indicates that the last VLC code word belongs necessarily to $\mathcal{C}_1$. Thus, $\eta_{\text{TB}}^1(\kappa,n)$ is the sum of the number of sequences of n bits, representing κ DCT coefficients, formed only with code words of $\mathcal{C}_0$ and ending with any code word $\mathbf{c}\in\mathcal{C}_1$. More formally, one may write

$$\eta_{\text{TB}}^1(\kappa,n)=\sum_{\mathbf{c}\in\mathcal{C}_1}\eta_{\text{TB}}^0(\kappa-\kappa(\mathbf{c}),n-\ell(\mathbf{c})). \tag{4.43}$$

The number $\eta_{\text{TB}}^{0}(\kappa,n)$ of sequences of n bits, representing κ DCT coefficients, and formed of elements of $\mathcal{C}_0$ can thus be computed recursively as indicated by (4.32)

$$\eta_{\text{TB}}^{0}(\kappa,n)=\sum_{\mathbf{c}\in\mathcal{C}_0}\eta_{\text{TB}}^{0}(\kappa-\kappa(\mathbf{c}),n-\ell(\mathbf{c})), \tag{4.44}$$

with $\eta_{\text{TB}}^{0}(0,0)=1$ and $\eta_{\text{TB}}^{0}(\kappa,n)=0$ whenever $\kappa<0$ or $n<0$. These limit conditions allow to take the knowledge of n and Constraint 4.24 into account.

In (4.42), all constraints (syntax, packetization, and semantic) have been taken into account. The global redundancy is, thus, evaluated from (4.42). To isolate the contribution of each constraint on the global redundancy, one may use simple FSMs representing the encoder or decoder as well as all involved constraints, see Sections 4.2.3 and 4.2.4.

Figure 4.13 shows the number of equivalent redundancy bits as a function of the size n of the encoded texture block. When considering all code words in $\mathcal{C}_0\cup\mathcal{C}_1$, the Kraft number is $K(\mathcal{C}_0\cup\mathcal{C}_1)=0.998$, indicating that the code is very close to a complete VLC. The relative redundancy due to the syntax which is about 5.10^{-4}. In contrast, the redundancy due to Constraint 4.25 is much more significant. For encoded packets of 100 bits, it is about 11 bits. The redundancy is that significant because when considering only $\mathcal{C}_0$ to perform encoding of the

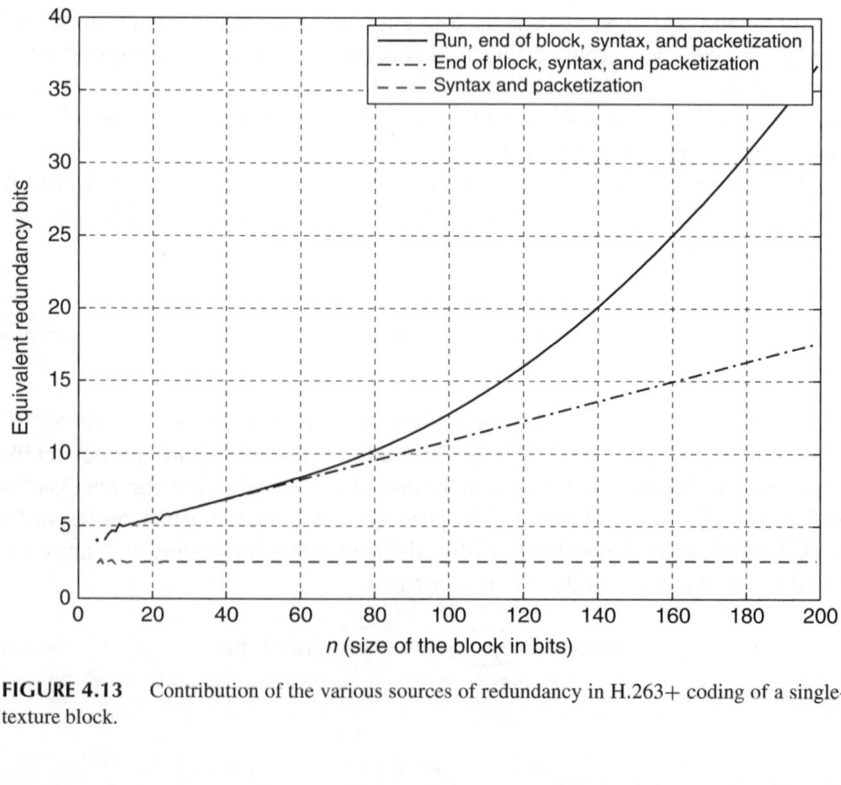

FIGURE 4.13 Contribution of the various sources of redundancy in H.263+ coding of a single-texture block.

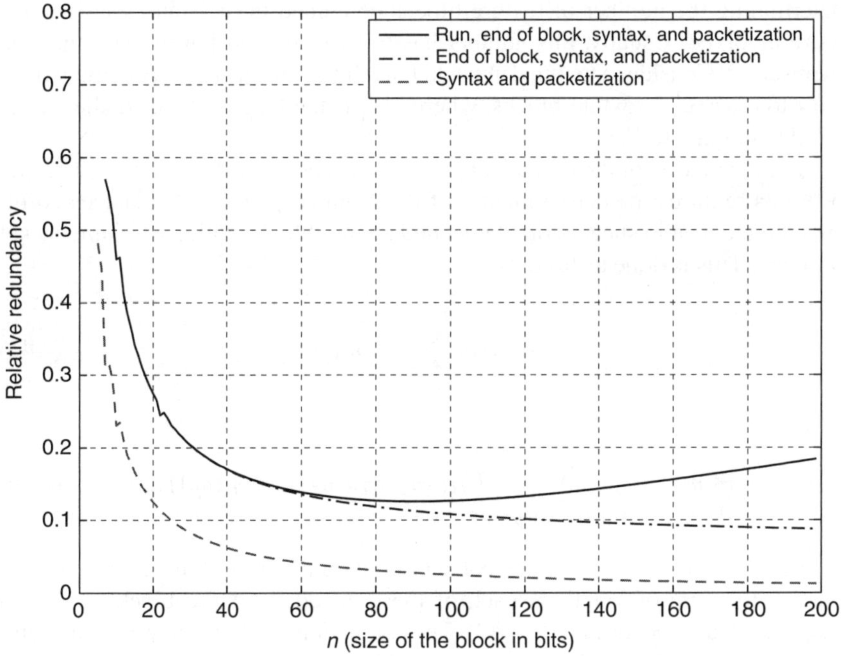

FIGURE 4.14 Relative redundancy as a function of the size n of an H.263+-encoded texture when taking various constraints into account.

first DCT coefficients of an image block, only 2/3 of the code words are used. Taking Constraint 4.24 into account results in an additional redundancy of about 2 bits, again for 100 bit packets. For larger values of n, Constraint 4.24 becomes more important. For example, for packets of $n = 160$ bits, it is about 15 bits. For large values of n, successions of very small VLC code words are no more possible, due to the limitation of the number of DCT coefficients of an image block.

All these results are confirmed by Figure 4.14 showing the evolution of the relative redundancy (i.e., the redundancy per bit of the encoded message) as a function of n when taking various constraints into account. For very short packets, n plays a very important role in the global relative redundancy. Its impact decreases to get minimal between 80 and 100 bits (around 0.12). It increases again afterward due to the constraint on the number of DCT coefficients, which are represented by the VLC code words.

Packet Containing Several Image Blocks

When receiving an H.263-encoded packet, assuming that data partitioning is activated, even in the presence of noise, it is possible to isolate the part of the packet dedicated to the coded texture blocks, see Chapter 9. Nevertheless, determining the number of texture blocks contained in the subpacket is much more involved. Usually, this number has to be estimated jointly with the data themselves. Consider a texture subpacket of n bits. This packet may contain from $t=1$ to $t=n/\ell^1_{\text{min}}$ texture blocks, where ℓ^1_{min} is the length of the smallest code word belonging to $\mathcal{C}_1$.

The evaluation of the redundancy left by the coder for the texture in a packet of n bits requires the determination of the number $\eta_{\text{TP}}(n)$ of sequences corresponding to a valid succession of encoded texture blocks within a texture packet of n bits. This is done as follows:

$$\eta_{\text{TP}}(n) = \sum_{t=1}^{n/\ell^1_{\text{min}}} \eta_{\text{TP}}(n,t), \tag{4.45}$$

with

$$\eta_{\text{TP}}(n,t) = \sum_{n_1+n_2+\cdots+n_t=n} \left(\eta_{\text{TB}}(n_1)\,\eta_{\text{TB}}(n_2)\ldots\eta_{\text{TB}}(n_p)\right), \tag{4.46}$$

where (4.45) considers texture packets containing from 1 to n/ℓ^1_{min} texture blocks; the sum of length of each of these encoded texture blocks is n, and η_{TB} is evaluated according to (4.42). The evaluation of (4.46) may become cumbersome, especially for large values of t. Nevertheless, (4.46) can be modified to allow an iterative evaluation of $\eta_{\text{TP}}(n,t)$ as follows

$$\begin{aligned}
\eta_{\text{TP}}(n,t) &= \sum_{n_t=1}^{n} \sum_{n_1+n_2+\cdots+n_{t-1}=n-n_t} \left(\eta_{\text{TB}}(n_1)\ldots\eta_{\text{TB}}(n_{t-1})\,\eta_{\text{TB}}(n_t)\right) \\
&= \sum_{n_t=1}^{n} \eta_{\text{TB}}(n_t) \sum_{n_1+n_2+\cdots+n_{t-1}=n-n_t} \left(\eta_{\text{TB}}(n_1)\ldots\eta_{\text{TB}}(n_{t-1})\right) \\
&= \sum_{n_t=1}^{n} \eta_{\text{TB}}(n_t)\,\eta_{\text{TP}}(n-n_t,t-1).
\end{aligned} \tag{4.47}$$

Iteration (4.47) is initialized with

$$\eta_{\text{TP}}(n,1) = \eta_{\text{TB}}(n), \tag{4.48}$$

evaluated using (4.42).

Figure 4.15 represents the residual redundancy in a texture packet of n bits

$$R_n^{\text{TP}}(\mathcal{C}_0 \cup \mathcal{C}_1, t) = n - \log_2 \eta_{\text{TP}}(n,t),$$

with n from 0 to 350 bits and containing t texture blocks for various values of t. As expected, for a given value of t, there exists some $\widetilde{n}_{\text{b}}(t)$ for which the redundancy is minimized. This length represents a kind of optimal (in terms of compression) packet size for t texture blocks.

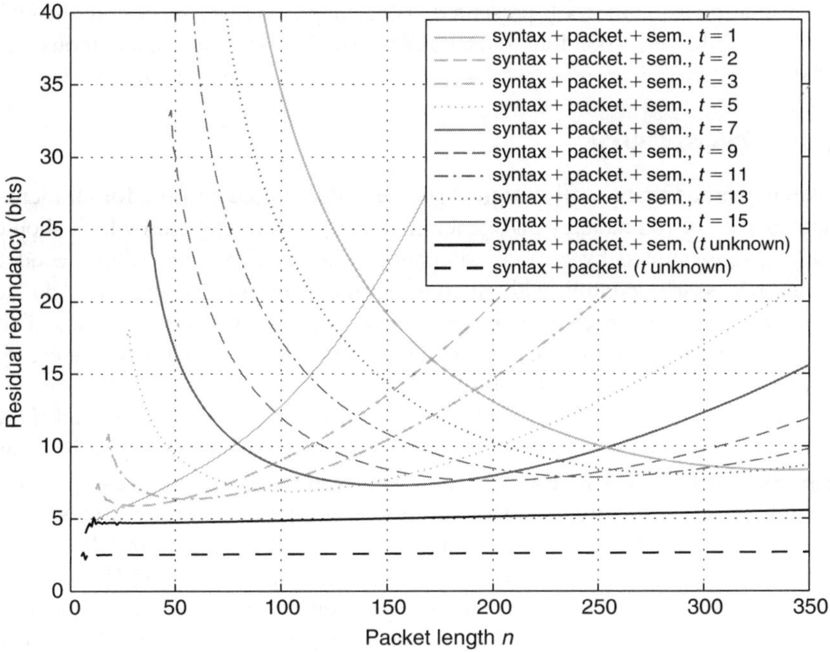

FIGURE 4.15 Redundancy in H.263+-encoded texture packets (unknown and known number of texture blocks).

Without any knowledge of the number of blocks within a texture packet, one obtains

$$R_n^{\mathrm{TP}}(\mathcal{C}_0 \cup \mathcal{C}_1) = n - \log_2 \eta_{\mathrm{TP}}(n)$$

represented in bold in Figure 4.15. $R_n^{\mathrm{TP}}(\mathcal{C}_0 \cup \mathcal{C}_1)$ is thus around 5 bits, among which, about 2.5 bits are due to the syntax of the code words. The additional redundancy provided by the knowledge of t is quite small when n is close to $\widetilde{n}_{\mathrm{b}}(t)$ but increases significantly as soon as n goes away from $\widetilde{n}_{\mathrm{b}}(t)$.

Discussion

The residual redundancy due to syntax, packetization, and semantic in subpackets containing H.263 compressed (data partitioned) texture information is about 5 bits. These bits correspond to the suboptimality of the entropy coder of the H.263 video coder. In H.264/AVC (ITU-T and ISO/IEC JTC 1, 2003) and in its scalable extension H.264/SVC, context-adaptive variable-length codes (CAVLC) or context-adaptive binary arithmetic codes (CABAC) have replaced the static Huffman-like VLC tables. These adaptive entropy coders, albeit more efficient, still produce packetized data. Redundancy due to semantic and packetization remains present in these more sophisticated source coders. Nevertheless, its evaluation using the tools presented in this chapter is much more complicated, since the FSM corresponding to the CABAC of H.264 would contain thousands of states.

4.4 CONCLUSION

One may argue that a well-designed source coder should account for all identified sources of redundancy and generate a compressed bitstream which should look as if generated by an i.i.d. symmetric binary source and when decoded would necessarily comply with all considered constraints. Nevertheless, due to complexity reasons, such ultimate source coder has not yet been designed. This opens many opportunities for the decoder to exploit the residual redundancy, see Shannon (1948), using JSCD tools.

However, it has to be noted that recent source coders take more and more into account such considerations and become more and more efficient in many aspects: They leave less residual redundancy (and CABAC makes a great job on this aspect), quite often the structuration in packets is carefully studied, and, more generally, even if some redundancy remains, it is much more difficult to identify. This evolution must be taken into account when studying JSCD.

Note also that all forms of redundancy described in this chapter were well structured, not too difficult to evaluate, and that efficient decoding schemes can be developed to exploit it, see Chapter 5.

Another source of redundancy that has not been described is due to the statistical dependence between symbols generated by some source. When the source is not memoryless, all successions of symbols are not equally likely. Intuitively, a decoder aware of the relation between source symbols would perform better than one working as if the source would be memoryless. The underlying redundancy relies on a source model which is not considered in what follows. In fact, such model would be highly dependent on the source realizations.

Chapter 5

Exploiting the Residual Redundancy

Chapter 4 identified some forms of stochastic and deterministic redundancy of source codes. In some cases, this redundancy could be evaluated based on a description in terms of finite-state machine (FSM). The aim of this chapter is to show how source-coded messages transmitted over communication channels may be efficiently recovered from noisy measurements obtained at the output of these channels. When source coders or decoders are described by FSMs, the decoding process is explained here in terms of *trellises*. Classical estimation algorithm, usually used to perform channel decoding are then adapted to perform joint source-channel decoding (JSCD) on these trellises. The resulting algorithms can thus be optimal. When the redundancy cannot be described by FSMs, or when the corresponding arithmetic complexity is too large, other types of algorithms (sequential decoders) can be used. They are described in Chapter 6.

Figure 5.1 represents the complete communication chain introduced in Chapter 1. As already mentioned, early results in JSCD do not consider the packetization process required to perform the transmission of multimedia contents over a realistic network. Channel outcomes or outputs of the channel decoders are assumed to be directly available to the source decoder. This assumption is equivalent to considering the part of the communication scheme contained in the shaded rectangle of Figure 5.1 as an idealized communication channel providing measurements of the bits at its input. Some of the tools required to get such

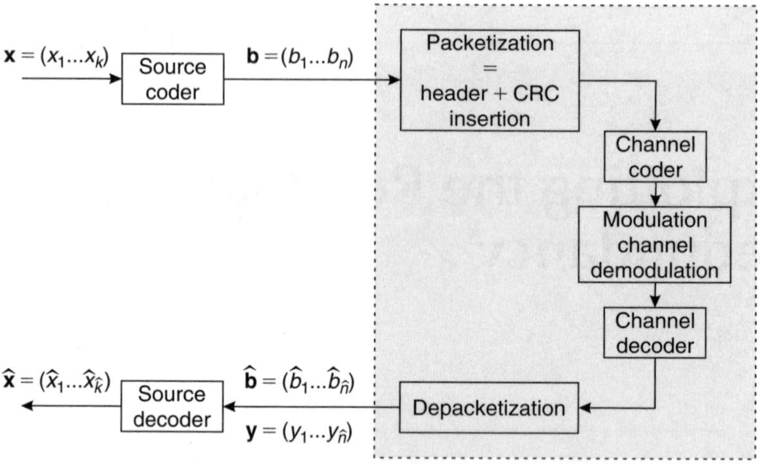

FIGURE 5.1 Complete communication model.

FIGURE 5.2 Simplified model of the communication system considered in this chapter.

idealized channel, e.g., robust packet defragmentation, reliable depacketization, raise in fact many issues, which will be considered in Chapter 8.

First, concentrate on the simplified communication system of Figure 5.2 where the source coder is only made of a variable-length code (VLC) encoder. A source $X_{1:k}$ generates a message $\mathbf{x}_{1:k} = (x_1, \ldots, x_k)$ belonging to $\mathcal{A}^k$. Message $\mathbf{x}_{1:k}$ passes through a binary source coder C, the output of which is represented by a random *variable-length* vector $\mathbf{B}_{1:N}$ producing a sequence of bits $\mathbf{b}_{1:n} = (b_1, \ldots, b_n) = C(\mathbf{x}_{1:k})$. For VLCs, n depends on $\mathbf{x}_{1:k}$ and is thus represented by the random variable N in $\mathbf{B}_{1:N}$. The sequence $\mathbf{b}_{1:n}$ is transmitted on a channel whose output is modeled by a random vector $\mathbf{Y}_{1:n}$ producing a sequence $\mathbf{y}_{1:n} = (y_1 \ldots y_n)$ of channel outcomes. The channel is characterized by its transition probability $P\left(\mathbf{Y}_{1:n} = \mathbf{y}_{1:n} | \mathbf{B}_{1:n} = \mathbf{b}_{1:n}\right)$, which simplifies to

$$P\left(\mathbf{Y}_{1:n} = \mathbf{y}_{1:n} | \mathbf{B}_{1:n} = \mathbf{b}_{1:n}\right) = \prod_{\nu=1}^{n} P\left(Y_\nu = y_\nu | B_\nu = b_\nu\right). \tag{5.1}$$

when the channel is memoryless.

In this chapter, for improved clarity of presentation, the number of bits n of the encoded sequence $\mathbf{b}_{1:n}$ is assumed to be known to the decoder. This assumption is seldom satisfied in practice, but some information is usually available. For example, when the encoded sequence is packetized before being processed by a protocol stack (see Figure 5.1), most standards include the length of packets in bytes in the packet header. As a result, the packets are *byte-aligned* (contain an integer number of bytes) but the actual packet may be of a slightly smaller number of bits. In this case, up to seven padding bits are appended to $\mathbf{b}_{1:n}$ to meet the packet length announced in the header. Therefore, in most practical situations, the algorithms explained in this chapter have to be modified to take this fact into account, at the cost of a slightly decreased performance and higher computational cost. The corresponding procedure is outlined at the corresponding places. Another solution would be to transmit the actual packet length (in bits) as side information but this would increase the required throughput.

The determination of the number of symbols k contained in the source sequence requires the decoding of $\mathbf{y}_{1:n}$. It may thus be associated with the realization of a random variable K, as it cannot be directly deduced from n. As a consequence, k has to be estimated from the received information or has to be transmitted as side information in a header. Both cases are addressed in this chapter.

The decoder aims at recovering the emitted data $\mathbf{x}_{1:k}$ or bits $\mathbf{b}_{1:n}$ from the received data. Several types of estimators for $\mathbf{x}_{1:k}$ or $\mathbf{b}_{1:n}$ are first presented in Section 5.1. Algorithms for computing these estimators are presented in Sections 5.2 and 5.3. A first application considering the estimation of the scale-factors of MPEG-4 AAC from noisy data is described in Section 5.4. Other more complex applications are described in Chapter 6. Section 5.5 gives some insight on taking into account other constraints than the first ones considered in Section 5.2.

5.1 ESTIMATORS

5.1.1 Element-by-Element Estimators

Using the knowledge of the channel output vector $\mathbf{y}_{1:n}$, the symbol maximum of posterior marginals (MPM) estimator

$$\widehat{x}_{\kappa}^{\mathrm{MPM}} = \arg\max_{x \in \mathcal{A}} P\left(X_{\kappa} = x | Y_{1:n} = \mathbf{y}_{1:n}\right) \tag{5.2}$$

allows to compute *symbol-by-symbol* maximum a posteriori (MAP) estimates. Reliability information for each symbol represented by the a posteriori probabilities (APPs) $P\left(X_{\kappa} = x | Y_{1:n} = \mathbf{y}_{1:n}\right)$, $x \in \mathcal{A}$ are also obtained as a by-product from the evaluation of (5.2). In what follows, the notation of probabilities may not show the random vectors (only its possible realization will be kept) when no confusion is possible. The BCJR algorithm (Bahl et al., 1974) is a popular algorithm to obtain such estimates in the case of convolutional codes, or fixed-length quantization indexes (Phamdo and Farvardin, 1994). In the case of VLCs, some adaptations have to be performed because each source symbol may be represented by a variable number of output bits, see Section 5.2.

Implementation of iterative decoders (see Section 9.2) requires the use of *bit-by-bit* reliability information. For that purpose, the bit MPM estimator

$$\widehat{b}_\nu^{\text{MPM}} = \arg\max_{b\in\{0,1\}} P\left(B_\nu = b|\mathbf{y}_{1:n}\right) \tag{5.3}$$

has to be considered in this case. Reliability information is often equivalently represented by the logarithm of the ratio of the APPs,

$$\Lambda_{\text{AP}}\left(\widehat{b}_\nu\right) = \log\frac{P\left(B_\nu = 0|\mathbf{y}_{1:n}\right)}{P\left(B_\nu = 1|\mathbf{y}_{1:n}\right)}, \tag{5.4}$$

which has the advantage, compared to $P\left(B_\nu = b|\mathbf{y}_{1:n}\right)$, to be of reduced dynamic range.

Using Bayes' rule, one may rewrite (5.4) as follows

$$\Lambda_{\text{AP}}\left(\widehat{b}_\nu\right) = \log\frac{P\left(\mathbf{y}_{1:n}|B_\nu = 0\right)}{P\left(\mathbf{y}_{1:n}|B_\nu = 1\right)} + \log\frac{P\left(B_\nu = 0\right)}{P\left(B_\nu = 1\right)} \tag{5.5}$$

$$= \Lambda\left(\widehat{b}_\nu\right) + \Lambda_{\text{P}}\left(\widehat{b}_\nu\right), \tag{5.6}$$

where

$$\Lambda_{\text{P}}\left(\widehat{b}_\nu\right) = \log\frac{P\left(B_\nu = 0\right)}{P\left(B_\nu = 1\right)} \tag{5.7}$$

is the logarithm of the ratio of the a priori probabilities and

$$\Lambda\left(\widehat{b}_\nu\right) = \Lambda_{\text{AP}}\left(\widehat{b}_\nu\right) - \Lambda_{\text{P}}\left(\widehat{b}_\nu\right) \tag{5.8}$$

$$= \log\frac{P\left(B_\nu = 0|\mathbf{y}_{1:n}\right)}{P\left(B_\nu = 1|\mathbf{y}_{1:n}\right)} \tag{5.9}$$

is the logarithm of the ratio of the likelihoods, usually called the *log-likelihood ratio*, which is often exchanged between decoders in an iterative context, see Chapter 9. In case of balanced a priori probabilities, $\Lambda\left(\widehat{b}_\nu\right) = \Lambda_{\text{AP}}\left(\widehat{b}_\nu\right)$.

From $\Lambda_{\text{AP}}\left(\widehat{b}_\nu\right)$, one may retrieve the APPs

$$P\left(B_\nu = 0|\mathbf{y}_{1:n}\right) = \frac{1}{1+\exp\left(-\Lambda_{\text{AP}}\left(\widehat{b}_\nu\right)\right)} \tag{5.10}$$

and

$$P\left(B_\nu = 1|\mathbf{y}_{1:n}\right) = \frac{1}{1+\exp\left(\Lambda_{\text{AP}}\left(\widehat{b}_\nu\right)\right)}. \tag{5.11}$$

The BCJR algorithm can also be used in this context to provide a solution for (5.3), see Section 5.2.

Since the BCJR algorithm can solve these two problems exactly, the resulting estimates are optimal in the sense of minimum bit error rate (BER) or symbol error rate (SER), depending on the chosen MPM criterion. It should be clear, however, that both estimators above can provide a succession of (symbol or bit) MPM estimates, which may not correspond to a valid sequence (i.e., a sequence that could have been generated by the source coder), since these estimates are only local even if the underlying trellis on which the BCJR algorithm is run describes only valid sequences of source symbols.

5.1.2 Sequence Estimators

With the sequence MAP estimator, the whole sequence of source outcomes is estimated from the knowledge of the channel output

$$\widehat{\mathbf{x}}_{1:\widehat{k}}^{\mathrm{MAP}} = \arg\max_{\mathbf{x}\in\Phi_n} P\left(\mathbf{x}|\mathbf{y}_{1:n}\right), \tag{5.12}$$

where

$$\Phi_n = \left\{\mathbf{x} \in \mathcal{A}^* | \ell\left(C\left(\mathbf{x}\right)\right) = n\right\}, \tag{5.13}$$

see Chapter 4. In (5.12), the number of symbols k is estimated jointly with $\mathbf{x}$. Using Bayes' rule, (5.12) may be written as

$$\begin{aligned}\widehat{\mathbf{x}}_{1:\widehat{k}}^{\mathrm{MAP}} &= \arg\max_{\mathbf{x}\in\Phi_n} \frac{P\left(\mathbf{y}_{1:n}|\mathbf{x}\right) P\left(\mathbf{x}\right)}{P\left(\mathbf{y}_{1:n}\right)} \\ &= \arg\max_{\mathbf{x}\in\Phi_n} P\left(\mathbf{y}_{1:n}|\mathbf{x}\right) P\left(\mathbf{x}\right)\end{aligned}$$

since $P\left(\mathbf{y}_{1:n}\right)$ does not depend on $\mathbf{x}$. Assuming further that $P(\mathbf{x})$ is constant for all $\mathbf{x} \in \Phi_n$ (which is reasonable since all $\mathbf{x} \in \Phi_n$ correspond to codes of the same length), one gets

$$\widehat{\mathbf{x}}_{1:\widehat{k}}^{\mathrm{MAP}} = \widehat{\mathbf{x}}_{1:\widehat{k}}^{\mathrm{ML}} = \arg\max_{\mathbf{x}\in\Phi_n} P\left(\mathbf{y}_{1:n}|\mathbf{x}\right), \tag{5.14}$$

the sequence maximum likelihood (ML) estimate of $\mathbf{x}_{1:k}$ based on the channel outcomes $\mathbf{y}_{1:n}$.

The ML estimate (i.e., the sequence MAP estimator under these assumptions) may be obtained from a variant of the Viterbi algorithm (VA) (Forney, 1973; Viterbi and Omura, 1979), which is detailed in Section 5.3.

Obviously, sequence estimates are of interest in our context since they provide sequences that meet the source decoder constraints. However, they do not provide bit or symbol APPs that can be useful, e.g., for iterative algorithms. The Soft-Output Viterbi Algorithm (SOVA) (Hagenauer and Hoeher, 1989) is a modification of the VA able to provide some reliability information in addition to the estimate $\widehat{\mathbf{x}}_{1:\widehat{k}}^{\mathrm{ML}}$, see Section 5.3.2. Nevertheless, only approximate symbol or bit APPs may be deduced. It is thus suboptimal in the sense of symbol error probability.

5.2 ELEMENT-BY-ELEMENT MAP ESTIMATION ALGORITHMS

Several estimation algorithms are now presented. Obviously, they differ depending on the available information characterizing the sequence to be estimated and on the quantities that have to be computed (bit or symbol APPs, log-likelihood ratios). These techniques assume that the source coder or decoder is represented by a FSM. For coders that do not satisfy this requirement, one may use suboptimal *sequential decoders* discussed in Chapter 6.

5.2.1 BCJR Algorithm for Bit-Clock FSMs

This section focuses on the solution of (5.3), where bit-by-bit estimates are to be obtained. First, only the number of bits n in $\mathbf{x}_{1:n}$ is assumed to be known. Second, the knowledge of the number of symbols k is taken into account as well.

Known Number of Bits in a Packet

A sequence $\mathbf{y}_{1:n}$ of noisy VLC-encoded symbols (without padding bits) has been received, n is thus assumed to be known, but the number of symbols k in $\mathbf{y}_{1:n}$ is not known a priori. For Huffman-based codes or integer arithmetic codes, Chapter 4 has shown that the source decoder may be described by a bit-clock FSM denoted by $\left(\{0,1\}, \mathcal{A}^*, \Sigma, 0, t, \omega, \mathcal{F}\right)$, with states in $\Sigma = \{0, 1, \dots, m\}$. Assume that at any time instant $\nu = 1, \dots, n$, the a priori transition probabilities $P\left(S_\nu = \mu | S_{\nu-1} = \mu'\right)$ between any pair of states $\left(\mu', \mu\right)$ of the FSM are available. Since we assume that the packet to be processed begins at the frontier of a variable-length encoded symbol, the initial state $S_0 = 0$ is fully determined. Moreover, since the FSM has a bit clock and the received sequence is of n bits, one has $S_n \in \mathcal{F}$, the set of admissible final states, and we are sure that the n bits of the sequence contain an integer number of encoded symbols, this is an additional knowledge that can be used in some cases. As an example, if VLCs are used, $\mathcal{F} = \{0\}$, indicating that $S_n = 0$. An example of such FSM is provided in Example 5.1.

Example 5.1: Consider a memoryless source X emitting symbols belonging to $\mathcal{A} = \{a, b, c, d, e\}$ according to the probability vector $\mathbf{p} = (1/4, 1/4, 1/4, 1/8, 1/8)$. This source is encoded with the *reversible* (it may be decoded from left to right or from right to left) VLC $\mathcal{C} = \{00, 11, 101, 010, 0110\}$ (Takishima et al., 1995). This code provides an average code length $E[\ell(X)] = 2.625$ bits/symbol, whereas the entropy of X is $H(X) = 2.25$ bits/symbol, resulting in a redundancy of 0.375 bits/symbol. This redundancy, if it were introduced by a channel code, would result in a coding rate

$$R_C = \frac{H(X)}{E[\ell(X)]} = 0.85. \tag{5.15}$$

When considering only the syntax of the code words, one gets, e.g., for a coded sequence of 200 bits a redundancy of about 25 bits, see Section 4.2.3. The FSM representing the coder has a simple structure similar to that in Figure 4.4. The FSM associated to the decoder is represented in Figure 5.3.

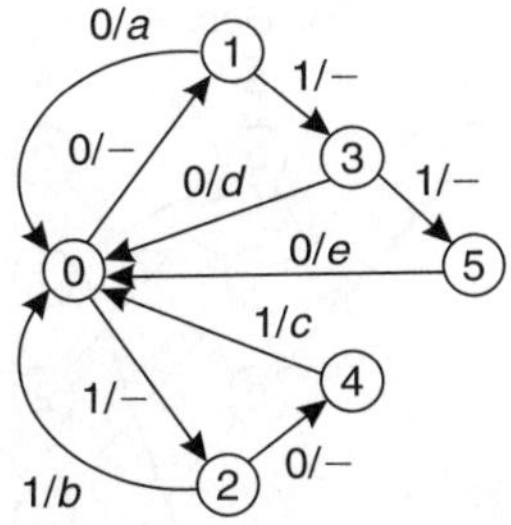

FIGURE 5.3 Bit-clock FSM of Example 5.1.

TABLE 5.1 State Indexes and Associated Sequences of Processed Bits for Example 5.1

State index	0	1	2	3	4	5
Processed bits	∅	0	1	01	10	011

Each value of the state of this FSM corresponds to a sequence of bits associated to a partially decoded symbol, see Table 5.1.

Another representation of a bit-clock FSM is provided by a trellis, which is the description of all possible evolutions of the state of the corresponding FSM with time (Bauer and Hagenauer, 1998). The states of the FSM are represented as nodes, which can take all possible values at each time, see Figure 5.4, where the five state values are indicated by numbered circles, the nodes of the trellis (vertical axis) and the bit time correspond to the horizontal axis. Nodes are connected by branches, representing the transitions between states (from each state of Figure 5.4, at most two possible states are feasible at next time, depending on the input bit). By construction, a path in the trellis corresponds to a succession of states taken by the FSM and fully describes the corresponding input and output sequences.

The trellises corresponding to Figure 5.3 are depicted in Figure 5.4 without the terminating constraint and in Figure 5.5 if the trellis is closed, i.e., the final state is known. However, the translation of this knowledge may not be so obvious, e.g., when arithmetic coding is used.

Example 5.2: The trellis of Figure 5.4 represents all successions of states that may be taken by the FSM of Example 5.1 for $\nu = 0, \ldots, 6$. The path in bold corresponds to the sequence of code words 00, 11, 00. For sufficiently large values of ν, the trellis becomes periodic. The steady state is attained here at time $\nu = 5$. This corresponds to some representation with time of Figure 5.3 taking into account the knowledge that the initial state is 0,

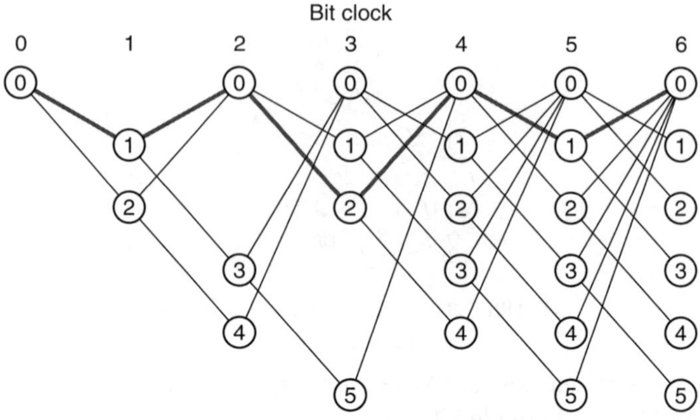

FIGURE 5.4 Trellis associated to the bit-clock FSM of Example 5.1, the bold path corresponds to the sequence 00, 11, 00.

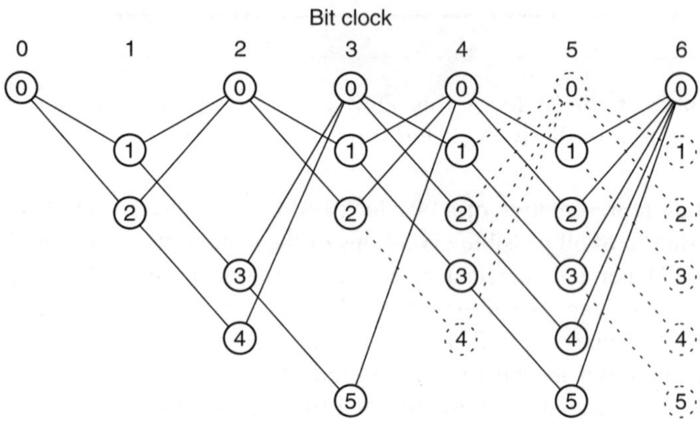

FIGURE 5.5 Trellis associated to the bit-clock FSM of Example 5.1, when a constraint on the number of bits is imposed.

meaning that the sequence does not begin in the middle of a symbol. Note also that this trellis does not take into account the fact that the sequence does not finish in the middle of a symbol.

Determining (5.3) for a bit-clock FSM such as that of Figure 5.3 may be formulated as

$$\begin{aligned}\widehat{b}_{\nu}^{\text{MPM}} &= \arg\max_{b\in\{0,1\}} P\left(B_{\nu}=b|\mathbf{y}_{1:n}\right) \\ &= \arg\max_{b\in\{0,1\}} \sum_{\mu'\in\Sigma} P\left(S_{\nu-1}=\mu', S_{\nu}=t\left(\mu',b\right)|\mathbf{y}_{1:n}\right) \qquad (5.16)\end{aligned}$$

and requires the evaluation of the APPs of the νth transition from μ' triggered by $b \in \{0, 1\}$. Here, $t(\mu', b)$ denotes the value taken by the next state when the previous one is μ' and the input bit is b.

The evaluation of probabilities of the form $P(S_{\nu-1} = \mu', S_\nu = \mu | \mathbf{y}_{1:n})$ is efficiently performed by the BCJR algorithm (Bahl et al., 1974), the principle of which is provided now, since this type of derivation is performed at many places in this book. The presentation closely follows that in Bahl et al. (1974) and Bauer and Hagenauer (1998) for VLCs.

The a posteriori probability (APP) $P(S_{\nu-1} = \mu', S_\nu = \mu | \mathbf{y}_{1:n})$ in (5.16) is obtained indirectly by first evaluating $\sigma_\nu(\mu', \mu, \mathbf{y}_{1:n})$ defined as follows:

$$\begin{aligned} \sigma_\nu(\mu', \mu, \mathbf{y}_{1:n}) &= P(S_{\nu-1} = \mu', S_\nu = \mu, \mathbf{y}_{1:n}) \\ &= P(S_{\nu-1} = \mu', S_\nu = \mu | \mathbf{y}_{1:n}) P(\mathbf{y}_{1:n}). \end{aligned} \tag{5.17}$$

In fact, since $P(\mathbf{y}_{1:n})$ does not depend on μ or μ', replacing $P(S_{\nu-1} = \mu', S_\nu = \mu | \mathbf{y}_{1:n})$ by $\sigma_\nu(\mu', \mu, \mathbf{y}_{1:n})$ in (5.16) provides the same result; they differ only by a constant factor.

Now, concentrate on the computation of $\sigma_\nu(\mu', \mu, \mathbf{y}_{1:n})$. Using the product rule, one may write (5.17) as

$$\begin{aligned} \sigma_\nu(\mu', \mu, \mathbf{y}_{1:n}) &= P(S_{\nu-1} = \mu', S_\nu = \mu, \mathbf{y}_{1:n}) \\ &= P(S_{\nu-1} = \mu', S_\nu = \mu, \mathbf{y}_{1:\nu-1}, \mathbf{y}_{\nu:n}) \\ &= P(S_{\nu-1} = \mu', \mathbf{y}_{1:\nu-1}) P(S_\nu = \mu, \mathbf{y}_{\nu:n} | S_{\nu-1} = \mu', \mathbf{y}_{1:\nu-1}) \\ &= P(S_{\nu-1} = \mu', \mathbf{y}_{1:\nu-1}) P(S_\nu = \mu, y_\nu | S_{\nu-1} = \mu', \mathbf{y}_{1:\nu-1}) \\ &\quad P(\mathbf{y}_{\nu+1:n} | S_\nu = \mu, S_{\nu-1} = \mu', \mathbf{y}_{1:\nu}). \end{aligned}$$

If the state of the FSM is known at time ν, any information before time ν does not bring more knowledge on the future outputs or states of the FSM (*Markov property* of the FSM). Thus, one may write

$$P(S_\nu = \mu, y_\nu | S_{\nu-1} = \mu', \mathbf{y}_{1:\nu-1}) = P(S_\nu = \mu, y_\nu | S_{\nu-1} = \mu')$$

and

$$P(\mathbf{y}_{\nu+1:n} | S_\nu = \mu, S_{\nu-1} = \mu', \mathbf{y}_{1:\nu}) = P(\mathbf{y}_{\nu+1:n} | S_\nu = \mu).$$

As a consequence, $\sigma_\nu(\mu', \mu, \mathbf{y}_{1:n})$ may be factorized as

$$\sigma_\nu(\mu', \mu, \mathbf{y}_{1:n}) = \alpha_{\nu-1}(\mu') \gamma_\nu(\mu', \mu) \beta_\nu(\mu) \tag{5.18}$$

with the following definitions

$$\alpha_{\nu-1}(\mu') = P(S_{\nu-1} = \mu', \mathbf{y}_{1:\nu-1}), \tag{5.19}$$

$$\beta_\nu(\mu) = P(\mathbf{y}_{\nu+1:n} | S_\nu = \mu), \tag{5.20}$$

and the probability of transition between state μ' and μ (i.e., the *branch probability*)

$$\gamma_\nu(\mu', \mu) = P(S_\nu = \mu, y_\nu | S_{\nu-1} = \mu'). \tag{5.21}$$

The quantities α_ν and β_ν may be recursively evaluated as follows:

$$\begin{aligned}\alpha_\nu(\mu) &= P(S_\nu = \mu, \mathbf{y}_{1:\nu}) = \sum_{\mu'} P(S_{\nu-1} = \mu', S_\nu = \mu, \mathbf{y}_{1:\nu}) \\ &= \sum_{\mu'} P(S_{\nu-1} = \mu', \mathbf{y}_{1:\nu-1}) P(S_\nu = \mu, y_\nu | S_{\nu-1} = \mu', \mathbf{y}_{1:\nu-1}) \\ &= \sum_{\mu'} \alpha_{\nu-1}(\mu') \gamma_\nu(\mu', \mu). \end{aligned} \tag{5.22}$$

Similarly,

$$\begin{aligned}\beta_\nu(\mu) &= \sum_{\mu'} P(S_\nu = \mu', \mathbf{y}_{\nu+1:n} | S_\nu = \mu) \\ &= \sum_{\mu'} P(S_{\nu+1} = \mu', y_{\nu+1} | S_\nu = \mu) P(\mathbf{y}_{\nu+2:n} | S_\nu = \mu, S_{\nu+1} = \mu', y_{\nu+1}) \\ &= \sum_{\mu'} \gamma_{\nu+1}(\mu, \mu') \beta_{\nu+1}(\mu'), \end{aligned} \tag{5.23}$$

using again in both cases the Markov property of the FSM.

To initialize α and β, one should consider the only allowed initial state 0 of the FSM and $\mathcal{F}$, the set of allowed final states. One may then get

$$\alpha_0(\mu) = \begin{cases} 1 & \text{if } \mu = 0 \\ 0 & \text{else.} \end{cases} \tag{5.24}$$

Assuming that all final states in $\mathcal{F}$ are equally likely, one gets

$$\beta_n(\mu) = \begin{cases} 1/|\mathcal{F}| & \text{if } \mu \in \mathcal{F} \\ 0 & \text{else.} \end{cases} \tag{5.25}$$

If more precise knowledge of the distributions of the possible final states is known, it can be used in (5.25).

Example 5.3: Assume that a sequence $\mathbf{x}_{1:k}$ of symbols generated by the source X of Example 5.1 has been generated, encoded using the VLC C, BPSK modulated, and transmitted on an additive white Gaussian noise (AWGN) channel with noise variance $\sigma_n^2 = N_0/2$. Here, one knows that the sequence terminates at bit n, which is the end of a VLC symbol. At channel output, a vector $\mathbf{y}_{1:6}$ of 6 real-valued channel outcomes is observed. The BCJR algorithm may be applied using the FSM in Figure 5.3. The initial state is 0 and the set of final states is $\mathcal{F} = \{0\}$. This last condition, which will be taken into account in the evaluation of β, leads to a modification of the trellis represented in Figure 5.4. When the bit clock approaches 6, some paths are no more allowed, as shown in Figure 5.5.

The branch probability is evaluated as

$$\gamma_\nu(\mu',\mu) = \sum_b P(B_\nu = b, S_\nu = \mu, y_\nu | S_{\nu-1} = \mu')$$
$$= \sum_b P(y_\nu | B_\nu = b) P(B_\nu = b | S_\nu = \mu, S_{\nu-1} = \mu')$$
$$P(S_\nu = \mu | S_{\nu-1} = \mu'). \tag{5.26}$$

In (5.26), $P(S_\nu = \mu | S_{\nu-1} = \mu')$ and $P(B_\nu = b | S_\nu = \mu, S_{\nu-1} = \mu')$ depend on the structure of the FSM (as well as on the probability vector $\mathbf{p}$ of the source) and are assumed to be known. For a bit-synchronized FSM representing a source decoder, there is at most one transition going from μ' to μ (triggered by 0 or 1), otherwise the code would not be uniquely decodable. Thus, when a transition between μ' and μ exists, it is triggered either by $b=0$ or by $b=1$ and $P(B_\nu = b | S_\nu = \mu, S_{\nu-1} = \mu')$ can only be equal to 0 or 1. Finally, $P(y_\nu | B_\nu = b)$ depends on the channel model. Example 5.4 illustrates the evaluation of the $P(S_\nu = \mu | S_{\nu-1} = \mu')$ and $P(B_\nu = b | S_\nu = \mu, S_{\nu-1} = \mu')$ for a VLC. Similar derivations can also be done for quasi-arithmetic codes.

Example 5.4 (Computing the a priori transition probabilities): Consider again the source and code $\mathcal{C}$ of Example 5.1, summarized in Table 5.2. The FSM associated to the decoder for $\mathcal{C}$ is represented in Figure 5.3. To evaluate the *a priori* transition probabilities $P(S_\nu = \mu | S_{\nu-1} = \mu')$, one has to sum up all events that would lead to this transition and then make sure that these transition probabilities sum up to 1. For example, from state 0, there are only two outgoing transitions $0 \to 1$ and $0 \to 2$. The transition $0 \to 1$ is realized by the decoder FSM if the source has emitted *a, d,* or *e* (the three symbols with code words starting with 0). The transition $0 \to 2$ is realized by the decoder FSM if the source has emitted *b* or *c* (the two symbols with code words starting with 1). Thus,

$$P(S_\nu = 1 | S_{\nu-1} = 0) = A(P(X=a) + P(X=d) + P(X=e))$$
$$= A/2$$
$$P(S_\nu = 2 | S_{\nu-1} = 0) = A(P(X=b) + P(X=c))$$
$$= A/2,$$

TABLE 5.2 Characteristics of the Source and of the Code Considered in Example 5.4

Symbol	a	b	c	d	e
p_i	1/4	1/4	1/4	1/8	1/8
Code	00	11	101	010	0110

where A is a normalization constant. Since

$$P(S_\nu=1|S_{\nu-1}=0)+P(S_\nu=2|S_{\nu-1}=0)=1,$$

one gets

$$P(S_\nu=1|S_{\nu-1}=0)=1/2 \quad \text{and} \quad P(S_\nu=2|S_{\nu-1}=0)=1/2.$$

Similar evaluations can be obtained for the other transitions of the FSM to get the transition matrix

$$\underline{\mathbf{P}}=\begin{pmatrix} 0 & 1/2 & 1/2 & 0 & 0 & 0 \\ 1/2 & 0 & 0 & 1/2 & 0 & 0 \\ 1/2 & 0 & 0 & 0 & 1/2 & 0 \\ 1/2 & 0 & 0 & 0 & 0 & 1/2 \\ 1 & 0 & 0 & 0 & 0 & 0 \\ 1 & 0 & 0 & 0 & 0 & 0 \end{pmatrix}. \tag{5.27}$$

The stationary probability vector $\boldsymbol{\pi}$ for the states of the FSM is the eigenvector of $\underline{\mathbf{P}}$ associated to the eigenvalue 1. Here,

$$\boldsymbol{\pi}=(0.38, 0.19, 0.19, 0.095, 0.095, 0.05),$$

thus, for example, $P(S_\nu=3)=0.095$. The probabilities $P(B_\nu=b|S_\nu=\mu, S_{\nu-1}=\mu')$ are directly deduced from the FSM. For example, $P(B_\nu=1|S_\nu=1, S_{\nu-1}=0)=0$ or $P(B_\nu=1|S_\nu=5, S_{\nu-1}=3)=1$.

An example of bit-by-bit MPM estimation with the BCJR algorithm and a bit-clock trellis is provided in Section 5.2.3.

Known Number of Bits and of Symbols

Assume now that the number k of encoded symbols within the received packet of n channel outcomes $\mathbf{y}_{1:n}$ is also known. A FSM or a trellis such as that of Figure 5.5 does not account for the number of symbols in the decoded sequence. To take k into account, the FSM representing the decoder has to keep track of the number of already decoded symbols, and the final state has to be associated to n processed bits and k decoded symbols. This may be done by supplementing the state of the bit-clock FSM with a symbol counter κ to get a 2D state (μ, κ), where μ is the state of the original bit-clock FSM and κ represents the number of already decoded symbols. The initial state of this extended FSM is $(0, 0)$ and the new set of final states is $\mathcal{F} \times \{k\}$.

Example 5.5: Consider again the source and the code $\mathcal{C}$ of Example 5.1. Figure 5.6 represents the extended FSM derived from that of Figure 5.3. The initial state is $S_0=(0, 0)$, and the final state at time n is necessarily $S_n=(0, k)$. One has to notice that not all trajectories (succession of states) within this new FSM are allowed, only trajectories starting from S_0 and

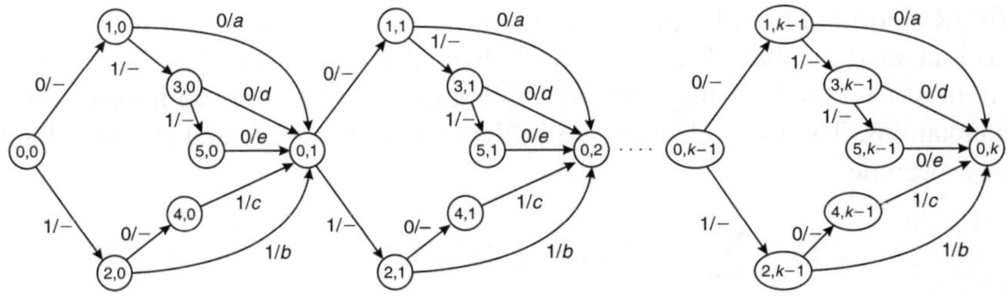

FIGURE 5.6 Bit-clock FSM taking into account the knowledge of the number of symbols contained in a packet.

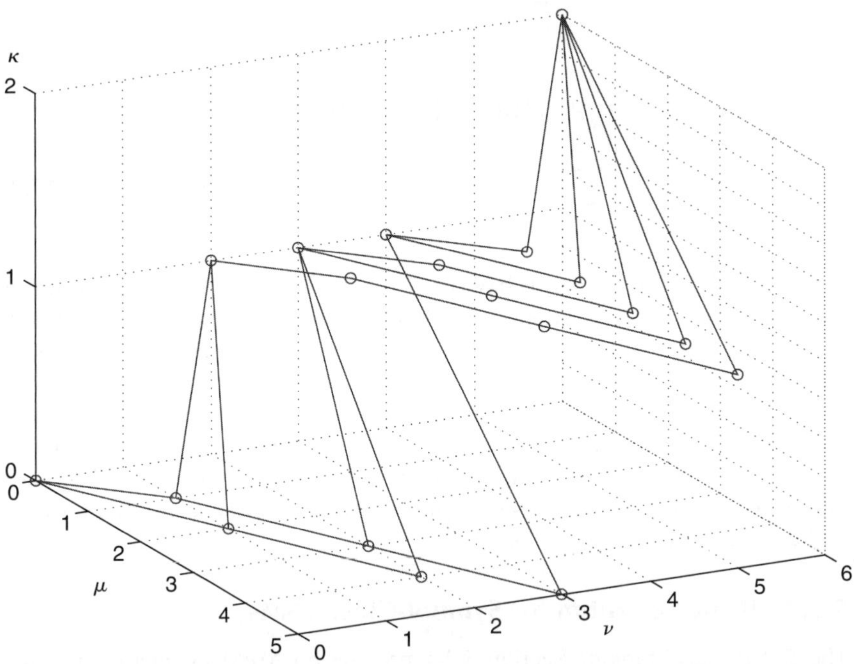

FIGURE 5.7 Three-dimensional bit/symbol trellis taking into account the knowledge of the length in bits of a packet and of the number of encoded symbols it contains.

ending in $S_n = (0, k)$, *i.e.*, *exactly* after n transitions, are allowed. These trajectories are more clearly represented by a three-dimensional trellis, such as that of Figure 5.7 for the case $n=6$ and $k=2$. In this trellis, the first dimension is the bit clock, the other dimensions correspond to the two dimensions of the state of the FSM.

Trellises such as that represented in Figure 5.7 are sometimes called *bit/symbol trellises*, see Malinowski et al. (2007). The "time clock" for the

BCJR algorithm is still represented by the accumulated number of bits, but the accumulated number of decoded symbols is also considered. The BCJR algorithm of Section 5.2.1 may then be applied to this new trellis with only minor adaptations. The main changes are in the recursions (5.22) and (5.23), which now become

$$\begin{aligned}\alpha_\nu(\mu,\kappa) &= \sum_{\mu'} \alpha_{\nu-1}(\mu',\kappa)\,\gamma_\nu\left((\mu',\kappa),(\mu,\kappa)\right) \\ &\quad + \sum_{\mu'} \alpha_{\nu-1}(\mu',\kappa-1)\,\gamma_\nu\left((\mu',\kappa-1),(\mu,\kappa)\right)\end{aligned} \tag{5.28}$$

and

$$\begin{aligned}\beta_\nu(\mu,\kappa) &= \sum_{\mu'} \gamma_{\nu+1}\left((\mu,\kappa),(\mu',\kappa)\right)\beta_{\nu+1}\left((\mu',\kappa)\right) \\ &\quad + \sum_{\mu'} \gamma_{\nu+1}\left((\mu,\kappa),(\mu',\kappa+1)\right)\beta_{\nu+1}\left((\mu',\kappa+1)\right).\end{aligned} \tag{5.29}$$

In (5.28) and (5.29), two terms have to be distinguished, depending on whether the considered transition outputs a symbol (the second terms in both equations) or not. The expression of $\gamma_\nu\left((\mu',\kappa-1),(\mu,\kappa)\right)$ and $\gamma_\nu\left((\mu',\kappa),(\mu,\kappa)\right)$ takes the same form as (5.26).

For a stationary source, the a priori transition probabilities $P(S_\nu=(\mu,\kappa)\,|\, S_{\nu-1}=(\mu',\kappa-1))$ and $P(S_\nu=(\mu,\kappa)\,|S_{\nu-1}=(\mu',\kappa))$ do not depend on the value of κ. Note, however, that bit/symbol trellises are well suited to take into account nonstationary sources, in which case, these a priori transition probabilities may change with κ.

5.2.2 BCJR Algorithm for Symbol-Clock FSMs

The BCJR algorithms of Section 5.2.1 provide bit MPM estimators. Getting symbol APPs from the bit APPs is possible but rather difficult. A better solution is to use a symbol-clock FSM $\left(\mathcal{A}, \{0,1\}^*, \Sigma, 0, t, \omega, \mathcal{F}\right)$ describing the decoder, such as that of Figures 4.4–4.6.

The main limitation of these FSMs comes from the fact that they do not account for the accumulated number of processed bits and symbols. Nevertheless, they are useful to build symbol-clock trellises (Bauer and Hagenauer, 2000c; Kaiser and Bystrom, 2000) such as that of Figure 5.8. For a simple Huffman-like code, as in this case, the FSM contains only a single state $\Sigma=\{0\}$. Here again, one has to define the time clock, which is now the accumulated number of decoded symbols. The state definition is adapted to take the decoding evolution into account; the values taken by the state S_κ at "time" κ are the number of bits ν which can correspond to κ symbols.

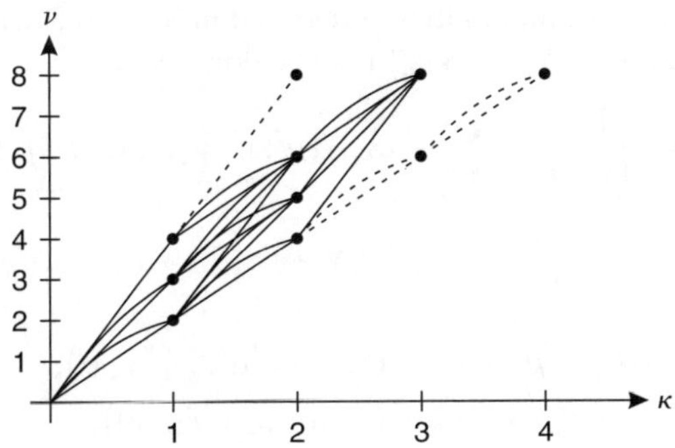

FIGURE 5.8 Symbol-clock trellis associated to the VLC of Example 5.6.

Example 5.6: Consider the same VLC $\mathcal{C} = \{00, 11, 101, 010, 0110\}$ as in Example 5.1. Figure 5.8 depicts the 2D symbol-clock trellis, with state $S_\kappa = \nu \in \{0, \ldots, 5\}$, κ being the cumulated number of symbols and ν the cumulated number of bits. The plain and dashed lines represent all successions of code words resulting in an encoded sequence of $n = 8$ bits. Parallel transitions correspond to code words with the same length. Knowing $n = 8$ does not give full knowledge on the number of symbols $k \in \{2, 3, 4\}$. If the number of symbols is known, e.g., $k = 6$, the trellis reduces to the transitions in plain lines. This trellis is the projection of the 3D one in Figure 5.7 onto the (κ, ν) plane.

For more general situations, corresponding to the FSMs taking into account constraints on the number of bits and symbols, such as those in Figures 4.5 or 4.6, all states of the FSM have to be taken into account to build a 3D symbol-clock trellis, which states S_κ at time κ are the pairs (μ, ν) indicating the state μ of the FSM and the number of processed bits ν to decode κ symbols.

Determining (5.2) using a 3D symbol-clock trellis derived from a symbol-clock FSM may be written as

$$\begin{aligned} \widehat{x}_\kappa^{\mathrm{MPM}} &= \arg\max_{x \in \mathcal{A}} P\left(X_\kappa = x | \mathbf{y}_{1:n}\right) \qquad (5.30) \\ &= \arg\max_{x \in \mathcal{A}} \sum_{(\mu', \nu'), (\mu, \nu)} P\left(S_{\kappa-1} = \left(\mu', \nu'\right), S_\kappa = (\mu, \nu), X_\kappa = x | \mathbf{y}_{1:n}\right) \end{aligned}$$

and now requires the evaluation of the APPs of the κth transition starting from $\left(\mu', \nu'\right)$ and triggered by some $x \in \mathcal{A}$. To perform decoding, again, we assume that probabilities like $P\left(S_\kappa = (\mu, \nu) | S_{\kappa-1} = \left(\mu', \nu'\right)\right)$ and $P\left(X_\kappa = x | S_{\kappa-1} = \left(\mu', \nu'\right), S_\kappa = (\mu, \nu)\right)$ are available for all $\kappa = 1, \ldots, k$, $\mu = 1, \ldots, m$ and $\nu = 1, \ldots, n$. Moreover, the channel is assumed to be memoryless, i.e., (5.1) is satisfied.

Using derivations similar to those presented in Section 5.2.1, one may show that the evaluation of $P\left(X_\kappa = x|\mathbf{y}_{1:n}\right)$ may be done as

$$P\left(X_\kappa = x|\mathbf{y}_{1:n}\right) = \frac{1}{P\left(\mathbf{y}_{1:n}\right)} \sum_{(\mu,\nu),(\mu',\nu')} \alpha_{\kappa-1}(\mu',\nu')\,\gamma_\kappa\left(x,(\mu',\nu'),(\mu,\nu)\right)\beta_\kappa(\mu,\nu), \tag{5.31}$$

where now,

$$\alpha_{\kappa-1}\left(\mu',\nu'\right) = P\left(S_{\kappa-1} = \left(\mu',\nu'\right), \mathbf{y}_{1:\nu'}\right), \tag{5.32}$$

$$\beta_\kappa(\mu,\nu) = P\left(\mathbf{y}_{\nu+1:n}|S_\kappa = (\mu,\nu)\right), \tag{5.33}$$

and the branch probability

$$\gamma_\kappa\left(x,\left(\mu',\nu'\right),(\mu,\nu)\right) = P\left(X_\kappa = x, S_\kappa = (\mu,\nu), \mathbf{y}_{\nu'+1:\nu}|S_{\kappa-1} = \left(\mu',\nu'\right)\right). \tag{5.34}$$

Again, α and β may be evaluated recursively,

$$\alpha_\kappa(\mu,\nu) = \sum_{\mu',\nu'}\sum_{x\in\mathcal{A}} \alpha_{\kappa-1}\left(\mu',\nu'\right)\gamma_\kappa\left(x,\left(\mu',\nu'\right),(\mu,\nu)\right),$$

$$\beta_\kappa(\mu,\nu) = \sum_{\mu',\nu'}\sum_{x\in\mathcal{A}} \beta_{\kappa+1}\left(\mu',\nu'\right)\gamma_{\kappa+1}\left(x,\left(\mu',\nu'\right),(\mu,\nu)\right),$$

with

$$\gamma_\kappa\left(x,\left(\mu',\nu'\right),(\mu,\nu)\right) = P\left(X_\kappa = x|S_{\kappa-1} = \left(\mu',\nu'\right), S_\kappa = (\mu,\nu)\right) \\ P\left(\mathbf{y}_{\nu'+1:\nu}|X_\kappa = x\right)P\left(S_\kappa = (\mu,\nu)\,|S_{\kappa-1} = \left(\mu',\nu'\right)\right). \tag{5.35}$$

In (5.35), $P\left(S_\kappa = (\mu,\nu)\,|S_{\kappa-1} = \left(\mu',\nu'\right)\right)$ and $P\left(X_\kappa = x|S_{\kappa-1} = \left(\mu',\nu'\right), S_\kappa = (\mu,\nu)\right)$ depend on the structure of the FSM describing the encoder and are assumed to be known. $P\left(X_\kappa = x|S_{\kappa-1} = \left(\mu',\nu'\right), S_\kappa = (\mu,\nu)\right) = 1$ when x fires the transition from $\left(\mu',\nu'\right)$ at time $\kappa-1$ to (μ,ν) at time κ, else it vanishes. $P\left(\mathbf{y}_{\nu'+1:\nu}|X_\kappa = x\right)$ is obtained from the characteristics of the channel.

The initialization depends on whether k is known to the receiver. One gets thus

$$\alpha_0(\mu,\nu) = \begin{cases} 1 & \text{if } \mu = 0 \text{ and } \nu = 0 \\ 0 & \text{else,} \end{cases}$$

and when k is known, one gets

$$\beta_k(\mu,\nu) = \begin{cases} 1/|\mathcal{F}| & \text{if } \mu \in \mathcal{F} \text{ and } \nu = n \\ 0 & \text{else,} \end{cases}$$

assuming again that all final states in $\mathcal{F}$ are equally likely. When k is unknown, the values for $\beta_\kappa(\mu,n)$ are more difficult to determine. For a Huffman-like VLC represented by the symbol-clock FSM $\left(\mathcal{A},\{0,1\}^*,\Sigma,0,t,\omega,\mathcal{F}\right)$, let

$$\ell_{\min} = \min_{\substack{\mu\in\Sigma,x\in\mathcal{A} \\ \omega(\mu,x)\neq\emptyset}} \ell\left(\omega(\mu,x)\right) \tag{5.36}$$

and

$$\ell_{\max} = \max_{\mu \in \Sigma, x \in \mathcal{A}} \ell(\omega(\mu, x)) \tag{5.37}$$

represent the lengths of the shortest and longest code words of the code. Then, necessarily, one has

$$k_{\min} \leqslant k \leqslant k_{\max},$$

with

$$k_{\max} = \lfloor n/\ell_{\min} \rfloor \text{ and } k_{\min} = \lceil n/\ell_{\max} \rceil .$$

Assuming that all final states in $\mathcal{F}$ are equally likely, one may then consider for $k_{\min} \leqslant \kappa \leqslant k_{\max}$,

$$\begin{aligned} \beta_\kappa(\mu, n) &= \frac{1}{|\mathcal{F}|(k_{\max} - k_{\min} + 1)} \text{ if } \mu \in \mathcal{F} \\ &= 0 \quad \text{else.} \end{aligned}$$

When k is not known to the receiver, it may be estimated from the quantities evaluated by the BCJR algorithm. For that purpose, one may use the fact that the FSM is only allowed to stop in a state belonging to $\mathcal{F}$ once n bits have been processed. The MAP estimate $\widehat{k}$ of k may thus be evaluated as follows:

$$\begin{aligned} \widehat{k} &= \arg\max_\kappa \left(\max_{\mu \in \mathcal{F}} P(S_\kappa = (\mu, n) | \mathbf{y}_{1:n}) \right) \\ &= \arg\max_\kappa \left(\max_{\mu \in \mathcal{F}} \frac{\alpha_\kappa(\mu, n)}{P(\mathbf{y}_{1:n})} \right) = \arg\max_\kappa \left(\max_{\mu \in \mathcal{F}} \alpha_\kappa(\mu, n) \right). \end{aligned} \tag{5.38}$$

Once the forward iteration of the BCJR algorithm has been performed, $\widehat{k}$ may be directly estimated by maximizing the $\alpha_\kappa(\mu, n)$s, with $\mu \in \mathcal{F}$.

5.2.3 Comparison

Consider again the VLC of Example 5.1. Random sequences of 100 symbols are generated, encoded, BPSK modulated, and transmitted over an AWGN channel. Three algorithms are compared in Figure 5.9: (i) a classical hard decoder, i.e., a decoder taking as input hard decisions on the channel output, (ii) the BCJR decoding algorithm working on a bit-clock trellis (BC-BCJR), without knowledge of k, and (iii) the BCJR decoding algorithm working on a symbol-clock trellis (SC-BCJR) knowing k. The BERs are plotted as a function of E_b/N_0, with $E_b = E_c/R_C$, and $E_c = 1$ the energy per code bit. For each point, enough sequences are generated to get at least 100 accumulated bits in error.

The BER is plotted, which raises some problems for the SC-BCJR algorithm, since in this case, only the symbol-by-symbol error probability is minimized. As a result, the sequence of symbols which is obtained does not always correspond to a sequence of the right number of bits: it may happen that two successive symbol estimates overlap (in terms of bits) or even leave some bits unestimated.

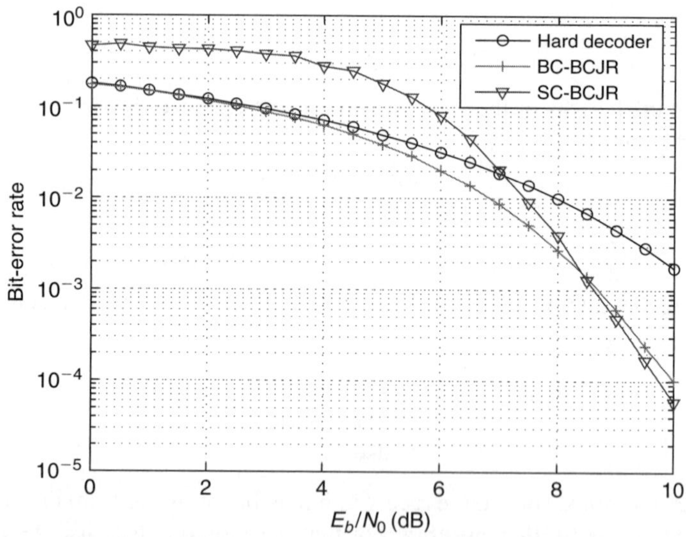

FIGURE 5.9 Bit-error rate comparison of a plain hard decoder, a BCJR decoder working on a bit-clock trellis and a BCJR algorithm working on a symbol-clock trellis.

To do a precise comparison, we have chosen to re-encode the sequence of symbol MPM estimates to get the sequences of bit estimates. When such a sequence is not long enough, all missing bits are considered as erroneous.

The performance gain provided by the BC-BCJR algorithm compared to a hard decoder increases significantly with E_b/N_0. At low values of E_b/N_0, the SC-BCJR algorithm performs worse than the hard decoder. This is mainly due to the fact that the SC-BCJR algorithm provides symbol MPM estimates while we are plotting bit MPM performance. However, it can be observed that at high values of E_b/N_0, SC-BCJR decoder improves over the BC-BCJR decoder. This is due to the fact that on top of the explicit knowledge of k, many quantities are evaluated on the underlying 3D trellis; therefore, the knowledge of the total number of bits is partly taken into account. Overall, this results in a higher deterministic redundancy when compared to the case of k unknown.

Figure 5.10 illustrates the symbol-error rate for the SC-BCJR and the BC-BCJR algorithms. Now, the SC-BCJR algorithm outperforms the BC-BCJR algorithm for all E_b/N_0 since the former is designed to get symbol MPM estimates, and benefits from the knowledge of the actual number of symbols.

5.2.4 Trellises

Several types of trellises have been considered in Sections 5.2.1–5.2.3. The natural question arises now about the choice of the trellis when trying to decode a given noisy sequence.

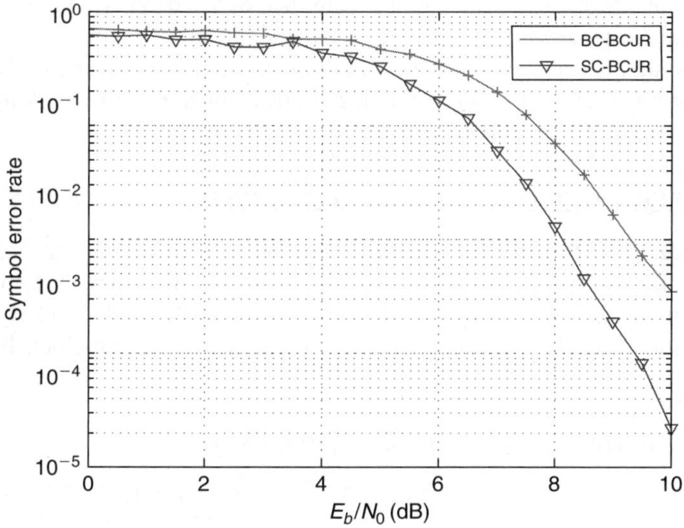

FIGURE 5.10 Symbol-error rate comparison of a BCJR decoder working on a bit-clock trellis and a BCJR algorithm working on a symbol-clock trellis.

All trellises differ in the dimensions they are considering (accumulated number of processed bits ν, accumulated number of decoded symbols κ, and state μ of the underlying FSM) and in their time clock. When bit a posteriori probabilities are required, bit-clock (ν, μ) or bit/symbol (ν, μ, κ) trellises facilitate significantly their evaluation. Now, 2D (κ, ν) or 3D symbol-clock (κ, ν, μ) trellises are better suited to the evaluation of symbol APPs.

For Huffman-based VLCs, and for a memoryless source, 2D and 3D symbol-clock trellises are equivalent, since there is only a single state in the encoding FSM. This is no more the case for arithmetic codes, which require 3D symbol/bit or bit/symbol trellises to perform decoding.

If no soft output is required, one may think that a 2D symbol-clock trellis may be less complex than a 3D bit/symbol trellis for decoding VLCs. This is wrong for what concerns computational complexity since the successions of valid sequences of encoded symbols are the same in both trellises. The computations are performed in a different way, but the amount of operations is the same. The main difference comes from the memory requirements, which are obviously larger for the bit/symbol trellis, to allow the evaluation of bit APPs.

In bit/symbol and 3D symbol-clock trellises, there are no parallel transitions between states, whereas they may appear in 2D symbol-clock trellises. Such parallel transitions may be efficiently processed simultaneously, as is detailed in Section 6.3, page 166.

Thus, it should be clear now that for Huffman-like VLCs, the symbol-clock BCJR algorithm is working implicitly on the 3D bit/symbol trellis which contains all possible information on the feasible paths in the closed trellis, taking all known information into account, but computes APPs only on the symbols.

It should also be clear that when all these constraints are taken into account in the bit-clock trellis, the corresponding BCJR algorithm is also working on the same 3D trellis but computes APPs only on the bits.

5.3 SEQUENCE ESTIMATION ALGORITHMS

As indicated in Section 5.1, the VA may be used to address the sequence MAP estimation problem (5.12). As for the BCJR algorithm, both bit-clock and symbol-clock FSM may be considered to build the trellis underlying the computations of the VA. Here, only computations on the bit-clock FSM are considered.

5.3.1 Viterbi Algorithm for Bit-Clock FSMs

Consider a bit-clock FSM $\left(\{0,1\}, \mathcal{A}^*, \Sigma, 0, t, \omega, \mathcal{F}\right)$ representing a source decoder. The structure of the FSM may account for the presence of some constraint imposed on sequences of source symbols. Assume that $\mathbf{y}_{1:n}$ is measured and correspond to k encoded code words (with k unknown). With this FSM, since the clock is bit synchronized, obtaining (5.12) is equivalent to determining the MAP estimate of the succession of states taken by the FSM, starting from $S_0 = 0$ and ending in some state $S_n \in \mathcal{F}$, after n transitions. Such sequences of states correspond to some path in the decoding trellis. Since k is not taken into account in our example, a bit-clock trellis is enough for the decoding.

Example 5.7: Consider the trellis of Figure 5.4, page 126. The sequence of states $\mathbf{s}^1 = (0,1,2,3,4,5,0)$ is not a path in the trellis since there is no transition from state 1 to state 2 between time $\nu = 1$ and $\nu = 2$. The sequence $\mathbf{s}^2 = (0,2,4,0,2,4,0)$ is a path in the trellis.

Only sequences of states corresponding to path in trellises are considered in what follows. To get (5.12), one is interested in obtaining the path in the trellis with the maximum APP,

$$\begin{aligned}
\widehat{\boldsymbol{\mu}}_{0:n} &= \arg \max_{\substack{\boldsymbol{\mu}_{0:n} \\ \mu_0 = 0, \mu_n \in \mathcal{F}}} P\left(\boldsymbol{\mu}_{0:n} | \mathbf{y}_{1:n}\right) \\
&= \arg \max_{\substack{\boldsymbol{\mu}_{0:n} \\ \mu_0 = 0, \mu_n \in \mathcal{F}}} \frac{P\left(\boldsymbol{\mu}_{0:n} | \mathbf{y}_{1:n}\right)}{P\left(\mathbf{y}_{1:n}\right)} \\
&= \arg \max_{\substack{\boldsymbol{\mu}_{0:n} \\ \mu_0 = 0, \mu_n \in \mathcal{F}}} \log P\left(\boldsymbol{\mu}_{0:n}, \mathbf{y}_{1:n}\right), \qquad (5.39)
\end{aligned}$$

where the maximization is taken over all paths of $n+1$ states $\boldsymbol{\mu}_{0:n}$. The term $\log P\left(\boldsymbol{\mu}_{0:n}, \mathbf{y}_{1:n}\right)$ in (5.39) is the *metric* of $\boldsymbol{\mu}_{0:n}$.

The VA evaluates (5.39) iteratively. For any $\nu = 1, \ldots, n$, consider

$$\Gamma_{\nu}(\mu) = \max_{\substack{\boldsymbol{\mu}_{0:\nu} \\ \mu_0 = 0, \mu_\nu = \mu}} \log P\left(\boldsymbol{\mu}_{0:\nu}, \mathbf{y}_{1:\nu}\right), \tag{5.40}$$

the largest *partial metric* at state μ and time ν, i.e., the largest value of $\log P\left(\boldsymbol{\mu}_{0:\nu}, \mathbf{y}_{1:\nu}\right)$ among all sequences of $\nu+1$ states starting from state 0 and ending at state μ and time ν. $\Gamma_{\nu}(\mu)$ is also called the *node metric* and the path $\widehat{\boldsymbol{\mu}}_{0:\nu}(\mu)$ corresponding to $\Gamma_{\nu}(\mu)$ is the *survivor* path at state μ and time ν.

One may write

$$P\left(\boldsymbol{\mu}_{0:\nu}, \mathbf{y}_{1:\nu}\right) = P\left(\boldsymbol{\mu}_{0:\nu-1}, \mathbf{y}_{1:\nu-1}\right) P\left(\mu_{\nu}, y_{\nu} | \boldsymbol{\mu}_{0:\nu-1}, \mathbf{y}_{1:\nu-1}\right) \tag{5.41}$$

The state of the FSM depends only on the previous state of the FSM (Markov property and memoryless channel); thus, (5.41) simplifies to

$$\begin{aligned} P\left(\boldsymbol{\mu}_{0:\nu}, \mathbf{y}_{1:\nu}\right) &= P\left(\boldsymbol{\mu}_{0:\nu-1}, \mathbf{y}_{1:\nu-1}\right) P(y_{\nu}, \mu_{\nu} | \mu_{\nu-1}) \\ &= P\left(\boldsymbol{\mu}_{0:\nu-1}, \mathbf{y}_{1:\nu-1}\right) P(y_{\nu} | \mu_{\nu}, \mu_{\nu-1}) P(\mu_{\nu} | \mu_{\nu-1}). \end{aligned} \tag{5.42}$$

Introducing (5.42) in (5.40), one gets

$$\begin{aligned} \Gamma_{\nu}(\mu) &= \max_{\substack{\boldsymbol{\mu}_{0:\nu} \\ \mu_0 = 0, \mu_\nu = \mu}} \left(\log P(\boldsymbol{\mu}_{0:\nu-1}, \mathbf{y}_{1:\nu-1}) + \log\left(P(y_{\nu} | \mu_{\nu}, \mu_{\nu-1}) P(\mu_{\nu} | \mu_{\nu-1})\right)\right) \\ &= \max_{\mu'} \left(\Gamma_{\nu-1}\left(\mu'\right) + \gamma_{\nu}\left(\mu', \mu\right)\right), \end{aligned} \tag{5.43}$$

where

$$\gamma_{\nu}\left(\mu', \mu\right) = \log P\left(y_{\nu} | S_{\nu} = \mu, S_{\nu-1} = \mu'\right) + \log P\left(S_{\nu} = \mu | S_{\nu-1} = \mu'\right) \tag{5.44}$$

is usually called the *branch metric* for a state transition from state μ' to state μ. Since 0 is again the only possible initial state when the packet begins by a full symbol, one gets

$$\Gamma_0(\mu) = \begin{cases} 0 & \text{if } \mu = 0 \\ -\infty & \text{else.} \end{cases}$$

Using (5.43), $\Gamma_{\nu}(\mu)$, $\mu \in \Sigma$ may thus be evaluated recursively from $\Gamma_{\nu-1}\left(\mu'\right)$, $\mu' \in \Sigma$ and with the evaluation of branch metrics $\gamma_{\nu}\left(\mu', \mu\right)$. This recursion has lead to the VA, a special type of dynamic programming algorithm (Busacker, 1965; Omura, 1969). One iteratively extends a set of up to $m = |\Sigma|$ survivor paths, each of which starts from $S_0 = 0$ and ends in $S_{\nu} = \mu \in \Sigma$ at time ν. The selection of the m survivors at time $\nu+1$ is done by extending the m survivors at time ν with all possible transitions, updating the node metric for all the obtained sequences of states, and for each state, keeping only one survivor sequence among the sequences arriving to that state. When $\nu = n$, only the survivors arriving at the allowed final states (i.e., belonging to $\mathcal{F}$) are considered, the estimate (5.39) and, as a consequence (5.12), correspond to the path with the largest node metric $\Gamma_n(\mu)$, $\mu \in \mathcal{F}$.

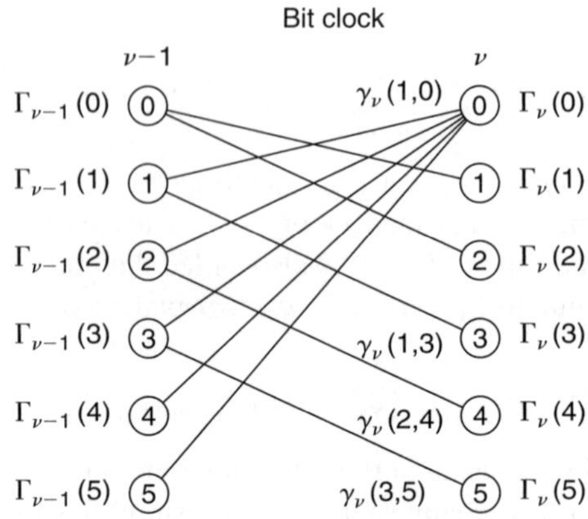

FIGURE 5.11 One iteration of the VA for the bit-clock trellis associated to the code of Example 5.1.

Example 5.8: Figure 5.11 illustrates one iteration of the VA as detailed in (5.43) for the bit-clock trellis corresponding to the code of Example 5.1.

Only some of the branch metrics have been represented, and one sees, for example, that

$$\Gamma_\nu(5) = \Gamma_{\nu-1}(3) + \gamma_\nu(3,5)$$

and that

$$\Gamma_\nu(0) = \max_{\mu'=1,\ldots5} \Gamma_{\nu-1}(\mu') + \gamma_\nu(\mu',0).$$

A single survivor path is kept at each node of the trellis.

The performance (in terms of BER as a function of E_b/N_0) of the VA, again on the same example, is provided in Figure 5.12, where the improvement over the hard decoder is clearly seen. This simulation also compares the symbol-by-symbol MAP estimate provided by the BCJR algorithm with the sequence ML estimate provided by the VA in terms of BER.

5.3.2 Soft-Output Viterbi Algorithm

The main drawback of the initial version of the VA is that it does not provide reliability information about the decoded bits or symbols, which prevents its use in iterative algorithms as developed in Chapter 9. Here, we focus on a modification of the VA allowing to obtain reliability information on decoded bits, assuming that a sequence of bits $\mathbf{b}_{1:n}$ has to be estimated from $\mathbf{y}_{1:n}$.

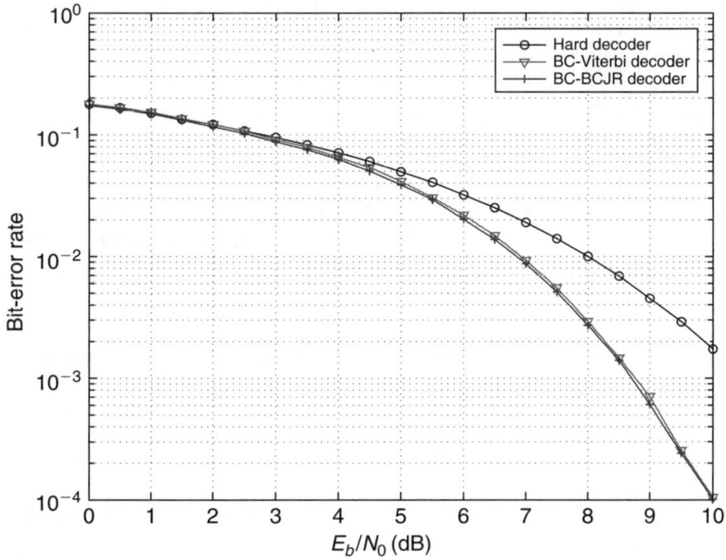

FIGURE 5.12 Viterbi algorithm decoding performance for the bit-clock trellis associated to the code of Example 5.1.

As introduced in Section 5.1, a measure of the confidence on the estimate $\widehat{b}_\nu$ is the logarithm of the ratio of the APPs (5.4), which may be rewritten as

$$\Lambda(\widehat{b}_\nu) = \log \frac{\sum_{\mathbf{b}_{1:n}, b_\nu = 0} P(\mathbf{b}_{1:n} | \mathbf{y}_{1:n})}{\sum_{\mathbf{b}_{1:n}, b_\nu = 1} P(\mathbf{b}_{1:n} | \mathbf{y}_{1:n})}. \tag{5.45}$$

Using the max-log approximation (Koch and Baier, 1990),

$$\log \sum_{i=1}^{n} a_i \approx \max_{i=1\ldots n} \log a_i, \tag{5.46}$$

one may write (5.45) as

$$\Lambda(\widehat{b}_\nu) = \max_{\mathbf{b}_{1:n}, b_\nu = 0} \log P(\mathbf{b}_{1:n} | \mathbf{y}_{1:n}) - \max_{\mathbf{b}_{1:n}, b_\nu = 1} \log P(\mathbf{b}_{1:n} | \mathbf{y}_{1:n}). \tag{5.47}$$

If $\widehat{\mathbf{b}}_{1:n}$ denotes the sequence MAP estimate, then

$$\begin{aligned} \Lambda(\widehat{b}_\nu) &= (1 - 2\widehat{b}_\nu) \left(\log P(\widehat{\mathbf{b}}_{1:n} | \mathbf{y}_{1:n}) - \max_{\mathbf{b}_{1:n}, b_\nu \neq \widehat{b}_\nu} \log P(\mathbf{b}_{1:n} | \mathbf{y}_{1:n}) \right) \\ &= (1 - 2\widehat{b}_\nu)\, \Delta_\nu, \end{aligned} \tag{5.48}$$

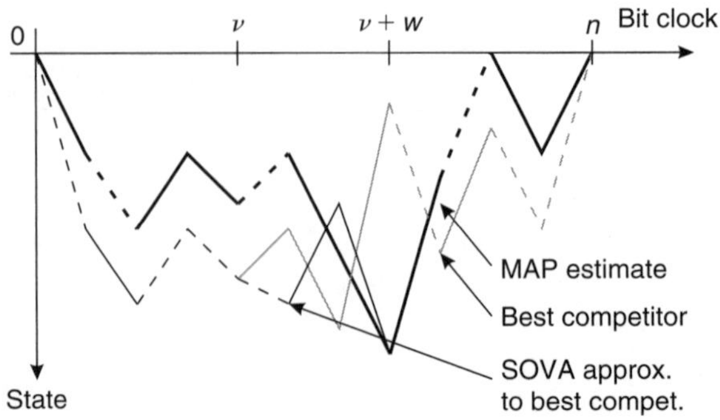

FIGURE 5.13 Several paths in a bit-clock trellis. Dashed branches correspond to zeros, and solid branches correspond to ones. The path corresponding to the MAP estimate is in bold. Its best competitor with $b_\nu \neq \widehat{b}_\nu$ (in gray) does not necessarily merge with the MAP path, whereas the SOVA approximate of the best competitor (in black) merges between ν and $\nu+w$.

where Δ_ν is the difference between the metric of the MAP estimate and the metric of the MAP estimate constrained to have $b_\nu \neq \widehat{b}_\nu$ (the corresponding sequence is called the *best competitor* to $\widehat{\mathbf{b}}_{1:n}$ with $b_\nu \neq \widehat{b}_\nu$). Nevertheless, for a bit-clock trellis, obtaining the best competitors for $\nu = 1, \ldots, n$, is not possible with the VA, since at each time instant ν and for each state μ, only a single survivor path is kept.

The SOVA (Hagenauer and Hoeher, 1989) is an extension of the VA providing an approximate evaluation of (5.48). The SOVA assumes that the best competitor to $\widehat{\mathbf{b}}_{1:n}$ with $b_\nu \neq \widehat{b}_\nu$ is found among the paths that merge with $\widehat{\mathbf{b}}_{1:n}$ between ν and $\nu+w$, where w denotes the size of a soft-output computation window, see Figure 5.13.

The main idea for that purpose is to consider for each state μ, the metric difference between the survivor path at time ν and the best nonsurvivor that can merge with the survivor path between times $\nu-w$ to ν, where w is chosen large enough so that all surviving paths at $\nu-w$ have merged together at time ν. Figure 5.14 shows a typical survivor path (in very bold) and the set of non-survivor paths (in black) that can merge with the survivor between time $\nu-4$ and ν.

To determine the best nonsurvivor for each bit location without storing all nonsurvivors in the trellis, the SOVA uses a simple update-and-discard procedure described in what follows. Let

$$\Delta^{(\mu,\nu)} = \left(\delta_{\nu-w}^{(\mu,\nu)}, \delta_{\nu-w+1}^{(\mu,\nu)}, \ldots, \delta_{\nu}^{(\mu,\nu)}\right)$$

be a vector of $w+1$ entries containing the metric difference $\delta_{\nu'}^{(\mu,\nu)}$ between the survivor path at state μ and time ν and the best nonsurvivor that can merge with it between time $\nu-w$ and ν, and has an opposite decision about bit $b_{\nu'}$ relative to the survivor as well. Let $\overline{\mu}$ be the incoming state associated with the

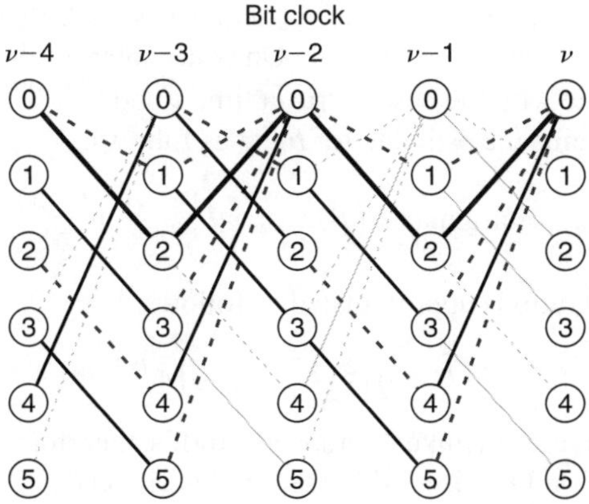

FIGURE 5.14 SOVA: A typical survivor path (bold) and a set of non-survivors (dark gray) that merge with the survivor between time $\nu-4$ and ν; transitions corresponding to an input 0 are in dashed lines, those corresponding to 1 are in solid lines.

survivor and $\underline{\mu}_1,\ldots,\underline{\mu}_q$, a set of q states associated with non-survivor paths (in Figure 5.14, $\overline{\mu}=2$ and $\underline{\mu}_1=1$, $\underline{\mu}_2=3$, $\underline{\mu}_3=4$, and $\underline{\mu}_4=5$). Let $\overline{\mathbf{b}}^{(\mu,\nu-1)}$ be the sequence of ν bits corresponding to the survivor path and $\underline{\mathbf{b}}_1^{(\mu,\nu-1)},\ldots,\underline{\mathbf{b}}_q^{(\mu,\nu-1)}$ be the set of q sequences corresponding to non-survivors merging the survivor path at time ν in state μ. Note that the set of nonsurvivors that can merge with path $\overline{\mathbf{b}}^{(\mu,\nu-1)}$, $\overline{\mathbf{b}}^{(\mu,\nu-2)}$, etc. can also merge with path $\overline{\mathbf{b}}^{(\mu,\nu)}$.

Assume that $\Delta^{(\overline{\mu},\nu-1)}$ for state $\overline{\mu}$ at time $\nu-1$ is known. Then the entry $\delta_{\nu'}^{(\overline{\mu},\nu-1)}$ of $\Delta^{(\overline{\mu},\nu-1)}$ represents the smallest metric difference between the survivor path of state $\overline{\mu}$ at time ν and its set of nonsurvivors, excluding the one merged at time ν, having an opposite decision about bit $b_{\nu'}$ relative to the survivor. To include the nonsurvivor paths at time ν, entries in $\Delta^{(\overline{\mu},\nu-1)}$ are compared to the metric difference between the survivor and the new nonsurvivor to form $\Delta^{(\mu,\nu)}$. Using this property, one may deduce the updating rule by evaluating first

$$d_{i,b}=\left(\Gamma_{\nu-1}(\overline{\mu})+\gamma_\nu(\overline{\mu},\mu,\overline{b}_\nu)\right)-\left(\Gamma_{\nu-1}\left(\underline{\mu}_q\right)+\gamma_\nu\left(\underline{\mu}_q,\mu,b\right)\right),\quad i=1\ldots q, b=0,1,$$

with $\overline{b}_\nu$ corresponding to the νth bit of the survivor path. Then,

$$\delta_\nu^{(\mu,\nu)}=\min_{i=1\ldots q} d_{i,1-\overline{b}_\nu}$$

and

$$\delta_{\nu'}^{(\mu,\nu)}=\min\left\{\delta_{\nu'}^{(\mu,\nu-1)},\min_{\substack{i=1\ldots q,b=0,1\\ \underline{b}_{i,\nu'}\neq\overline{b}_{\nu'}}} d_{i,b}\right\}.$$

The minimization is performed over all sequences such that the bit ν' differs between the survivor and a nonsurvivor. One thus obtains $\Delta^{(\mu,\nu)}$ from $\Delta^{(\overline{\mu},\nu-1)}$.

If μ^* is the node with the best metric at time ν and $\overline{\mathbf{b}}^{(\mu^*,\nu)}$ the corresponding path, then one obtains the estimate for $b_{\nu-w}$ as follows:

$$\widehat{b}_{\nu-w} = \arg\max_{b=0,1} P\left(B_{\nu-w} = b | \overline{b}_{\nu-w}^{(\mu^*,\nu)}, \overline{b}_{\nu-w-1}^{(\mu^*,\nu)}\right),$$

and the likelihood ratio is approximated as follows:

$$\Lambda(\widehat{b}_{\nu-w}) \simeq (1 - 2\widehat{b}_{\nu-w})\, \delta_{\nu-w}^{(\mu,\nu)}. \tag{5.49}$$

Such modifications are not linked to a specific trellis; therefore, similar extensions can be obtained for all trellises addressed in this chapter.

5.4 EXAMPLE: DECODING MPEG-4 AAC SCALE FACTORS

In Section 4.3.1, the redundancy present in the scale factors field of MPEG-4 AAC (ISO/IEC, 1997) has been evaluated. Here, the estimation techniques of Section 5.1 are applied to the robust decoding of the content of such scale factor fields when transmitting MPEG-4 AAC encoded data over noisy channels.

5.4.1 Context

We consider a monophonic audio signal compressed with the *Low-Complexity* profile of MPEG-4 AAC. Encoded data are put into packets, the structure of which has been briefly presented in Section 4.3.1. Assume that packets are BPSK modulated, transmitted over an AWGN channel, and that headers have been correctly received (this issue will be considered in Chapter 8), so that the beginning of the scale factor field and the number of scale factors are perfectly known. The main difficulty in estimating the scale factors from noisy data comes from the fact that the length of the scale factor field is not easily obtained when transmission errors impair the content of the field, since other data (*spectral data* and optional *data stream elements*) are stored just after the scale factors in MPEG-4 AAC packets. As indicated in Section 4.3.1, this length may be transmitted as side information, at the price of a reduced coding efficiency. As will be seen in what follows, this knowledge is not necessary to improve significantly the decoding performance.

Scale factors are encoded using a binary VLC $\mathcal{C} = \{\mathbf{c}_1, \mathbf{c}_2 \ldots \mathbf{c}_J\}$ of $J = 121$ code words. To help decoding, one may use the fact that three bits (*Pulse data present*, *Temporal noise shaping present*, and *Gain control present*) separate the scale factors from the remaining data of the MPEG-4 AAC packet. These three bits can be determined a priori depending on the encoding profile and are assumed to be known. Here, they are considered to be set to 000. One may thus introduce an additional code word $\mathbf{c}_{J+1} = 000$ to $\mathcal{C}$, interpreted as a mark indicating the end of the sequence of scale factors and playing the role of a basic *synchronization marker*, which will facilitate the decoding. A new code $\mathcal{C}'$ with $J+1$ code words is thus obtained.

Assume that $\mathbf{y}_{1:n} = (y_1, \ldots, y_n)$ represents the channel outcomes corresponding to the considered packet starting with the scale factor field and ending at the end of that packet. This packet contains $k-1$ scale factors, which may be modeled as a realization $\mathbf{x}_{1:k-1}$ of a source vector $X_{1:k-1}$ encoded with $\mathcal{C}$, followed by the code word $\mathbf{c}_{J+1}$ (the kth symbol is no more random) and by the other data. Thus, the beginning of $\mathbf{y}_{1:n}$ corresponds to k VLC code words $\mathbf{c}_{1:k} = \left(\mathbf{c}_{i_1} \ldots \mathbf{c}_{i_k}\right)$, with $\mathbf{c}_{i_k} = \mathbf{c}_{J+1}$. Necessarily, one has

$$\sum_{\kappa=1}^{k} \ell\left(\mathbf{c}_{i_\kappa}\right) \leqslant n.$$

The coder may be represented by a symbol-clock FSM of the same kind as that of Figure 4.6, page 99, with the main difference that the last code word must be $\mathbf{c}_{J+1}$; the transition between states $k-1$ and k is uniquely determined. If bit-level reliability information is not required, the BCJR algorithm for symbol-clock FSMs of Section 5.2.2 is a good choice to perform the evaluation of the MPM estimates of the emitted code words. The structure of the trellis, which may be deduced from the FSM representing the encoder, is provided in Figure 5.15, with a codebook $\mathcal{C}'$ limited to four code words.

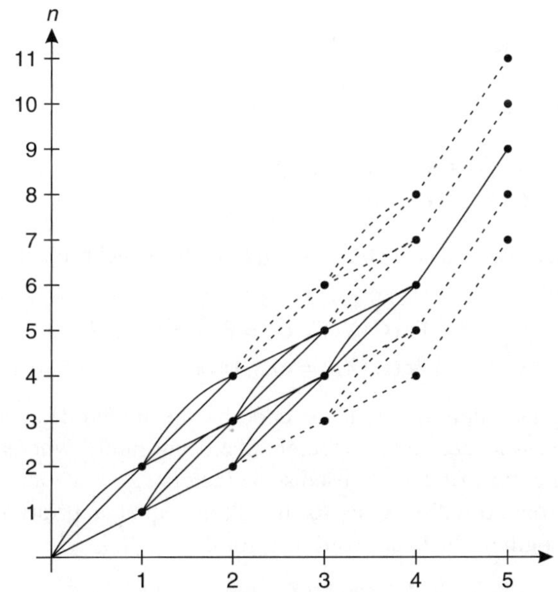

FIGURE 5.15 Structure of the bit-clock trellis used to perform the decoding of MPEG AAC scalefactors.

This trellis is a 2D symbol-clock trellis: the first dimension is the symbol clock κ and the second dimension is ν, the accumulated number of bits. The branches at the end of the trellis correspond to the code word $\mathbf{c}_{J+1}$. When the number of bits n associated with the scale-factor field is known, one gets a *closed trellis* represented in solid lines in Figure 5.15. The scale-factor error rate (SER) performance of both cases is depicted in Section 5.4.3.

5.4.2 Decoding Algorithm

The BCJR algorithm of Section 5.2.2 directly applies to get the MPM estimates of the κth scale factor from $\mathbf{y}_{1:n}$,

$$\widehat{x}_{\kappa}^{\text{MPM}} = \arg \max_{x=1...J+1} p\big(X_{\kappa} = x | \mathbf{y}_{1:n}\big). \tag{5.50}$$

The main difficulty is to determine probabilities like $P\big(S_{\kappa} = \nu | S_{\kappa-1} = \nu'\big)$ and $P\big(X_{\kappa} = x | S_{\kappa-1} = \nu', S_{\kappa} = \nu\big)$, as required by the BCJR algorithm from the a priori probabilities of the code words in $\mathcal{C}'$, which are usually not known to the receiver.

A technique estimating jointly these a priori probabilities and the transmitted sequence for noisy measurements has been proposed by Mannoni et al. (2006). Using an idea reminiscent from the turbo principle, a first code word decoding is performed with, e.g., uniform a priori probabilities. Using this result, better estimates for the a priori probabilities of the code words are obtained, which in turn help in a second code word decoding step, until convergence has been reached. This iterative process is not implemented here, and only the initialization of this decoder is considered. From approximate estimate of the a priori probabilities described below, one obtains estimates of the transmitted scale-factor sequence.

A first approximation of the a priori probabilities consists in assuming that they are function of the length of the code words, i.e., for $\kappa = 1, \ldots, k-1$

$$p_{i,\kappa} = \begin{cases} \Pr\left(C\left(X_{\kappa}\right) = \mathbf{c}_i\right) = 2^{-\ell(\mathbf{c}_i)} / \sum_{j=1}^{J-1} 2^{-\ell(\mathbf{c}_j)}, \ i = 1, \ldots, J \\ \Pr\left(C\left(X_{\kappa}\right) = \mathbf{c}_{J+1}\right) = 0, \end{cases} \tag{5.51}$$

except for the last symbol, which is known to be the synchronization marker, as described above.

$$p_{i,k} = \begin{cases} \Pr\left(C\left(X_k\right) = \mathbf{c}_i\right) = 0, \ i = 1, \ldots, J \\ \Pr\left(C\left(X_k\right) = \mathbf{c}_{J+1}\right) = 1. \end{cases} \tag{5.52}$$

The underlying assumption is that the code has been well designed, i.e., long code words are associated to scale factors that occur rarely, whereas short code words are for the most frequently used scale factors.

A second approximation consists in taking equal a priori probabilities, whatever the length of the code word, i.e., for $\kappa = 1, \ldots, k-1$

$$p_{i,\kappa} = \Pr\left(C\left(X_{\kappa}\right) = \mathbf{c}_i\right) = 1/J \tag{5.53}$$

and $p_{i,k}$ as in (5.52).

Now, considering one of these two a priori probability models, the a priori transition probabilities may be derived as follows:

$$P\left(S_\kappa = \nu | S_{\kappa-1} = \nu'\right) = \sum_{i,\ell(\mathbf{c}_i)=\nu-\nu'+1} p_{i,\kappa}$$

and

$$P\left(X_\kappa = x_i | S_{\kappa-1} = \nu', S_\kappa = \nu\right) = \begin{cases} 0, & \text{if } \ell(\mathbf{c}_i) \neq \nu - \nu' + 1, \\ p_{i,\kappa} / \sum_{i,\ell(\mathbf{c}_i)=\nu-\nu'+1} p_{i,\kappa}, & \text{else.} \end{cases}$$

The second difficulty comes from the initialization of β. Let $\mathcal{N}_k$ be the set of admissible accumulated number of bits at time k. If n_{SF}, the number of bits of the scale-factor field, is known, $\mathcal{N}_k = \{n_{\text{SF}}\}$. If this is not the case, using the minimum and maximum lengths ($\ell_{\min}$ and $\ell_{\max}$) of the code words in $\mathcal{C}$ and the size n of the packet, one may get

$$\mathcal{N}_k = \{(k-1)\,\ell_{\min} + 3, (k-1)\,\ell_{\min} + 4, \ldots, \min(n, (k-1)\,\ell_{\max} + 3)\},$$

see the end of Section 5.2.2 (the length of the kth code word is 3). Again, several a priori probabilities can be defined over $\mathcal{N}_k$. The most accurate a priori probability model would be obtained by considering the code word a priori probabilities (5.51). Nevertheless, when they are not perfectly known, one may consider that all values in $\mathcal{N}_k$ are equally likely, i.e.,

$$\beta_k(\nu) = \begin{cases} 1/|\mathcal{N}_k| & \text{if } \nu \in \mathcal{N}_k, \\ 0 & \text{else.} \end{cases} \tag{5.54}$$

5.4.3 Simulation Results

To evaluate decoding schemes based on a BCJR algorithm working on a symbol-clock trellis, MPEG-4 AAC encoded data are packetized into frames, BPSK-modulated, and sent onto an AWGN channel. Three decoding methods are applied:hard decoding, BCJR with known n_{SF} (*informed decoder*), and BCJR with unknown n_{SF} (*uninformed decoder*). The initialization of the a priori probabilities is done using (5.53) and the initialization for β is done according to (5.54). Performance is measured in terms of SER and objective perceptual quality at the decoder output, by running the PEMO-Q algorithm (Huber and Kollmeier, 2006). PEMO-Q gives a reliable prediction of subjective quality evaluations. The quality level is given by the Objective Differential Grade (ODG). An ODG of 0 means that the decoded signal is perceptually identical to the reference signal (unprocessed). An ODG of -4 means a maximum perceptual distortion. The audio material chosen for the tests is *Tom's Diner* by Suzanne Vega (first 5 s, sample rate 48 kHz), which is extensively used for evaluating audio codecs. The signal is encoded at 64 kbits/s with an MPEG-4 AAC Low-Complexity profile.

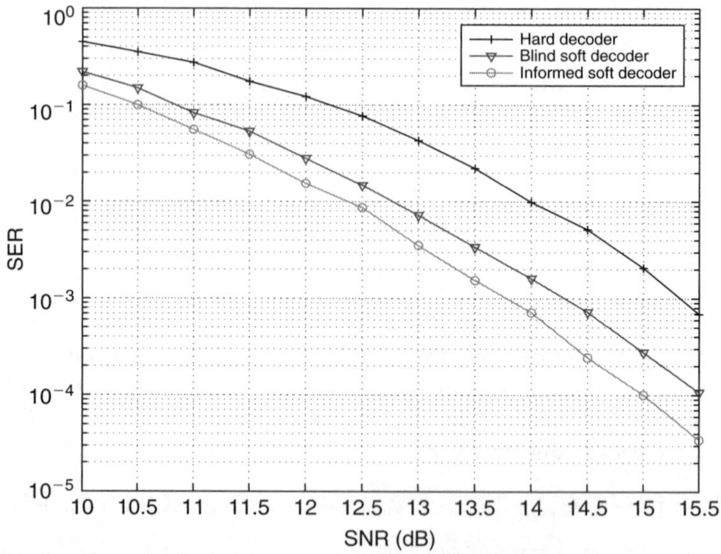

FIGURE 5.16 Scale-factor error rate for different values of the SNR on *Suzanne Vega*.

No correlation between successive scale factors has been taken into account here. Since errors on critical data is usually considered equivalent to a frame loss, critical data are assumed to be received without error. Spectral data are decoded and an error concealment algorithm is used if necessary, see Herre and Shulz (1998). The noiseless decoding of scale factors, with errors affecting the spectral data, represents the reference. This is denoted as "noiseless decoder" (meaning noiseless on the scale factors, but noisy estimates of spectral data, compensated by error concealment if necessary).

Figure 5.16 shows the SER for different values of SNR. For an SER of 10^{-3}, approximately 1.5 dB and 1.0 dB in SNR are gained with the informed and the uninformed decoders respectively. Figure 5.17 represents the ODG for different values of SNR. To get reasonably smooth plots, 10 noise realizations have been generated, and the average ODG values have been plotted.

With the hard decoder, the audio quality falls down when the SNR gets below 16 dB. The soft decoders (semiblind and informed) both have almost the same performance, which is about 0.5 dB better than the hard decoder. The SNR/ODG slope is almost the same for the soft decoders and the noiseless decoder. With the noiseless decoder, the gain is about 1 dB compared to the hard decoder. In other words, the soft decoders save half the performance loss introduced by transmission errors in the scale factors.

Note that the improvement of the soft decoders with respect to the hard decoder is lower in terms of subjective quality (0.5 dB) than in terms of SER (1 dB) because here, the quality criterion combines the effect of noise on both

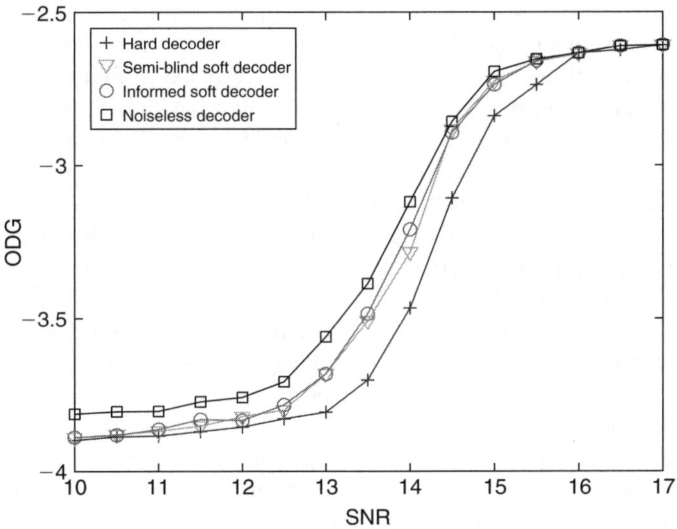

FIGURE 5.17 Objective quality evaluation for different SNR values on *Suzanne Vega*.

the scale factors and the spectral data. Below 13 dB, soft-decoding methods are close to the hard decoder.

5.5 POSSIBLE EXTENSIONS

Estimators and estimation algorithms presented in this chapter are of quite general purpose. Nevertheless, their application to perform joint source-channel decoding in practical situations usually requires some adaptations, as illustrated in Section 5.4 for the decoding of MPEG-4 AAC scale factors. The main difficulty consists in identifying the various sources of redundancy and to structure them, following the ideas presented in Chapter 4. For the situations considered in Section 5.2, only basic constraints have been taken into account (redundancy due to the syntax of the code words, to some basic semantic rules, to constraints on the number of symbols in a packet, or to the presence of some synchronization markers). Other constraints, provided that they allow a FSM representation of the source coder or decoder, may readily be considered, with minor adaptation of the considered estimation algorithms. Two examples are provided in the next chapters to show that there is no theoretical difficulty, but only possible practical difficulties.

Another possible extension is linked to sources with memory. Until now, sources were assumed memoryless. Taking the memory of a source into account is easily done using a more complex FSM, derived from the FSM representing the memoryless source, as illustrated by Example 4.6. Memory increases the number of states of the FSM and that of the corresponding decoding trellises.

For example, for a first-order Markov source, an additional dimension has to be incorporated in any (bit-clock or symbol-clock) trellis used for the decoding. For example, with the symbol-clock trellis represented in Figure 5.8, it is not possible to know, during decoding, which symbol has been decoded between $\kappa - 1$ and κ. The idea proposed by Thobanen and Kliewer (2003) and Jeanne et al. (2005) is to supplement this trellis with a new dimension storing the information corresponding to the previously decoded symbol.

In another context, the redundancy due to the semantic of the source, to the packetization of the encoded data, as detailed for the texture data of H.263+ in Section 4.3.2, can again be taken into account by a FSM. In Lee et al. (2005), it is shown that two dimensions have to be added to the symbol-clock trellis not accounting for source semantic and packetization. The main problem is that the resulting 4D trellis is not usable to perform efficient decoding in a reasonable amount of time.

These two examples illustrate that taking more complex situations or constraints into account often results in extended definitions of the state in the trellis, the dimension of which can become very large. In such case, or when the constraints imposed on the source or on the source coder are difficult to model by a simple FSM, one may switch to suboptimal estimation algorithms, which are presented in Chapter 6.

Chapter 6

Toward Practical Implementations

Chapter 5 has introduced several estimators and optimal estimation algorithms. All of them require a description with a finite state machine (FSM) of the coder or decoder and are the most efficient when the number of bits and the number of symbols of the sequence to be decoded are known. It was shown that with this knowledge, all successions of valid code words can be represented with trellises.

However, some current image and video coding standards, such as JPEG (ITU, 1992), H263+ (ITU-T, 2005), or the baseline profile of H.264/AVC (ITU-T/SG16/VCEG and ISO/JTC1/SC29/WG11, 2002), involve large variable-length codebooks, and the complexity of the trellises describing such VLCs may become prohibitive. Moreover, in most examples of Chapter 5, only simple constraints involving the syntax of the source codes, the number of bits of the encoded sequence, or the number of symbols it contains have been considered. Accounting for more specific constraints, e.g., related to the semantic of the source, may even add more dimensions to the "basic" bit/symbol decoding trellis, as will be seen in Section 6.2. Dealing with four or higher dimensional trellis with codes resulting in FSM with many states is clearly intractable when the aim is to perform real-time decoding of received bitstream.

On another side, source coders become more and more efficient. Structuring the residual redundancy with the help of FSM and trellises is quite easy for source coders such as JPEG, H263+, or MPEG4-AAC. For more recent (and more efficient) coders, this task is much more involved. It becomes almost impossible for compression techniques involving universal codes, such as gzip (Deutsch, 1996), or arithmetic codes with many contexts, such as H.264/AVC in its main profile, since the number of states required to describe such source coders would be unmanageable. In some other situations, it may even happen that the trellis representation does not even exist.

Does this mean that for most practical situations, joint source-channel decoding (JSCD) is useless? The answer is that in many situations, even if a straightforward application of the optimal decoding algorithms as presented in Chapter 5 may well be out of scope, two possible remedies may exist. In a first situation one may design trellises of reduced number of states from the full complexity ones, resulting in suboptimal algorithms that may implement the estimators of Chapter 5 with a controlled degree of performance degradation. In a second situation, when the redundancy of source coders is not that well structured or when the first strategy does not result in practically usable algorithms, sequential decoders may be the solution. This chapter describes both approaches.

This chapter first introduces several suboptimal decoding algorithms that have been proposed to improve the applicability of source-channel decoding techniques. *Aggregated trellises* (Jegou et al., 2005; Malinowski et al., 2007) are introduced in Section 6.1 to reduce the number of states of a trellis by replacing the constraint on the number of symbols of the sequence to be decoded by a constraint on the number of symbols modulo K_a. In Section 6.2, some constraints are simply neglected. As a consequence, a suboptimal decoder working on a small-dimensional trellis is obtained without accounting for all constraints imposed by the source. This small-dimensional trellis is a projection of the complete trellis on some dimension of the complete state. These projected trellises have usually many parallel transitions between states, e.g., corresponding to code words of the same length. Section 6.3 shows how such parallel transitions can be grouped in classes to get a reduced number of transitions and to significantly reduce the amount of computations performed during decoding. Finally, Section 6.4 introduces sequential decoders; the aim of which is to explore only part of the decoding trellises in an efficient way. Such decoders are usable even when it is difficult to structure the residual redundancy with a FSM or trellis.

6.1 STATE AGGREGATION

When the number k of symbols of a sequence $\mathbf{x}_{1:k}$ to be decoded from measurements $\mathbf{y}_{1:n}$ is known at decoder side, one may use a 2D or 3D symbol-clock trellis or a 3D bit/symbol trellis, such as that introduced in Section 5.2.1. The number of nodes of such bit/symbol trellis is $O(mkn)$, where m is the number of states of the FSM representing the decoder. The decoding complexity is thus of the order of the number of nodes in the trellis $O(mkn)$.

The aim of *state aggregation* techniques (Jegou et al., 2005; Malinowski et al., 2007) is to reduce the decoding complexity without degrading too much the decoding performance in terms of bit or symbol error rate. Aggregated trellises realize a compromise between low-complexity bit trellises, which do not take the knowledge of k into account, and high-complexity bit/symbol trellises which do.

6.1.1 Description of the Aggregated Trellis

An aggregated trellis is a bit/symbol trellis where the state $\overline{S}_\nu$ at time ν is the pair $(\mu, \overline{\kappa})$, with μ, the state of the FSM describing the decoder, and $\overline{\kappa} = \kappa \% K_a$, the remainder of the division of the number of already decoded symbols κ by a constant K_a. The termination constraint involving the total number of symbols k in the bit/symbol trellis is replaced by a constraint on $\overline{k}$, the number of symbols k modulo K_a; see Example 6.1.

Example 6.1: Consider the variable-length code (VLC) $\mathcal{C} = \{0, 10, 11\}$ associated to a source with alphabet $\mathcal{A} = \{a, b, c\}$. Figure 6.1 represents all successions of code words resulting in an encoded sequence of $n = 8$ bits. When the number of symbols k is also known, all successions of code words are best represented by the bit/symbol trellis of Figure 6.2, here with $k = 5$. Taking, for example, $K_a = 2$, one gets the aggregated trellis of Figure 6.3.

When $K_a = 1$, one gets a bit-clock trellis, and when $K_a = k$, the aggregated trellis is simply the bit/symbol trellis. Aggregated trellises keep track of the number of already decoded symbols modulo K_a. The number of states of the aggregated trellis is $O(mK_a n)$. Even if the number of states of the bit/symbol trellis of Figure 6.2 is smaller than that of the aggregated trellis of Figure 6.3, for sufficiently large values of n, the complexity of aggregated trellises becomes much lower than that of bit/symbol trellises. Moreover, the transmission of $\overline{k}$ to the decoder costs usually less than that of k, reducing thus the amount of side information to be transmitted to improve the efficiency of JSCD techniques.

The BCJR and Viterbi algorithms of Chapter 5 obviously apply on aggregated trellises, provided that one is able to evaluate the a priori state transition

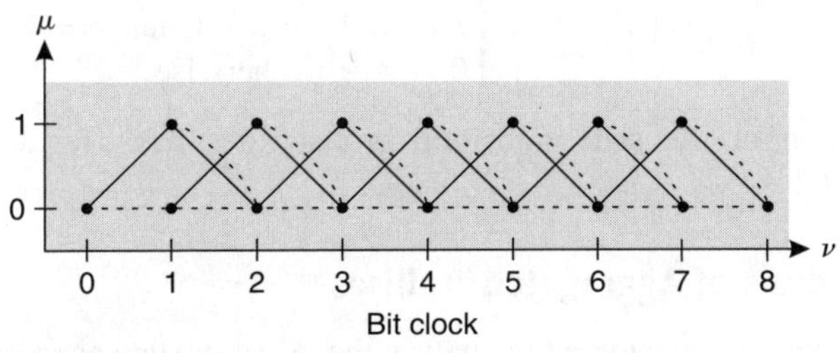

FIGURE 6.1 Bit-clock trellis for Example 6.1, dotted transitions correspond to 0s and plain transitions to 1s.

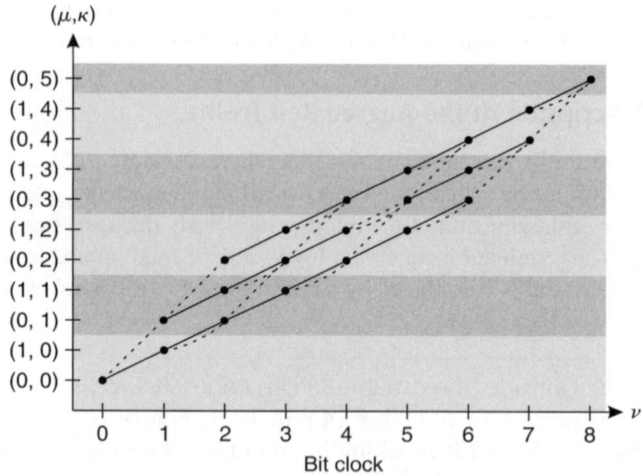

FIGURE 6.2 2D view of a bit/symbol trellis for Example 6.1, the actual trellis is 3D.

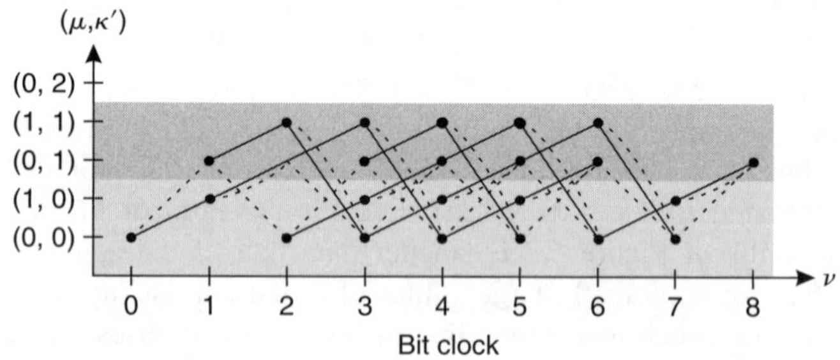

FIGURE 6.3 Aggregated trellis with $K_a = 2$ for Example 6.1.

probabilities for the aggregated trellis. They may be directly derived from the state transition probabilities $P(\mu|\mu')$ of the FSM representing the decoder

$$P\left(\overline{S}_\nu = (\mu, \overline{\kappa})\,|\overline{S}_{\nu-1} = \left(\mu', \overline{\kappa}'\right)\right) = \begin{cases} P\left(\mu|\mu'\right) & \text{if } \mu \neq 0 \text{ and } \overline{\kappa} = \overline{\kappa}' \\ P\left(\mu|\mu'\right) & \text{if } \mu = 0 \text{ and } \overline{\kappa} = \left(\overline{\kappa}' + 1\right) \% K_a. \\ 0 & \text{otherwise.} \end{cases}$$

The evaluation of the state transition probabilities has been explained in Section 5.2.1.

6.1.2 Analysis of Aggregated Trellises

Two approaches are considered to analyze the effect of state aggregation. First, one evaluates the deterministic redundancy in a sequence when its number of symbols is only known modulo K_a. A second approach analyzes the number of inserted/removed symbols when decoding an encoded bitstream corrupted by noise and evaluates the information provided by the knowledge of the number of symbols modulo K_a.

Deterministic Redundancy

The impact of using an aggregated trellis in place of a complete bit/symbol trellis can be evaluated by comparing the deterministic redundancy of a source code $\mathcal{C}$ when the number of symbols k of an encoded sequence of n bits is known and when this number is only known modulo K_a.

Consider the symbol-clock FSM $(\mathcal{A}, \{0,1\}^*, \Sigma, \sigma_0, t, \omega, \mathcal{F})$ describing $\mathcal{C}$. For any $\mathbf{x} \in \mathcal{A}^*$, let

$$\xi_k(\mathbf{x}) = \begin{cases} 1 & \text{if } \ell(\mathbf{x}) = k \\ 0 & \text{else} \end{cases} \tag{6.1}$$

be the constraint indicating that the source sequence should be of k symbols. Then, the amount of sequences of n bits corresponding to k symbols encoded with $\mathcal{C}$ is

$$|\Omega_n(\xi_k)| = \eta(n, k, \sigma_0) \tag{6.2}$$

and can be evaluated recursively using (4.41). The deterministic redundancy is then

$$R_n(\mathcal{C}, \xi_k) = n - \log_2 |\Omega_n(\xi_k)|. \tag{6.3}$$

Now, the constraint

$$\xi^{\mathrm{m}}_{k',K_\mathrm{a}}(\mathbf{x}) = \begin{cases} 1 & \text{if } \ell(\mathbf{x}) \% K_\mathrm{a} = k' \\ 0 & \text{else} \end{cases} \tag{6.4}$$

indicates that the number of symbols of a sequence $\mathbf{x}$ modulo K_a is k'. The number of sequences $|\Omega_n(\xi^{\mathrm{m}}_{k',K_\mathrm{a}})|$ of n bits encoded with $\mathcal{C}$ and satisfying this constraint involves again $\eta(n, k, \sigma_0)$ as follows:

$$\left|\Omega_n\left(\xi^{\mathrm{m}}_{k',K_\mathrm{a}}\right)\right| = \sum_{\kappa, \kappa\% K_\mathrm{a} = k'} \eta(n, \kappa, \sigma_0), \tag{6.5}$$

and the deterministic redundancy is now

$$R_n\left(\mathcal{C}, \xi^{\mathrm{m}}_{k',K_\mathrm{a}}\right) = n - \log_2 \left|\Omega_n\left(\xi^{\mathrm{m}}_{k',K_\mathrm{a}}\right)\right|. \tag{6.6}$$

Example 6.2 evaluates the residual redundancy of a coded sequence as a function of K_a.

Example 6.2: Consider a source X emitting symbols in the alphabet $\mathcal{A} = \{a, b, c, d, e\}$ and the code $\mathcal{C}_5 = \{01, 00, 11, 100, 101\}$ from Zhou and Zhang (2002). Figure 6.4 shows the evolution of the deterministic redundancy (6.6) and of the number of nodes in the trellis as a function of K_a for sequences

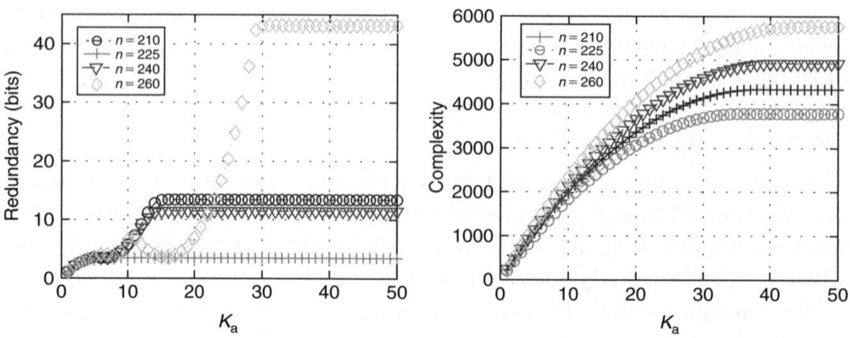

FIGURE 6.4 Residual redundancy (left) and upper bound for the number of nodes in the trellis (right) as a function of K_a for $k = 100$ and several values of n.

of $k = 100$ symbols and several values of n. The redundancy has been evaluated using (4.41) and (6.5), considering in all cases that n is known by the decoder.

When $K_\text{a} = k$, the aggregated trellis is exactly the bit/symbol trellis. In this case, as already noticed in Example 4.23 the residual redundancy depends on the value of n. Here, $R_n(\mathcal{C}, \xi_{100})$ is minimum for $n = 225$, which corresponds to $100\overline{\ell}$, where $\overline{\ell}$ is the average length of the code words under the assumption of dyadic probabilities. When K_a is very small, $R_n(\mathcal{C}, \xi^{m}_{k', K_\text{a}})$ hardly depends on n: the residual redundancy is mainly due to the knowledge of n and to the fact that the encoded sequence contains an integer number of symbols. When K_a increases, the convergence of $R_n(\mathcal{C}, \xi^{\text{m}}_{k', K_\text{a}})$ to $R_n(\mathcal{C}, \xi_{100})$ is very fast when $R_n(\mathcal{C}, \xi_{100})$ is small and requires larger values of K_a when $R_n(\mathcal{C}, \xi_{100})$ is large.

This observation is confirmed by Figure 6.5, showing the evolution of the redundancy as a function of the complexity. One observes that for encoded sequences of $n = 225$ bits, which present the least deterministic redundancy, this redundancy is obtained for small values of K_a, allowing the number of states in the trellis to be divided by 4. For sequences with more deterministic redundancy, the complexity may be halved without any degradation in the deterministic redundancy. Finally, for sequences with high deterministic redundancy, i.e., for which the knowledge of k brings a lot of information, small values of K_a result in a significant degradation of the redundancy.

Example 6.2 shows that even when $R_n(\mathcal{C}, \xi_{100})$ is quite large, K_a may be chosen significantly less than k, thus leading to an aggregated trellis of reduced complexity, without losing too much information (measured in terms of deterministic redundancy).

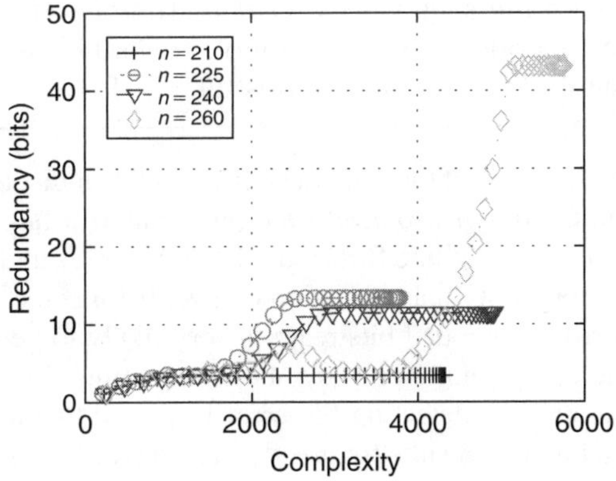

FIGURE 6.5 Residual redundancy as a function of the number of nodes in the trellis parametrized with K_a for $k=100$ and for several values of n.

Error State Diagram

The error state diagram (Malinowski et al., 2007) has been proposed to study the difference ΔK (called *gain/loss*), caused by a single bit error between the number of emitted and decoded symbols, after hard decoding of the received bitstream. On top of this target, this tool brings an interesting point of view on the usefulness of integrating the knowledge of the number of symbols into a decoding procedure, as pointed out below.

As illustrated by Example 6.3, ΔK depends on the sequence and on the location of the error. It is thus considered as a random variable taking values in $\mathbb{Z}$, the set of relative integers.

Example 6.3: Consider again the source and code of Example 6.2. The sequence of code words associated to $\mathbf{x}=(ebcbba)$ is

$$\mathbf{b}=(1010011000001).$$

Assume that during transmission over a binary symmetric channel, the second bit has been flipped to get

$$\mathbf{y}_1=(1\mathbf{1}10011000001).$$

The decoded message is then $\mathbf{x}_1=(cdcbba)$, the decoder has *resynchronized* after the second symbol, i.e., it is able to decode correctly the bitstream after the second erroneous symbol has been decoded. Since the length of $\mathbf{x}$ is the same as that of $\mathbf{x}_1$, in this case, $\Delta K=0$. Assume now that the fourth bit has been flipped

$$\mathbf{y}_1=(101\mathbf{1}011000001)$$

so that the decoded message is $\mathbf{x}_2 = (eedba)$. The decoder has resynchronized once it has decoded the third symbol. Nevertheless, in the decoded sequence one symbol is missing and thus $\Delta K = -1$.

The *error state diagram* of the code, which is an FSM indicating the evolution of a decoder from the *desynchronized* state (after an error has occurred) to the resynchronized state (Maxted and Robinson, 1985; Swaszek and DiCicco, 1995), allows to get the probability mass function of ΔK. If for some code $P(\Delta K = 0)$ is high, a bit in error in a coded bitstream is very likely to generate a decoded sequence with the correct number of symbols. Assuming that the number k of source symbols is known, detecting the error is very difficult. For codes with low $P(\Delta K = 0)$, a bit error results frequently in a gain or loss of symbols, which allow an easy detection of the error.

The variance $\sigma^2_{\Delta K}$ of ΔK or its entropy $H(\Delta K)$ are more accurate figures of merit for the resynchronization property of a code. Using codes with poor resynchronization properties, i.e., with high $H(\Delta K)$, is thus very interesting in the use of decoding techniques involving symbol-clock or bit/symbol trellises (see Chapter 5). Indeed, in such trellises, erroneous sequences that do not have the correct number of symbols are not considered. $H(\Delta K)$ is a measure of the "information" brought by the knowledge of the length k of a sequence when performing JSCD with a symbol-clock or a bit/symbol trellis. In other words, if a single error likely results in an erroneous number of symbols, a decoding algorithm using the correct number of symbols as a constraint will likely correct it.

Example 6.4: Using the error state diagram of $\mathcal{C}_5$, one gets $P(\Delta K = -1) = 0.1023$, from Malinowski et al. (2007), $P(\Delta K = 0) = 0.8352$, and $P(\Delta K = 1) = 0.0625$, which leads to $\sigma^2_{\Delta K} = 0.1632$ and $H(\Delta K) = 0.8035$ bits/symbol.

When several errors occur, the symbol gains and losses may compensate. Assuming that when a bit error occurs, the decoder has already resynchronized, with possibly some gain/loss of source symbols, Malinowski et al. (2007) extend the results of Swaszek and DiCicco (1995) by considering the effect on ΔK of several bit errors resulting from the transmission of a sequence of length k over an additive white Gaussian noise (AWGN) channel followed by a hard decision on the bits. In such case, the probability mass function of ΔK depends on k and on the signal-to-noise ratio (SNR) E_b/N_0 on the channel. Even for long sequences, large absolute values for ΔK are rather unlikely; thus, for any $\eta > 0$, one may find some positive integer δk_η such that

$$\delta k_\eta = \arg\min_{\delta k > 0} P(\Delta K < -\delta k) + P(\Delta K > \delta k) \leqslant \eta. \tag{6.7}$$

Example 6.5: For sequences of $k=100$ symbols and $E_b/N_0=6$ dB, one gets for the code $\mathcal{C}_5$ (from Malinowski et al. (2007))

$$\begin{cases} P(\Delta K \leqslant -4) & = 0.0000002 \\ P(\Delta K = -3) & = 0.0000235 \\ P(\Delta K = -2) & = 0.0013201 \\ P(\Delta K = -1) & = 0.0493389 \\ P(\Delta K = 0) & = 0.9186664 \\ P(\Delta K = 1) & = 0.0301524 \\ P(\Delta K = 2) & = 0.0004930 \\ P(\Delta K = 3) & = 0.0000053 \\ P(\Delta K \geqslant 4) & = 0.0000001 \end{cases}$$

For $\eta = 10^{-6}$, one has $\delta k_\eta = 3$.

The parameter δk_η is important, since it allows the selection of a value for K_a that does not significantly modify the information brought by the knowledge of k. Indeed, with $K_a = 2\delta k_\eta + 1$, Malinowski et al. (2007) show that

$$H(\Delta K) - (k + 2\delta k_\eta + 1)\eta \log \eta \leqslant H(\Delta K \% (2\delta k_\eta + 1)) \leqslant H(\Delta K), \qquad (6.8)$$

meaning that with η sufficiently small, and thus $K_a = 2\delta k_\eta + 1$ large enough, there is no information loss in using an aggregated trellis.

Example 6.6: Consider again the code $\mathcal{C}_5$ with encoded sequences of $k = 100$ symbols (from Malinowski et al. (2007)). Table 6.1 shows the evaluation of the frame error rate (FER) for E_b/N_0 between 3 and 7 dB and for various K_a from 1 to 100. As expected, the FER decreases when E_b/N_0 increases but reaches a floor for a value of K_a depending on E_b/N_0. For example, for $E_b/N_0 = 6$ dB, this floor is reached for $K_a = 5$, which is close to the value of 7 deduced from Example 6.5 and (6.8).

6.1.3 Comparison

The two previously considered analysis techniques evaluate the impact of K_a on the deterministic redundancy for the first approach and on the "information" provided by the number of symbols in a coded sequence.

The first technique is mainly based on the lengths of the code words, whereas in the second approach, the length and (binary) representation of each code word is important to predict the performance with different K_a (see also Swaszek and DiCicco, 1995). The effect of the channel noise is also better caught by the second approach.

TABLE 6.1 Evolution of $H(\Delta K)$ as a Function of K_a and E_b/N_0 (Taken from Malinowski et al. (2007))

E_b/N_0	3	4	5	6	7
$K_a=1$	0.99120	0.92330	0.70464	0.38774	0.14558
$K_a=2$	0.98805	0.90368	0.66193	0.34633	0.12452
$K_a=3$	0.98698	0.89901	0.65527	0.34313	0.12388
$K_a=4$	0.98665	0.89795	0.65457	0.34298	0.12386
$K_a=5$	0.98652	0.89782	0.65449	0.34296	⋮
$K_a=10$	0.98651	0.89780	0.65488	⋮	⋮
$K_a=100$	0.98651	0.89780	0.65448	0.34296	0.12386

Both methods demonstrate that a noticeable reduction of complexity is feasible without too much loss in terms of performance. Finally, the study of resynchronization properties provides some insight on the usefulness of building higher dimension trellis incorporating the number of symbols.

6.2 PROJECTED TRELLISES

The 3D bit/symbol trellis considered in Chapter 5 allows to easily account for the knowledge (i) of the number of bits to process and (ii) of the number of symbols to decoder. Its main advantage is that there are no parallel transitions between states. The evaluation of bit a posteriori probabilities is then easy; provided that a BCJR algorithm or a soft output Viterbi algorithm (SOVA) is used to perform decoding; see Section 5.2.1. The drawback of bit/symbol trellises and other high-dimensional trellises is the size of the memory they require. Here, $O(nkm)$ data have to be stored, due to the three dimensions (ν, κ, μ), where $\nu \in \{1, \ldots, n\}$ and $\kappa \in \{1, \ldots, k\}$ are, respectively, the number of already decoded symbols and processed bits, and $\mu \in \{1, \ldots, m\}$ is the state of the bit-clock FSM associated to the decoder.

Some additional constraints, e.g., on the semantic of the source, are taken into account by adding new dimensions to a trellis, as discussed in Section 6.2.2. The number of possible path in such high-dimensional trellis is usually smaller than the number of path in a "simple" bit/symbol trellis, since the constraint eliminates paths. The complexity is thus not directly related to the dimensionality of the trellis, but more by the number of sequences it may describe.

The aim of this section is to show that projecting high-dimensional trellises onto subspaces formed by several components of the state may reduce the memory requirements. Combined with code-word grouping techniques, see Section 6.3, projected trellises may also lead to significant computational savings.

6.2.1 Projections of the Bit/Symbol Trellis

To reduce the memory requirements and the number of operations needed to perform decoding, one may use a trellis with a reduced number of dimensions by projecting, e.g., the bit/symbol trellis onto two of the dimensions of the state. The price to be paid is some potential suboptimality in the associated decoding algorithms.

When performing a projection of the bit/symbol trellis on the (κ, ν)-plane, one obtains a symbol-clock trellis such as that of Figure 5.8, page 133. Many parallel transitions between states now appear due to the projection. Such transitions correspond in the case of a Huffman-like VLC to code words of the same length. The memory requirements are now reduced to $O(kn)$. Nevertheless, for each transition between states, branch metrics involve code words of at most $m + 1$ bits; thus, a better evaluation of the decoding complexity would lead to $O(nkm)$.

With a projection on the (ν, μ)-plane, one obtains a bit-clock trellis, such as that represented in Figure 5.5, page 126. The constraint on the number of symbols the sequence contains cannot be taken into account. When this information is available, such bit-clock trellis is thus suboptimal. The bit-clock trellis allows an easy computation of bit-level a posteriori probabilities. Nevertheless, symbol-level a posteriori probabilities are much more difficult to obtain. The memory requirements and decoding complexity are now $O(nm)$.

A further projection can be done on the ν-plane for Huffman-based VLCs to get a ID trellis. As for the symbol-clock trellis, many parallel transitions appear now. The complexity remains the same, but the memory requirements are limited to $O(n)$.

6.2.2 Projections of Higher-Dimensional Trellises

The idea of projecting high-dimensional trellises onto a smaller number of dimensions takes its full power when applied to situations where many constraints are taken into account.

In fact, one way to take additional constraints (e.g., corresponding to the semantic of the source) is to add new dimensions to the bit/symbol trellis. The resulting high-dimensional trellis, even if optimal in terms of consideration of the constraints, becomes quite difficult to exploit due to the resulting computational complexity. A projection onto some of the components of the state allows to focus on the most important properties of the source while making the complexity tractable. This section is illustrated by the decoding of texture blocks, i.e., motion-compensation residuals (see Section 3.4), encoded with the H.263+ encoder (ITU-T, 2005), but the methodology introduced may be directly translated to many other situations involving high-dimensional decoding trellises.

5D Trellis for Decoding H.263+ Texture Blocks

The way texture blocks are encoded in the H.263+ standard has been presented in Section 4.3.2 and is briefly recalled here. After motion-compensation, each texture block of 8×8 pixels is transformed and zig-zag scanned in order to get a vector of 64 transformed coefficients, which are then encoded with triples (*run*, *level*, *last*). For each triple, *run* indicates the number of null coefficients before the next nonzero coefficient, *level* indicates the amplitude of this coefficient, and *last* indicates whether it is the last nonzero coefficient of the block. Each triple is then encoded using a VLC code $\mathcal{C}_{\text{Text-H.263}}$ containing 204 code words with lengths between 3 and 13 bits. $\mathcal{C}_{\text{Text-H.263}}$ also contains a set of 22-bit fixed-length code words starting with *escape* (000 0011), containing the *last* bit, a run value ranging from (000 000) to (111 111), and a 8-bit coded level between -127 and 127 (the values -128 (1000 0000) and 0 (0000 0000) are not used).

Consider a vector $\mathbf{b}$ of n bits corresponding to an H.263+ encoded texture block. The residual redundancy present in $\mathbf{b}$ has been evaluated in Section 4.3.2. A 3D bit/symbol trellis is not able to fully exploit this redundancy since with such trellis, the decoding could result in sequences corresponding to more than 64 transformed coefficients, which is not feasible. Thus, a 4D bit/symbol/transformed coefficients (BSTC) trellis with state (κ, ν, μ, τ) has to be considered, where τ is the number of already recovered transformed coefficients. This trellis allows to account for the number of bits and encoded symbols, as well as for the semantic of the source (number of transformed coefficients in a texture block limited to 64).

Assume now that a vector $\mathbf{b}_\text{p}$ of n_p bits contains several H.263+ encoded texture blocks. To decode such vector transmitted over a noisy channel, the preceding 4D trellis may be used several times. The main difficulty comes from the fact that the number of bits of each subvector associated to a single-texture block is unknown. Moreover, when the number of texture blocks is known precisely, there is no way to take this into account. The solution is again to supplement the state of the 4D BSTC trellis with an additional state component θ, indicating the number of already decoded texture blocks. The resulting trellis is now 5D, with state vector $(\kappa, \nu, \mu, \tau, \theta)$. The memory required to store this trellis is now $O(knmt_{\text{TC}}t_{\text{TB}})$, where $t_{\text{TC}} = 64$ is the number of transformed coefficients and t_{TB} is the number of texture blocks of the vector. The decoding complexity is of the same order.

Projections of the 5D Trellis

Such trellis presents too many states to allow a manageable real-time decoder to be implemented. A suboptimal decoding solution for the texture blocks has been proposed by Lee et al. (2005). When receiving a packet containing several texture blocks, neither their location nor the number of code words within each subvector corresponding to a single-texture block is known. However, the number of texture blocks may be deduced from packet headers, assumed to be perfectly decoded;

see Chapter 8 for more details on reliable header decoding. The main idea is to determine the location of each encoded texture block in a measured vector $\mathbf{y}_\mathrm{P}$ and then to perform the decoding of each subvector of $\mathbf{y}_\mathrm{P}$, taking only into account the useful constraints.

For that purpose, a projection onto the (ν, θ)-plane is considered first, which allows to estimate the likeliest locations of the frontiers between texture blocks. The resulting 2D bit-clock trellis allows to take into account the constraints on the number of bits and on the number of texture blocks (when available) contained in a code vector. The total number of code words being unknown, it is useless to keep κ. The two other components μ and τ are dropped to get a trellis as simple as possible: the FSM describing the decoder would consist of about 2^{14} states due to the presence of fixed-length code words in the code used to represent the triples (run, level, and last). The value of θ is incremented each time a VLC corresponding to *last* = 1 is encountered, indicating a last nonzero transformed coefficient of a texture block. The location ν_θ of the end of the θth texture block is estimated by looking at the block locations of the maximum-likelihood (ML) sequence estimate (computed with the Viterbi algorithm) consistent with the 2D trellis.

If necessary, this trellis allows to estimate the number of texture blocks using the same technique as that used to estimate the number of symbols in a code vector (see Section 5.2.2). The main drawback of this trellis is that it does not account for the constraint on the number of transformed coefficients in each block, leading to a suboptimal estimation algorithm.

Once the locations of each texture block have been obtained, the 5D trellis may be projected onto the (κ, ν, τ)-plane to get a symbol-clock trellis taking into account the number of transformed coefficients of each texture block. The decoding with this 3D trellis is then performed in a similar way as the decoding with the symbol-clock trellis, where a constraint on the number of transformed coefficients has to be taken into account; see Section 5.2.1 for more details.

For both trellises, the number of parallel transitions may be very large, since 2^{15} fixed-length code words have to be taken into account when computing the various branch metrics. To reduce the computational effort, parallel branches have been grouped into classes (see Section 6.3).

Experimental Results

The performance of the texture block decoding technique, which principle has been described in the previous section, is now evaluated. The 101 first frames of `foreman.qcif` are considered with a quantization step tuned to $Q_\mathrm{P} = 16$, leading to a bitrate of about 0.2 bpp. Compressed data are put in packets of at most 1024 bits and sent over an AWGN channel. A soft-input Viterbi algorithm is used to localize the texture blocks in packets containing INTER-encoded texture blocks and in packets containing INTRA-encoded texture blocks. A 2D bit-clock trellis with dimensions (ν, θ) is used for that purpose. Only the number of texture blocks and the length in bits of each packet are assumed to be known.

Once texture blocks have been localized, a 3D trellis and several decoders are used to decode each individual block. A classical hard input decoder (HD) serves as reference. It is compared with hard-input Viterbi algorithm (Const-HD), two soft-input Viterbi algorithms, one implementing an ML, one implementing an estimator, and a Viterbi algorithm implementing an MAP estimator to which the actual block locations have been provided (Known loc. MAP). The a priori probability of each code word $\mathbf{c}$ is taken as proportional to $2^{\ell(\mathbf{c})}$, where $\ell(\mathbf{c})$ is the length of $\mathbf{c}$.

Three scenarios have been considered. For scenario A, all frames are INTRA encoded. This results in an average number of blocks per packet, which is about 16 and a mean number of bits allocated to the texture which is 758. For scenario B, only the first frame is INTRA encoded, and all other frames are INTER encoded. The average texture size is now 602 bits, and each packet contains about 29 blocks. The results of the two-step decoding technique are represented in Figure 6.6.

For INTRA-encoded frames (scenario A), as soon as the SNR is larger than 8 dB, the results provided by MAP are very similar to that obtained by Known loc. MAP, which requires the correct transmission of the locations of each texture block. For INTER-encoded frames (B), the performances of MAP are between 0.5 and 1 dB worse than those obtained by Known loc. MAP. This is mainly due to the number of blocks contained in an INTER packet, which is twice that of an INTRA packet for this value of Q_P. Reducing the maximum packet length may help to reduce the discrepancy between these two techniques.

Scenario C consists in putting one INTRA-encoded frame every 10 encoded frames. The PSNR of the luminance is evaluated on each frame after using HD,

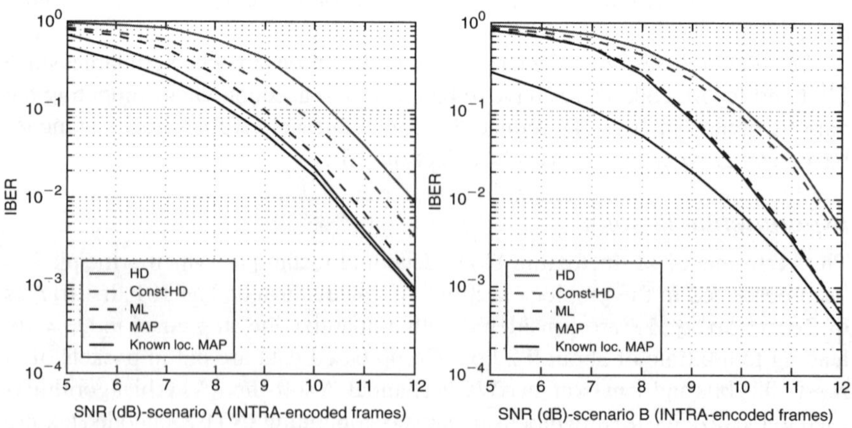

FIGURE 6.6 IBER for scenario A (INTRA-encoded frames) and B (INTER-encoded frames).

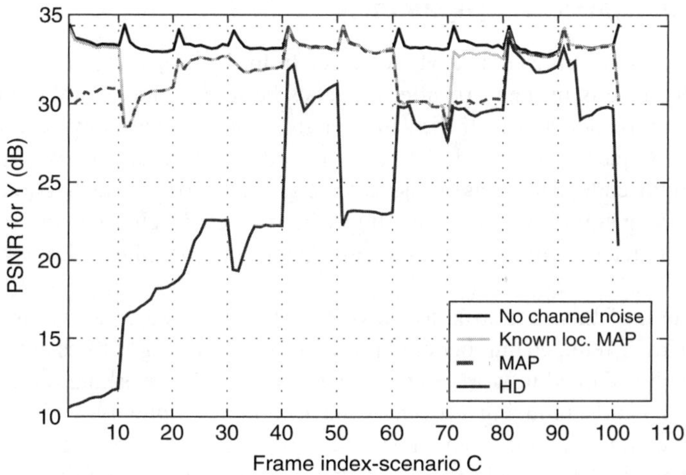

FIGURE 6.7 Scenario C, evolution of the peak signal-to-noise ratio (PSNR) of `Foreman.cif` encoded with H.264/AVC and decoded with various decoders, channel SNR of 11 dB.

MAP, and Known loc. MAP at an SNR of 11 dB. Figure 6.7 shows that again, the performances of MAP, and Known loc. MAP are quite similar, illustrating the efficiency of the proposed two-step algorithm.

6.2.3 Iterative Processing

Further improvements may be obtained in the projection techniques presented in the previous section, especially when two different projections may be considered.

First, assume that to decode a sequence **y**, a complete trellis $\mathcal{T}$ could be used. Nevertheless, because of complexity constraints, two projected trellises $\mathcal{T}_1$ and $\mathcal{T}_2$ are considered, on which bit-level a posteriori probabilities can be obtained. The a posteriori probabilities obtained from $\mathcal{T}_1$ can then be used for the decoding with $\mathcal{T}_2$. An iterative decoding between both trellises may even be considered (see Chapter 9). Nevertheless, contrary to turbo codes, where interleavers are present between the two constituent coders so that the decoding errors of one decoder look as spread as possible to the other decoder, here, decoding errors are bursty, and the turbo effect may be much less significant.

Second, as has been briefly mentioned, when projecting a trellis on an hyperplane which does not contain all states of the FSM describing the decoder, one gets several parallel transitions between states. These parallel transitions are taken into account when computing the branch metric in the various decoding algorithms presented in Chapter 5. This metric evaluation may be simplified by grouping code words, as is detailed in Section 6.3.

6.3 GROUPING CODE WORDS

Building a small-dimensional symbol-clock trellis by projecting, e.g., a 3D symbol trellis, results in many parallel branches between states. Parallel branches correspond to code words of the same length. Their number may be very large when the sequence to decode has been encoded with a codebook containing fixed-length code words, as is the case, e.g., in H.263+ (see Section 6.2.2). The decoding complexity when dealing with projected trellises is thus not only dependent on the number of states but also on the number of transitions between states.

The aim of this section is to show that Huffman-like code words may be grouped into a reduced number of classes. The decoding algorithms of Chapter 5 may then be adapted to work on these classes instead of working on the whole set of code words, reducing drastically the decoding complexity. Variable-length code reduction techniques have been used previously. In the study by Nguyen and Duhamel (2003), code words, at the cost of a reduced coding efficiency. Other authors also group code words, despite the fact that this procedure has seldom been explicitly written.

This section introduces first the main idea of classes, defined as groups of code words. Modifications of the decoding algorithms to benefit from these classes are then presented. Situations where replacing a group of code words leads to some suboptimality are identified. The way code words are grouped into classes to get a minimum number of classes is then sketched, before showing an example on localizing the limits of encoded macroblocks in H.263+ texture data.

6.3.1 Introduction

Consider the subset of 16 code words of 8 bits taken from the set of code words used to encode the texture in H.263+,

$$\mathcal{C}_8 = \{00100000,\ 00100001, \ldots, 00101110,\ 00101111\}. \qquad (6.9)$$

If a symbol-clock trellis with state ν, such as that of Figure 5.8, page 133, is used to decode an H.263+ encoded sequence, the code words of $\mathcal{C}_8$ lead to 16 parallel transitions between any pair $(\nu, \nu+8)$ of values of the states. The evaluation of a branch probability, in the case of the BCJR or Viterbi algorithm, involves 16 independent evaluations of the likelihood of the code words.

All code words of $\mathcal{C}_8$ share the same 4-bit prefix 0010, and all other bits take all possible values. Thus, this set of code words may be compactly represented by the class 0010XXXX, where X is either 0 or 1 (Mohammad-Khani et al., 2006). In the symbol-clock trellis, 16 parallel branches may be replaced by a single branch corresponding to 0010XXXX. The next section introduces the way branch transition probabilities of classical decoding algorithms may be simplified using classes, which are first defined more formally.

6.3.2 Classes and Branch Transition Probabilities

As will be seen in this section, when the a priori probability of each grouped code word is the same or when ML decoding is performed, this simplification is harmless and decoding techniques inspired from those for punctured convolutional codes (Cain et al., 1979) may be put at work. When there are discrepancies between a priori probabilities, MAP decoding may become suboptimal, as will be detailed in what follows.

Definitions

Consider the set of binary numbers $\mathbb{B}=\{0,1\}$ supplemented with undetermined X corresponding to 0 or 1, to obtain the set of *extended* binary numbers

$$\overline{\mathbb{B}}=\{0,1,\mathrm{X}\}. \tag{6.10}$$

The *Hamming distance* for elements of $\overline{\mathbb{B}}$ is defined as usual

$$\forall x\in\overline{\mathbb{B}},\forall y\in\overline{\mathbb{B}},\ d_{\mathrm{H}}(x,y)=\begin{cases}0 & \text{if } x=y,\\ 1 & \text{if } x\neq y.\end{cases}$$

$\mathbb{B}^L$ and $\overline{\mathbb{B}}^L$ denote, respectively, the set of L-dimensional vectors of binary and extended binary numbers.

Definition 6.7: *A* class $\underline{\mathbf{c}}$ *is an element of* $\overline{\mathbb{B}}^L$. ■

A class $\underline{\mathbf{c}}$ is also a subset of $\mathbb{B}^L$, denoted by $\mathcal{S}(\underline{\mathbf{c}})$.

Consider now a memoryless source X with probability vector $\mathbf{p}$, taking values in $\mathcal{A}$ that is encoded with a binary VLC

$$\mathcal{C}=\mathcal{C}_{\ell_{\min}}\cup\cdots\cup\mathcal{C}_{\ell_{\max}}, \tag{6.11}$$

where $\mathcal{C}_\ell$, $\ell_{\min}\leqslant\ell\leqslant\ell_{\max}$, is the subset of $\mathcal{C}$ containing only code words of ℓ bits.

Definition 6.8: *A* classification *for a set* $\mathcal{C}_\ell\subset\mathbb{B}^\ell$ *is a partition for* $\mathcal{C}_\ell$ *with* nonoverlapping *classes*

$$\mathcal{C}_\ell=\mathcal{S}\left(\underline{\mathbf{c}}_{\ell,1}\right)\cup\cdots\cup\mathcal{S}\left(\underline{\mathbf{c}}_{\ell,n_\ell}\right) \tag{6.12}$$

such that $\mathcal{S}(\underline{\mathbf{c}}_{\ell,i})\cap\mathcal{S}(\underline{\mathbf{c}}_{\ell,j})=\emptyset$ *for any* $i\neq j$. ■

Several classifications may be considered for a given set $\mathcal{C}_\ell$. Any set $\mathcal{C}_\ell$ may be represented by a classification containing at least one single class, and at most $|\mathcal{C}_\ell|$ classes, each of which being a code word of $\mathcal{C}_\ell$. A classification with a minimum number of code words is called a *minimum classification* (Mohammad-Khani et al., 2006). Such minimum classification allows to represent a set of code words of the same length with a minimum number of classes.

Branch Transition Probability for the BCJR Algorithm

Assume that a sequence $\mathbf{x}_{1:k}$ of k source outcomes has been encoded using $\mathcal{C}$ to get $\mathbf{b}_{1:n}$, which is transmitted over a noisy channel to get finally $\mathbf{y}_{1:n}$. The forward recursion of the BCJR algorithm for estimating the symbol-by-symbol MAP estimate for $\mathbf{x}_{1:k}$ may be written as

$$\alpha_\kappa(\nu) = \sum_{\nu'} \alpha_{\kappa-1}\left(\nu'\right) \sum_{x\in\mathcal{A}} \gamma_\kappa\left(x, \nu', \nu\right), \tag{6.13}$$

with

$$\begin{aligned}\gamma_\kappa\left(x, \nu', \nu\right) = P\left(X_\kappa = x|S_{\kappa-1} = \nu', S_\kappa = \nu\right) \\ P\left(\mathbf{y}_{\nu'+1:\nu}|X_\kappa = x\right) P\left(S_\kappa = \nu|S_{\kappa-1} = \nu'\right).\end{aligned} \tag{6.14}$$

Combining (6.13) and (6.14), one gets

$$\alpha_\kappa(\nu) = \sum_{\nu'} \alpha_{\kappa-1}\left(\nu'\right) P\left(S_\kappa = \nu|S_{\kappa-1} = \nu'\right) \underline{\gamma}_\kappa\left(\nu', \nu\right), \tag{6.15}$$

with

$$\underline{\gamma}_\kappa\left(\nu', \nu\right) = \sum_{x\in\mathcal{A}} P\left(X_\kappa = x|S_{\kappa-1} = \nu', S_\kappa = \nu\right) P\left(\mathbf{y}_{\nu'+1:\nu}|X_\kappa = x\right). \tag{6.16}$$

Assume that in (6.16), $\nu' = \nu - \ell$; then, $P(X_\kappa = x|S_{\kappa-1} = \nu - \ell, S_\kappa = \nu)$ is nonzero only for source symbols corresponding to code words of ℓ bits, i.e., belonging to $\mathcal{C}_\ell$. Thus, there is a one-to-one mapping between the source symbols and the code and

$$\underline{\gamma}_\kappa\left(\nu', \nu\right) = \sum_{\mathbf{c}\in\mathcal{C}_\ell} P\left(C\left(X_\kappa\right) = \mathbf{c}|S_{\kappa-1} = \nu - \ell, S_\kappa = \nu\right) P\left(\mathbf{y}_{\nu-\ell+1:\nu}|C\left(X_\kappa\right) = \mathbf{c}\right). \tag{6.17}$$

The sum in $\underline{\gamma}_\kappa(\nu - \ell, \nu)$ is due to the parallel transitions between $\nu - \ell$ and ν. It involves as many terms as code words in $\mathcal{C}_\ell$. Using a classification $(\underline{\mathbf{c}}_{\ell,1}, \ldots, \underline{\mathbf{c}}_{\ell,n_\ell})$ for $\mathcal{C}_\ell$, (6.17) may be rewritten as

$$\begin{aligned}\underline{\gamma}_\kappa(\nu - \ell, \nu) = \sum_{i=1}^{n_\ell} \sum_{\mathbf{c}\in\mathcal{S}(\underline{\mathbf{c}}_{\ell,i})} P\left(C\left(X_\kappa\right) = \mathbf{c}|S_{\kappa-1} = \nu - \ell, S_\kappa = \nu\right) \\ P\left(\mathbf{y}_{\nu-\ell+1:\nu}|C\left(X_\kappa\right) = \mathbf{c}\right).\end{aligned} \tag{6.18}$$

To simplify (6.18) further, one has to assume that the a priori probabilities for all code words of the class $\underline{\mathbf{c}}_{\ell,i}$ are all equal, i.e.,

$$P\left(C\left(X_\kappa\right) = \mathbf{c}|S_{\kappa-1} = \nu - \ell, S_\kappa = \nu\right) = p_{\ell,i}. \tag{6.19}$$

This is quite reasonable since all these code words have the same length. If this is not the case, another classification may be considered, or some approximation error (and suboptimality) has to be tolerated. With this assumption, one gets

$$\underline{\gamma}_\kappa(\nu - \ell, \nu) = \sum_{i=1}^{n_\ell} p_{\ell,i} d_{\mathrm{B}}\left(\mathbf{y}_{\nu-\ell+1:\nu}, \underline{\mathbf{c}}_{\ell,i}\right), \tag{6.20}$$

with

$$d_{\mathrm{B}}\left(\mathbf{y}_{\nu-\ell+1:\nu},\underline{\mathbf{c}}_{\ell,i}\right)=\sum_{\mathbf{c}\in\mathcal{S}\left(\underline{\mathbf{c}}_{\ell,i}\right)}P\left(\mathbf{y}_{\nu-\ell+1:\nu}|C\left(X_\kappa\right)=\mathbf{c}\right). \tag{6.21}$$

The term $d_{\mathrm{B}}(\mathbf{y}_{\nu-\ell+1:\nu},\underline{\mathbf{c}}_{\ell,i})$ is the sum of the likelihoods of $\mathbf{c}\in\mathcal{S}(\underline{\mathbf{c}}_{\ell,i})$. Assume that the channel is AWGN with variance σ^2; then, (6.21) may be simplified as

$$d_{\mathrm{B}}\left(\mathbf{y}_{\nu-\ell+1:\nu},\underline{\mathbf{c}}\right)=\prod_{\lambda=1}^{\ell}d_{\mathrm{B}}\left(y_{\nu-\ell+\lambda},\underline{c}_\lambda\right), \tag{6.22}$$

where $\underline{c}_\lambda$ is the λth entry of the class $\underline{\mathbf{c}}$. In (6.22), one has

$$d_{\mathrm{B}}\left(y,\underline{c}\right)=\begin{cases}\left(2\pi\sigma^2\right)^{-1}\exp\left(-\left(y-m\left(\underline{c}\right)\right)^2/2\sigma^2\right) & \text{if } \underline{c}\in\{0,1\},\\ d_{\mathrm{B}}\left(y,0\right)+d_{\mathrm{B}}\left(y,1\right)=2\left(2\pi\sigma^2\right)^{-1}\cosh\left(y/\sigma^2\right)\exp\left(-\left(y^2+1\right)/2\sigma^2\right) & \text{if } \underline{c}=\mathrm{X},\end{cases} \tag{6.23}$$

where m stands for the formatting function used by the modulation scheme.

Using a classification of n_ℓ classes, the evaluation of (6.17), instead of requiring $|\mathcal{C}_\ell|$ evaluations of likelihood functions, requires only to evaluate n_ℓ terms (6.23), which have a computational complexity similar to that of the evaluation of a likelihood.

Branch Transition Probability for the Viterbi Algorithm

In the case of Viterbi algorithm used with the 2D symbol-clock trellis for the VLC $\mathcal{C}$, one gets the following recursion for the evaluation of the node metrics

$$\Gamma_\kappa(\nu)=\max_{x_\kappa}\left(\Gamma_{\kappa-1}\left(\nu-\ell\left(C\left(x_\kappa\right)\right)\right)+\gamma_\kappa\left(x_\kappa,\nu-\ell\left(C\left(x_\kappa\right)\right),\nu\right)\right), \tag{6.24}$$

with

$$\gamma_\kappa\left(x_\kappa,\nu-\ell\left(C\left(x_\kappa\right)\right),\nu\right)=\log P\left(\mathbf{y}_{\nu-\ell(C(x_\kappa))+1:\nu}|x_\kappa\right)+\log P\left(x_\kappa\right). \tag{6.25}$$

Using the decomposition (6.11), one may rewrite (6.24) as follows:

$$\begin{aligned}\Gamma_\kappa(\nu)&=\max_{\ell_{\min}\leqslant\ell\leqslant\ell_{\max}}\max_{\mathbf{c}\in\mathcal{C}_\ell}\left(\Gamma_{\kappa-1}(\nu-\ell)+\gamma_\kappa\left(C^{-1}(\mathbf{c}),\nu-\ell,\nu\right)\right)\\&=\max_{\ell_{\min}\leqslant\ell\leqslant\ell_{\max}}\left(\Gamma_{\kappa-1}(\nu-\ell)+\underline{\gamma}_\kappa(\nu-\ell,\nu)\right),\end{aligned} \tag{6.26}$$

with

$$\underline{\gamma}_\kappa(\nu-\ell,\nu)=\max_{\mathbf{c}\in\mathcal{C}_\ell}\gamma_\kappa\left(C^{-1}(\mathbf{c}),\nu-\ell,\nu\right). \tag{6.27}$$

The evaluation of $\Gamma_\kappa(\nu)$ thus requires to perform in (6.27) a maximization of the branch metrics for all possible codes of the same length ℓ, and then, at most, $\ell_{\max}-\ell_{\min}+1$ terms have to be compared. The most time-consuming step is the first maximization, since for each ℓ, $|\mathcal{C}_\ell|$ terms have to be evaluated and compared. Again, classes allow to simplify the evaluation of (6.27), which may be written as follows:

$$\underline{\gamma}_{\kappa}(\nu-\ell,\nu)=\max_{i=1,\ldots,n_\ell}\max_{\mathbf{c}\in\mathcal{S}(\underline{\mathbf{c}}_{\ell,i})}\gamma_\kappa\left(C^{-1}(\mathbf{c}),\nu-\ell,\nu\right). \tag{6.28}$$

Assuming again that all code words in the same class have an identical a priori probability $p_{\ell,i}$ and that the channel is AWGN with variance σ^2, one gets

$$\underline{\gamma}_{\kappa}(\nu-\ell,\nu)=\max_{i=1,\ldots,n_\ell}\left(\log p_{\ell,i}-\ell\log\left(2\pi\sigma^2\right)+d_{\mathrm{V}}\left(\mathbf{y}_{\nu+1:\nu+\ell},\underline{\mathbf{c}}_{\ell,i}\right)\right), \tag{6.29}$$

where

$$d_{\mathrm{V}}\left(\mathbf{y}_{\nu+1:\nu+\ell},\underline{\mathbf{c}}\right)=\sum_{\lambda=1}\ell d_{\mathrm{V}}\left(y_{\nu+\lambda},\underline{c}_\lambda\right) \tag{6.30}$$

and

$$d_{\mathrm{V}}(y,\underline{c})=\begin{cases}-(y-m(\underline{c}))^2/2\sigma^2 & \text{if } \underline{c}\in\{0,1\},\\ \max\{d_{\mathrm{V}}(y,0),d_{\mathrm{V}}(y,1)\} & \text{if } \underline{c}=\mathrm{X}.\end{cases} \tag{6.31}$$

In (6.31), m is still the formatting function used by the considered modulation scheme.

Instead of comparing $|\mathcal{C}_\ell|$ terms, the complexity of each of which is that of computing the likelihood of a code word; (6.27) only requires the evaluation of n_ℓ terms, each of which has a complexity comparable to computing the likelihood of a code word.

6.3.3 Minimum Classification

This section briefly describes the way a minimum classification $\underline{\mathcal{C}}_\ell$ may be obtained for a set of vectors $\mathcal{C}_\ell$ of ℓ bits. For more details, see Mohammad-Khani et al. (2006).

The classification $\underline{\mathcal{C}}_\ell$ is built iteratively starting from the empty set. First, all pairs of code words in $\mathcal{C}_\ell$ are considered. If for a pair $(\mathbf{c}_i,\mathbf{c}_j)$ of code words, $d_{\mathrm{H}}(\mathbf{c}_i,\mathbf{c}_j)=1$; these two code words may be "grouped" together to form a class with a single X at the place where $\mathbf{c}_i$ and $\mathbf{c}_j$ differ. One obtains a first set of classes $\underline{\mathcal{C}}_\ell^{(1)}$. Any code word that was not grouped with another code word is put into $\underline{\mathcal{C}}_\ell$.

Now, all pairs of classes of $\underline{\mathcal{C}}_\ell^{(1)}$ are considered. If for a pair $(\underline{\mathbf{c}}_i,\underline{\mathbf{c}}_j)$ of classes, $d_{\mathrm{H}}(\underline{\mathbf{c}}_i,\underline{\mathbf{c}}_j)=1$; these two classes may again be grouped together to form a new class with a second X at the place where they differ. A second set of classes $\underline{\mathcal{C}}_\ell^{(2)}$ is obtained, each of which having two X in its entries. Assume that the set of classes that was not grouped with another class is

$$\underline{\mathcal{C}}_\ell^{(1')}=\{\underline{\mathbf{c}}_1,\ldots,\underline{\mathbf{c}}_n\};$$

then, for some pairs $(\underline{\mathbf{c}}_i,\underline{\mathbf{c}}_j)$, one may have $\underline{\mathbf{c}}_i\neq\underline{\mathbf{c}}_j$ and $\mathcal{S}(\underline{\mathbf{c}}_i)\cap\mathcal{S}(\underline{\mathbf{c}}_j)\neq\emptyset$. Since a classification is a partition of $\mathcal{C}_\ell$, only a subset of nonoverlapping classes of

$\underline{\mathcal{C}}_\ell^{(1')}$ has to be put into $\underline{\mathcal{C}}_\ell$ so that the union of the code words of these selected classes contains all code words contained in $\underline{\mathcal{C}}_\ell^{(1')}$. This step is combinatorial in the number of classes of $\underline{\mathcal{C}}_\ell^{(1')}$.

This process is repeated with $\underline{\mathcal{C}}_\ell^{(2)}$ until it is no more possible to group classes from $\underline{\mathcal{C}}_\ell^{(I)}$ together. A subset of classes of $\underline{\mathcal{C}}_\ell^{(I)}$ is then added to $\underline{\mathcal{C}}_\ell$ so that all classes form a partition of $\mathcal{C}_\ell$ (see Example 6.9).

Example 6.9: Consider the set $\mathcal{D}_7$ of 24 code words of length $\ell = 7$ bits

$$\begin{aligned}\mathcal{D}_7 = \{&0000000, 0000001, \ldots, 0001010, 0001011,\\ &0010100, 0010101, \ldots, 0010111,\\ &0101000, 0101001, \ldots, 0101011,\\ &1010100, 1010101, \ldots, 1010111\}.\end{aligned}$$

One wants to build a minimum classification for $\mathcal{D}_7$. After one iteration, one has

$$\begin{aligned}\underline{\mathcal{D}}_7^{(1)} = \{&000000\text{X}, 00000\text{X}0, 00000\text{X}1, 0000\text{X}00, \ldots, 0000\text{X}11, 000\text{X}000,\\ &000\text{X}000, \ldots, 000\text{X}011, 000100\text{X}, 000101\text{X}, 00010\text{X}0, 00010\text{X}1,\\ &001010\text{X}, 00101\text{X}0, 001011\text{X}, 00101\text{X}1,\\ &010100\text{X}, 0101001\text{X}, 01010\text{X}0, 01010\text{X}1,\\ &00\text{X}0100, \ldots, 00\text{X}0111, \text{X}010100, \ldots, \text{X}010111,\\ &101010\text{X}, 101011\text{X}, 10101\text{X}0, 10101\text{X}1\},\end{aligned}$$

and $\underline{\mathcal{C}}_\ell^{(1')} = \emptyset$. After a second iteration, one obtains

$$\begin{aligned}\underline{\mathcal{D}}_7^{(2)} = \{&00000\text{XX}, 00001\text{XX}, 00010\text{XX}, 0000\text{XX}0, 0000\text{XX}1, 000\text{X}0\text{X}0, 000\text{X}0\text{X}1\\ &00\text{X}010\text{X}, 00\text{X}011\text{X}, 00\text{X}01\text{X}0, 00\text{X}01\text{X}1,\\ &0\text{X}0100\text{X}, 0\text{X}0101\text{X}, 0\text{X}010\text{X}0, 0\text{X}010\text{X}1,\\ &\text{X}01010\text{X}, \text{X}01011\text{X}, \text{X}0101\text{X}0, \text{X}0101\text{X}1,\\ &00101\text{XX}, 01010\text{XX}, 10101\text{XX}\},\end{aligned}$$

and still $\underline{\mathcal{C}}_\ell^{(2')} = \emptyset$. After three iterations, one gets

$$\underline{\mathcal{D}}_7^{(3)} = \{0000\text{XXX}, 000\text{X}0\text{XX}, 00\text{X}01\text{XX}, 0\text{X}010\text{XX}, \text{X}0101\text{XX}\},$$

and again $\underline{\mathcal{C}}_\ell^{(3')} = \emptyset$. The classes in $\underline{\mathcal{D}}_7^{(3)}$ may no more be grouped. Moreover, some of them are overlapping. For example, $\mathcal{S}(0000\text{XXX}) \cap \mathcal{S}(000\text{X}0\text{XX}) \neq \emptyset$. With a combinatorial search, one gets a minimum classification

$$\underline{\mathcal{D}}_7 = \{0000\text{XXX}, 0\text{X}010\text{XX}, \text{X}0101\text{XX}\};$$

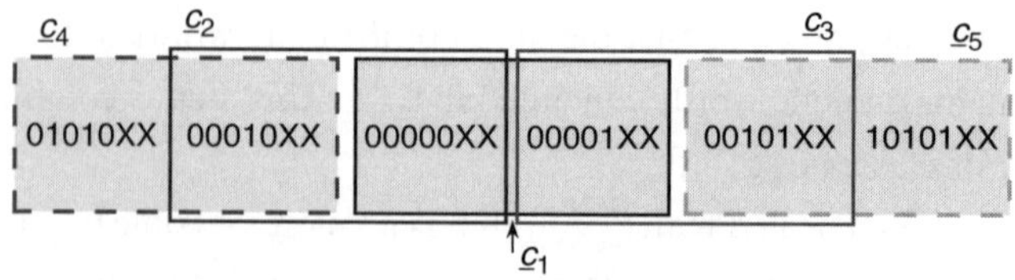

FIGURE 6.8 Classes belonging to $\underline{\mathcal{D}}_7^{(3)}$.

see Figure 6.8. As a result, 24 code words are represented by only three classes.

In practice, even if this classification algorithm takes some time, it is performed only once and depends only on the considered code.

6.3.4 Application

Section 6.2.2 introduces an H.263+ texture decoding example. The proposed two-step decoding algorithm estimates first the boundaries of each encoded texture block and then estimates the content of each of them. A bit-clock trellis accounting for the number of already localized blocks is used to localize each encoded texture block. When grouping code words, the information on whether the code word is the last one of a texture block (EOB) has to be considered. Thus, for each code-word length, two classifications have to be built.

The VLC table used in H.263+ has two components: a truly variable-length one, containing 204 code words, and a very large fixed-length part representing about 2^{15} code words. Both components can be worked out as explained above to reduce the number of equivalent code words (classes).

Table 6.2 describes the minimum classification containing 34 classes obtained using the technique presented in Section 6.3.3 for the 204 VLCs used to encode the texture in H.263+. Table 6.3 presents the minimum classification for the fixed-length codes. Now, only 14 classes may represent about 2^{15} code words.

The performance of the texture block localization algorithm is now compared using two VLC tables: the original H263+ VLCs and the classified VLCs of Tables 6.2 and 6.3. Data have been put in packets of about 1024 bits and sent over an AWGN channel. A Viterbi algorithm is used to localize the texture blocks in packets containing INTER-encoded texture blocks and in packets containing INTRA-encoded texture blocks. Performance is evaluated in terms of block location error rate in Figure 6.9 and in terms of average computing time in Table 6.4.

Using a simplified VLC table allows to reduce the computing time by a factor of 3 compared with computations based on the original VLC table. Except for the decoder using a MAP metric for INTRA blocks, where a small loss of 0.2 dB is obtained, this speedup is obtained with negligible performance loss.

TABLE 6.2 Classification of the Code Words Involved in the Texture Encoding by the H263+ Standard

ℓ	EOB	CW	Minimal Classifications	Number of Classes
3	0	2	10X	1
4	0	2	110X	1
5	0	4	111XX	1
	1	2	0111X	1
6	0	6	0110XX, 01011X	2
7	0	12	01010XX, 0100XXX	2
	1	8	0011XXX	1
8	0	8	00101XXX	1
	1	8	00100XXX	1
9	0	10	000111XXX, 00011011X	2
	1	16	00011010X, 0001100XX, 000101XXX, 00010011X	4
10	0	24	00010010XX, 0001000XXX, 0000111XXX, 00001101XX	4
	1	18	00001100XX, 0000101XXX, 00001001XX, 000010001X	4
11	0	20	000010000XX, 0000001XXXX	2
	1	8	00000001XXX	1
12	0	12	0000000011XX, 000001000XXX	2
	1	12	0000000010XX, 000001001XXX	2
13	0	16	000001010XXXX	1
	1	16	000001011XXXX	1
Total		204		34

Code words indicating the end of a texture block (EOB) are distinguished from the others.

6.3.5 Summary

The simplification of VLC tables generally results in a large decrease of the number of branches in the associated decoding trellis, thus in a much reduced computational complexity. The ML decoder has the same properties as the one working on the original VLC table. For the MAP algorithm, provided that the code words that are grouped in a class have the same a priori probabilities, and the

TABLE 6.3 Minimal Classification without Intersection Obtained for the 22-Bit FLC Code Words for Texture Encoding in H263+

Minimal Classification	
Not EOB	**EOB**
000 0011 0 XXXXXX X1XX XXXX	000 0011 1 XXXXXX X1XX XXXX
000 0011 0 XXXXXX X01X XXXX	000 0011 1 XXXXXX X01X XXXX
000 0011 0 XXXXXX X001 XXXX	000 0011 1 XXXXXX X001 XXXX
000 0011 0 XXXXXX X000 1XXX	000 0011 1 XXXXXX X000 1XXX
000 0011 0 XXXXXX X000 01XX	000 0011 1 XXXXXX X000 01XX
000 0011 0 XXXXXX X000 001X	000 0011 1 XXXXXX X000 001X
000 0011 0 XXXXXX X000 0001	000 0011 1 XXXXXX X000 0001

TABLE 6.4 Mean Decoding Time (ms/block)

	ML		MAP	
	Init. Table	Simpl. Table	Init. Table	Simpl. Table
INTRA	165.77	62.53 (−62%)	242.01	92.41 (−62%)
INTER	356.18	108.46 (−70%)	445.76	131.03 (−71%)

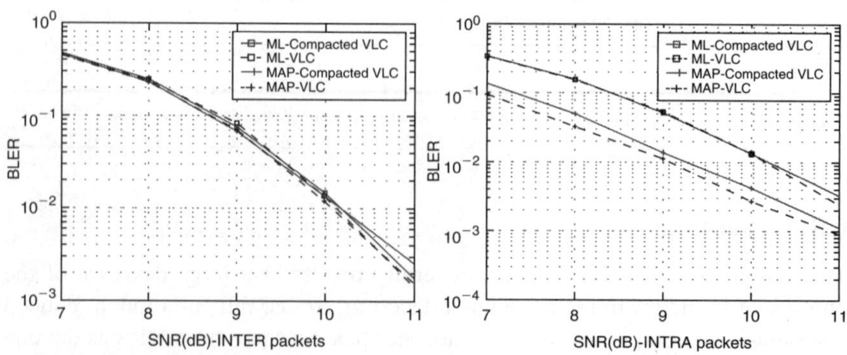

FIGURE 6.9 Block error rate for INTER packets (left) and INTRA packets (right).

decoder has again the same properties as if it would be working on the original VLC table. This is likely to be the case, as the code words that are merged in a given class have the same length.

6.4 SEQUENTIAL DECODERS

A source X with alphabet $\mathcal{A}$ and probability vector $\mathbf{p}$ emits a sequence of k symbols $\mathbf{x}_{1:k}$. These symbols are encoded with some binary encoding function $C:\mathcal{A}^*\rightarrow\{0,1\}^*$ to get $\mathbf{b}_{1:n}=C(\mathbf{x}_{1:k})$, where n is a function of $\mathbf{x}_{1:k}$. Then, $\mathbf{b}_{1:n}$ is sent over a channel, and the output of which is $\mathbf{y}_{1:n}$. The sequence MAP estimates $\widehat{\mathbf{b}}_{1:n}^{\text{MAP}}$ for $\mathbf{b}_{1:n}$ knowing $\mathbf{y}_{1:n}$ is

$$\widehat{\mathbf{b}}_{1:n}^{\text{MAP}} = \arg\max_{\mathbf{b}\in\Omega_n} P\left(\mathbf{b}|\mathbf{y}_{1:n}\right) \tag{6.32}$$

$$= \arg\max_{\mathbf{b}\in\Omega_n} \frac{P\left(\mathbf{y}_{1:n}|\mathbf{b}\right)P(\mathbf{b})}{P\left(\mathbf{y}_{1:n}\right)}, \tag{6.33}$$

where

$$\Omega_n=\{\mathbf{b}\in\{0,1\}^n \mid \exists\mathbf{x}\in\mathcal{A}^*, C(\mathbf{x})=\mathbf{b}\}. \tag{6.34}$$

When the coder or the decoder may be represented by a FSM, Chapter 5 provides efficient decoding techniques involving trellises to get (6.33). The tools presented in the previous sections allow to reduce the decoding complexity if it is too high. This section addresses the situations where the FSM describing the coder or the decoder is too complex to allow an efficient use of BCJR or Viterbi-like decoding techniques, even after simplification.

Thus, only the availability of the encoding C and decoding C^{-1} functions is assumed. In such a case, a brute-force estimation algorithm would consider all sequences $\mathbf{b}$ of n bits, evaluate whether $\mathbf{x}=C^{-1}(\mathbf{b})$ exists, and determine (6.33). Alternatively, the set $\mathcal{A}^*$ could be explored. These explorations may be better organized by taking into account the fact that all sequences of n bits or all sequences of $\mathcal{A}^*$ may be represented with a *tree* (see Figure 6.10). A sequence of n bits may be described by a path of length n starting from the root of the tree. Each branch stemming from a node of the tree represents a given symbol or bit value. Since some sequences of bits do not belong to Ω_n, some path does not belong to the tree. An optimal estimator has thus to explore at most 2^n paths of the tree to determine (6.33). Since this may take quite a lot of time for large values of n, the purpose of sequential decoders (Anderson and Mohan, 1991a) is to find the best path, according to a chosen metric, without examining too many branches of the tree. The most popular sequential decoding algorithms are the stack algorithm (SA) (Jelinek, 1969; Zigangirov, 1966) and the M-algorithm (MA) (Anderson and Mohan, 1991a). These algorithms perform an iterative (sequential) bit-by-bit or symbol-by-symbol exploration of the tree representing $\{0,1\}^n$ or $\mathcal{A}^*$, focusing on one or several paths in the trellis, which are the most likely to correspond

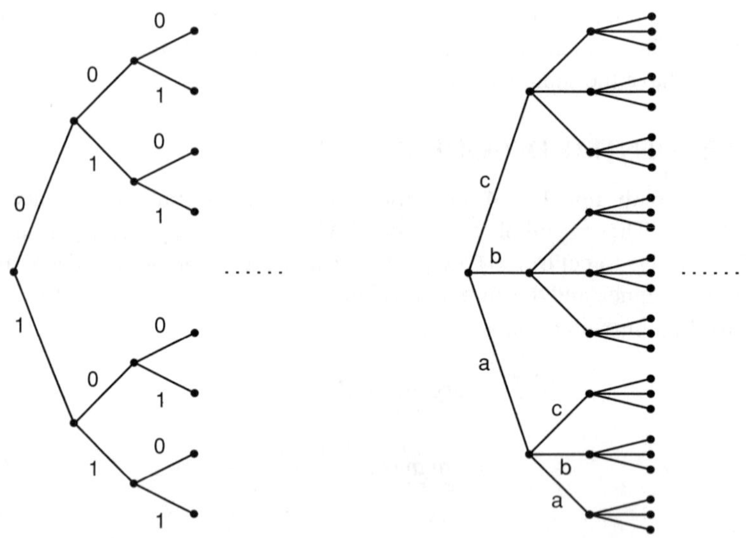

FIGURE 6.10 Trees used to perform sequential decoding, bit-by-bit (left) and symbol-by-symbol (right).

to the prefix of the MAP estimate. These branches are then further explored in different ways.

6.4.1 Metric for Sequential Decoding

Using a sequential decoding algorithm, a decoding metric has to be considered to perform an iterative bit-by-bit (respectively, symbol-by-symbol) construction of the sequence estimate for $\mathbf{b}_{1:n} = C(\mathbf{x}_{1:k})$ (respectively, for $\mathbf{x}_{1:k}$) with the knowledge of $\mathbf{y}_{1:n}$. A popular decoding metric, see, e.g., Grangetto et al. (2005), Park and Miller (2000), Perros-Meilhac and Lamy (2002), and Wen and Villasenor (1999), is derived from the MAP sequence estimator

$$\mathcal{M}\left(\mathbf{b}_{1:\nu}, \mathbf{y}_{1:\nu}\right) = \log P\left(\mathbf{b}_{1:\nu} | \mathbf{y}_{1:\nu}\right), \tag{6.35}$$

with $\nu = 1, \ldots, n$ and

$$\widehat{\mathbf{b}}_{1:n}^{\mathrm{MAP}} = \arg\max_{\mathbf{b} \in \Omega_n} \mathcal{M}\left(\mathbf{b}_{1:\nu}, \mathbf{y}_{1:\nu}\right).$$

Using Bayes' rule, (6.35) may be rewritten as

$$\mathcal{M}\left(\mathbf{b}_{1:\nu}, \mathbf{y}_{1:\nu}\right) = \log P\left(\mathbf{y}_{1:\nu} | \mathbf{b}_{1:\nu}\right) + \log P\left(\mathbf{b}_{1:\nu}\right) - \log P\left(\mathbf{y}_{1:\nu}\right). \tag{6.36}$$

The first term in (6.36) is the likelihood of $\mathbf{b}_{1:\nu}$ determined from the considered channel model. The a priori probability $P(\mathbf{b}_{1:\nu})$ vanishes if $\mathbf{b}_{1:\nu}$ is not the prefix of a sequence belonging to Ω_n. The evaluation of the last term $P(\mathbf{y}_{1:\nu})$ is detailed in what follows.

When sequences of the same length are compared to find the best candidate prefixes of ν bits for $\widehat{\mathbf{b}}_{1:n}^{\text{MAP}}$, as is the case, for example, of the MA, the term $\log P(\mathbf{y}_{1:\nu})$ is common to all candidates. There is thus no need to evaluate it. If prefixes of a varying number of bits have to be compared, as is the case of the SA, $P(\mathbf{y}_{1:\nu})$ has to be evaluated. An exact evaluation is quite complex, since

$$P\left(\mathbf{y}_{1:\nu}\right) = \sum P\left(\mathbf{y}_{1:\nu}|\mathbf{b}_{1:\nu}\right) P\left(\mathbf{b}_{1:\nu}\right), \tag{6.37}$$

where the sum is over all ν bit prefixes of sequences in Ω_n. Park and Miller (2000) propose the simple but very coarse approximation

$$P\left(\mathbf{y}_{1:\nu}\right) \approx 2^{-\nu}. \tag{6.38}$$

A better approximation has been suggested by Fano (1963) and adapted to VLCs by Massey (1972). The main idea is to assume that all channel outputs are independent

$$P\left(\mathbf{y}_{1:\nu}\right) \approx P_0\left(\mathbf{y}_{1:\nu}\right) = \prod_{i=1}^{\nu} P\left(y_i\right) \tag{6.39}$$

with

$$P\left(y_i\right) = \sum_{j=0,1} P\left(y_i|B_i = j\right) P\left(B_i = j\right). \tag{6.40}$$

This Fano–Massey approximation is usually much closer to $P(\mathbf{y}_{1:\nu})$ than (6.38).

6.4.2 M-Algorithm

The MA is a *breadth first* search, the breadth being M. The decoding tree exploration is performed by iteratively incrementing the depth of the paths, keeping in a list $\mathcal{L}$ only the best M paths according to the metric (6.36). Since all compared sequences have the same length, there is no need to evaluate $\log P(\mathbf{y}_{1:\nu})$. Algorithm 6.1 outlines the various steps of the MA.

Algorithm 6.1: M-Algorithm (input: $\mathbf{y}_{1:n}$, M; output: $\widehat{\mathbf{b}}_{1:n}$ and $\widehat{\mathbf{x}}_{1:k}$)

---

1. Initialize $\mathcal{L}$ with the root of the decoding tree to which a null metric is assigned.
2. Extend all paths in $\mathcal{L}$ to the following branches, creating new paths from each stored path.
3. Keep only the M best paths in terms of metric (6.36), i.e., the path with the largest metric.
4. Stop as soon as one of the M paths is n bits long or if $\mathcal{L}$ is empty. Otherwise, go to 2.
5. If $\mathcal{L}$ is not empty, select the path of n bits with the largest metric to get the estimates $\widehat{\mathbf{b}}_{1:n}$ and $\widehat{\mathbf{x}}_{1:k}$.

At each iteration, $2M$ paths have to be examined if the exploration is done bit by bit and $|\mathcal{A}|M$ paths if it is done symbol by symbol. Only M sequences are considered at each iteration; none of them may correspond to the actual prefix of the sequence to estimate if the metric of the latter is not among the M best metrics. After n iterations, the MA provides in the best case the sequence $\widehat{\mathbf{b}}_{1:n}^{\text{MAP}}$ a Viterbi algorithm would have obtained and in the worst case is not able to give any sequence belonging to Ω_n when none of the M candidates is in Ω_n. The decoding performance and complexity depend thus mainly on M. Increasing M allows the decoding algorithm to perform better ($\mathcal{L}$ is less likely to get empty), the price to be paid is an increasing decoding complexity, which is well controlled as it is $O(Mn)$.

6.4.3 Stack Algorithm

The stack algorithm (SA) (Jelinek, 1969; Zigangirov, 1966) is a *metric first* search performed iteratively: an ordered stack $\mathcal{S}$ containing previously examined paths (of different lengths in bits and in symbols) is maintained. The best path (located in the top of the stack, i.e., with the largest metric) is extracted from $\mathcal{S}$ and expanded by the exploration of the two branches following the current node. The two new paths are put in the stack, which is then reordered by decreasing metric.

Since all partially estimated bit sequences are of various lengths, $\log P(\mathbf{y}_{1:\nu})$ in (6.36) has to be approximated using, e.g., (6.38) or (6.39). Algorithm 6.2 describes the basic steps involved in the SA.

Algorithm 6.2: Stack Algorithm (input: $\mathbf{y}_{1:n}$, output: $\widehat{\mathbf{b}}_{1:n}$ and $\widehat{\mathbf{x}}_{1:k}$)

1. Initialize $\mathcal{S}$ with the root of the decoding tree to which a null metric is assigned.
2. Extend the top path in $\mathcal{S}$ by exploring all branches stemming from the node ending the top path.
3. Store the created paths in $\mathcal{S}$ with their corresponding metric.
4. Sort the stack according to (6.36), paths with the largest metric are on the top of $\mathcal{S}$.
5. Stop as soon as the path at the top of $\mathcal{S}$ is of n bits long. Otherwise, go to 2.
6. From the top path in $\mathcal{S}$, get the estimates $\widehat{\mathbf{b}}_{1:n}$ and $\widehat{\mathbf{x}}_{1:k}$.

At each iteration, two paths are generated if the exploration is done bit by bit and $|\mathcal{A}|$ paths if it is done symbol by symbol. With good channel conditions, the prefixes of the actual sequence always remain near to the top of the stack during the iterations, and in the best case, the estimates $\widehat{\mathbf{b}}_{1:n}$ and $\widehat{\mathbf{x}}_{1:k}$ are obtained after n iterations. When the channel quality worsens, contrary to the MA, there is no risk to eliminate the actual sequence; however, the memory required to store all partially decoded sequences may become very important. The complexity of the algorithm is thus rather difficult to estimate in advance; it depends on the channel conditions. Usually, the number of iterations and the number of lines kept in $\mathcal{S}$ are limited to upper-bound the decoding time and memory requirements, and the price to be paid is a risk of not being able to get any estimate, as for the MA.

Many variations of Algorithm 6.2 have been proposed in the literature, such as the generalized SA (Haccoun and Ferguson, 1975), which extends the first $M \geqslant 2$ paths in the stack instead of only the first one. This allows to reduce the decoding failures especially when the beginning of the received sequence is strongly disturbed.

Accurate analyses of the MA and SA are provided in the study by Anderson and Mohan (1991).

6.4.4 Getting Soft Estimates from the MA and the SA

The algorithms presented in Sections 6.4.2 and 6.4.3 provide only hard estimates, and no reliability information on the obtained sequences of bits or symbols is available. The aim of this section is to show how the basic MA and SA may be modified to get these reliability information in terms of log a posteriori ratio, as introduced in (5.4),

$$\begin{aligned}\Lambda_{\mathrm{AP}}\left(\widehat{b}_\nu\right) &= \log \frac{P\left(B_\nu = 0 | \mathbf{y}_{1:n}\right)}{P\left(B_\nu = 1 | \mathbf{y}_{1:n}\right)} \\ &= \log \sum_{\mathbf{b}_{1:n},\, b_\nu = 0} P\left(\mathbf{b}_{1:n} | \mathbf{y}_{1:n}\right) - \log \sum_{\mathbf{b}_{1:n},\, b_\nu = 1} P\left(\mathbf{b}_{1:n} | \mathbf{y}_{1:n}\right). \end{aligned} \tag{6.41}$$

Using the max-log approximation, (6.41) becomes

$$\widetilde{\Lambda}_{\mathrm{AP}}\left(\widehat{b}_\nu\right) = \log \max_{\mathbf{b}_{1:n},\, b_\nu = 0} P\left(\mathbf{b}_{1:n} | \mathbf{y}_{1:n}\right) - \log \max_{\mathbf{b}_{1:n},\, b_\nu = 1} P\left(\mathbf{b}_{1:n} | \mathbf{y}_{1:n}\right) \tag{6.42}$$

$$= \left(1 - 2\widehat{b}_\nu\right)\left(\log P\left(\widehat{\mathbf{b}}_{1:n} | \mathbf{y}_{1:n}\right) - \log \max_{\mathbf{b}_{1:n},\, b_\nu \neq \widehat{b}_\nu} P\left(\mathbf{b}_{1:n} | \mathbf{y}_{1:n}\right)\right). \tag{6.43}$$

These approximations have to be evaluated in the case of the SA and of the MA.

Soft Output Stack Algorithm

Assume that Algorithm 6.2, working bit by bit, has been successfully run on a sequence of channel outputs $\mathbf{y}_{1:n}$. Upon termination, the stack $\mathcal{S}$ contains several paths of various length on the decoding tree. They are denoted by $\mathbf{b}^1_{1:n_1}, \ldots, \mathbf{b}^t_{1:n_t}$ with associated metrics satisfying

$$\mathcal{M}^* = \mathcal{M}\left(\mathbf{b}^1_{1:n_1}, \mathbf{y}_{1:n_1}\right) \geqslant \mathcal{M}\left(\mathbf{b}^2_{1:n_2}, \mathbf{y}_{1:n_2}\right) \geqslant \cdots \geqslant \mathcal{M}\left(\mathbf{b}^t_{1:n_t}, \mathbf{y}_{1:n_t}\right). \tag{6.44}$$

Thus, $\widehat{\mathbf{b}}_{1:n} = \mathbf{b}^1_{1:n_1}$ corresponds to the estimate provided by the SA and necessarily $n_1 = n$.

A solution to get soft bit estimates proposed by Lamy and Perros-Meilhac (2003) is inspired by the bidirectional SOVA presented by Vucetic and Yuan

(2000). It considers a log-max approximation of each term in (6.41) to get

$$\widetilde{\Lambda}_{\text{AP},1}\left(\widehat{b}_\nu\right)=\max_{\substack{\tau=1\ldots t\\ b_\nu=0}}\mathcal{M}\left(\mathbf{b}^{\tau}_{1:n_\tau},\mathbf{y}_{1:n_\tau}\right)-\max_{\substack{\tau=1\ldots t\\ b_\nu=1}}\mathcal{M}\left(\mathbf{b}^{\tau}_{1:n_\tau},\mathbf{y}_{1:n_\tau}\right), \tag{6.45}$$

corresponding to the difference of the metric of the best path with $b_\nu=0$ and the best path with $b_\nu=1$. If $\widehat{b}_\nu=0$, then the first term on the right-hand side of the equality in (6.45) is $\mathcal{M}^*$; else, the last term in (6.45) is $\mathcal{M}^*$.

The second solution proposed by Lamy and Perros-Meilhac (2003) considers all path with $b_\nu=0$ and all path with $b_\nu=1$ to approximate (6.41). One then gets

$$\widetilde{\Lambda}_{\text{AP},2}\left(\widehat{b}_\nu\right)=\sum_{\substack{\tau=1\ldots t\\ b_\nu=0}}\mathcal{M}\left(\mathbf{b}^{\tau}_{1:n_\tau},\mathbf{y}_{1:n_\tau}\right)-\sum_{\substack{\tau=1\ldots t\\ b_\nu=1}}\mathcal{M}\left(\mathbf{b}^{\tau}_{1:n_\tau},\mathbf{y}_{1:n_\tau}\right). \tag{6.46}$$

In (6.46), only paths of more than ν bits are considered when evaluating $\widetilde{\Lambda}_{\text{AP},2}(\widehat{b}_\nu)$.

To provide good quality estimates for $\Lambda_{\text{AP}}(\widehat{b}_\nu)$, the stack $\mathcal{S}$ has to contain many sequences. The approximation of $\Lambda_{\text{AP}}(\widehat{b}_\nu)$ provided by (6.46) is very good for all $\nu\leqslant\min_{\tau=1,\ldots,t}n_\tau$. Nevertheless, for these bits

$$\widetilde{\Lambda}_{\text{AP},2}\left(\widehat{b}_\nu\right)\neq\Lambda_{\text{AP}}\left(\widehat{b}_\nu\right)$$

because the metrics are only approximate values of the a posteriori probabilities involved in (6.41). For ν close to n, the approximations (6.45) and (6.46) are quite similar.

Experimental results provided by Lamy and Perros-Meilhac (2003) do not show significant differences between both approximations.

Soft Output M-Algorithm

Assume that Algorithm 6.1, also working bit by bit, has decoded a sequence of channel outputs $\mathbf{y}_{1:n}$. At a given iteration ν, the working list $\mathcal{L}$ contains at most $2M$ partially decoded sequences denoted by $\mathbf{b}^{1,\nu}_{1:\nu},\ldots,\mathbf{b}^{M_\nu,\nu}_{1:\nu}$, with $M_\nu\leqslant 2M$ and with associated metrics satisfying

$$\mathcal{M}\left(\mathbf{b}^{1,\nu}_{1:\nu},\mathbf{y}_{1:\nu}\right)\geqslant\mathcal{M}\left(\mathbf{b}^{2,\nu}_{1:\nu},\mathbf{y}_{1:\nu}\right)\geqslant\cdots\geqslant\mathcal{M}\left(\mathbf{b}^{M_\nu,\nu}_{1:\nu},\mathbf{y}_{1:\nu}\right). \tag{6.47}$$

Now, after n iterations, if the list $\mathcal{L}$ is not empty, one gets $\widehat{\mathbf{b}}_{1:n}=\mathbf{b}^{1,n}_{1:n}$ corresponding to the estimate provided by the MA.

The main difference between the SA and the MA is that upon completion, the MA contains at most $2M$ sequences, with which it may be very difficult to obtain soft estimates using expressions similar to (6.45) or (6.46). For example, if $\widehat{b}_\nu=1$, if M is low, there is a high probability that there is no sequence in $\mathcal{L}$ with $b_\nu=0$. No log a posteriori ratio may then be evaluated for the νth bit.

The soft output M algorithm (SOMA) proposed by Wong (2006) aims at evaluating an approximation for (6.41) during the iterations of the MA. For that purpose, two vectors $\boldsymbol{\delta}^0_{1:n}$ and $\boldsymbol{\delta}^1_{1:n}$ are introduced to determine the smallest difference of metrics between the best partially decoded sequence and other decoded sequences. The content of the vectors $\delta^0_{1:n}$ and $\delta^1_{1:n}$ is updated at each iteration and initialized with $-\infty$.

At iteration ν, just before the elimination of the $M_\nu - M$ sequences with the largest metric, the possible contributions of the dropped sequences to $\Lambda_{\text{AP}}(\widehat{b}_\nu)$ are used to update $\delta^0_{1:n}$ and $\delta^1_{1:n}$ according to

$$\delta^b_\nu = \min\left(\delta^b_\nu, \min_{\substack{\tau=M+1\ldots M_\nu \\ b^{1,\nu}_\nu = b^{\tau,\nu}_\nu = b}} \left(\mathcal{M}\left(\mathbf{b}^{1,\nu}_{1:\nu}, \mathbf{y}_{1:\nu}\right) - \mathcal{M}\left(\mathbf{b}^{\tau,\nu}_{1:\nu}, \mathbf{y}_{1:\nu}\right)\right)\right), b=0,1. \tag{6.48}$$

Upon completion, at iteration n, a final update of $\delta^0_{1:n}$ and $\delta^1_{1:n}$ is performed according to

$$\delta^b_\nu = \min\left(\delta^b_\nu, \min_{\substack{\tau=2\ldots M_\nu \\ b^{1,\nu}_\nu = b^{\tau,\nu}_\nu = b}} \left(\mathcal{M}\left(\mathbf{b}^{1,\nu}_{1:\nu}, \mathbf{y}_{1:\nu}\right) - \mathcal{M}\left(\mathbf{b}^{\tau,\nu}_{1:\nu}, \mathbf{y}_{1:\nu}\right)\right)\right), b=0,1. \tag{6.49}$$

Now all sequences are involved in the update of $\delta^0_{1:n}$ and $\delta^1_{1:n}$.

Finally, the log a posteriori ratio is approximated as

$$\widetilde{\Lambda}_{\text{AP},3}\left(\widehat{b}_\nu\right) = \left(1 - 2\widehat{b}_\nu\right)\delta^{1-\widehat{b}_\nu}_{1:n} \tag{6.50}$$

Thus, even sequences that will be dropped are taken into consideration. Moreover, better estimates for $\Lambda_{\text{AP}}(\widehat{b}_\nu)$ are obtained when M is large, at the cost of an increased complexity.

6.4.5 Objective Adjustment of the Efficiency-Complexity Tradeoff

In sequential algorithms, the decoding performance in terms of error resilience efficiency is improved when more paths are explored in the decoding tree; however, this improvement goes with increased decoding complexity. To adjust a tradeoff between decoding complexity and efficiency, this section introduces some test to decide whether a path is worth being expanded by the MA or SA.

Consider a path of $\nu - 1$ bits corresponding to $\mathbf{b}_{1:\nu-1}$. In a bit-by-bit exploration, this path may be extended with $b_\nu = 0$ and $b_\nu = 1$. Assume that both extended paths correspond to prefixes of sequences in Ω_n. A hard decoder discards systematically one of them and, thus, explores only one path in the decoding tree. The idea is to derive a test allowing to decide if both of the two choices deserve being considered or if a hard decision on the current bit is sufficient.

Focusing on the bit-by-bit exploration, Pettijohn et al. (2001) proposed to determine if

$$\Lambda_\nu = \log \frac{P\left(B_\nu = 0 | y_\nu\right)}{P\left(B_\nu = 1 | y_\nu\right)} \tag{6.51}$$

belongs to $[-\Delta, \Delta]$, in which case the two paths are followed. Otherwise, if $\Lambda_\nu > \Delta$, only the path with $b_\nu = 0$ is kept, and if $\Lambda_\nu < -\Delta$, only that with $b_\nu = 1$ is kept. This idea has been refined by Ben-Jamaa et al. (2009), where Δ is adjusted depending on the channel conditions and on some parameter α.

Assuming that $P(B_\nu = 0) = P(B_\nu = 1)$, the probability of losing the correct path P_e is

$$P_e = \frac{1}{2}\left(P(\Lambda_\nu < -\Delta | B_\nu = 0) + P(\Lambda_\nu > \Delta | B_\nu = 1)\right). \tag{6.52}$$

In what follows, the b_νs are assumed to be mapped into a BPSK symmetric signaling $\pm\sqrt{E_b}$ and transmitted over an AWGN channel with noise variance $N_0/2$. Then,

$$P_e = \frac{1}{2}\left(1 - \text{erf}\left(\sqrt{\frac{E_b}{N_0}}\left(\frac{\Delta N_0}{4E_b} + 1\right)\right)\right), \tag{6.53}$$

where

$$\text{erf}(z) = \frac{2}{\sqrt{\pi}}\int_0^z e^{-t^2}\,dt. \tag{6.54}$$

The threshold Δ may then be expressed as a function of E_b/N_0 and P_e as

$$\Delta(E_b/N_0, P_e) = \max\left\{4\sqrt{\frac{E_b}{N_0}}\left(\left(\text{erf}^{-1}(1 - 2P_e)\right) - \sqrt{\frac{E_b}{N_0}}\right), 0\right\}. \tag{6.55}$$

Several choices may then be considered for P_e. For example, one may take

$$P_e(E_b/N_0) = \alpha P_{\text{h}}(E_b/N_0),\text{with } \alpha < 1, \tag{6.56}$$

where $P_{\text{h}}(E_b/N_0)$ is the probability that the hard decoder locally fails, expressed in the AWGN case as

$$P_{\text{h}}(E_b/N_0) = \frac{1}{2}\left(1 - \text{erf}\left(\sqrt{\frac{E_b}{N_0}}\right)\right); \tag{6.57}$$

see Proakis (2001). The parameter α allows one to adjust the decoding complexity-efficiency tradeoff, as discussed in Section 6.4.6.

Figure 6.11 represents the evolution of Δ for several values of P_e as a function of E_b/N_0. The values of E_b/N_0 at which $\Delta = 0$ are those at which a hard decoder produces a probability of error equal to P_e.

6.4.6 Examples

First, the performance of various MA and SA is compared, on the decoding of data encoded with the Context adaptive binary arithmetic coding (CABAC) of H.264/AVC. Then, whole H.264/AVC Network Abstraction Layer units (NALUS) are decoded using the MA. If getting a description of the source coder

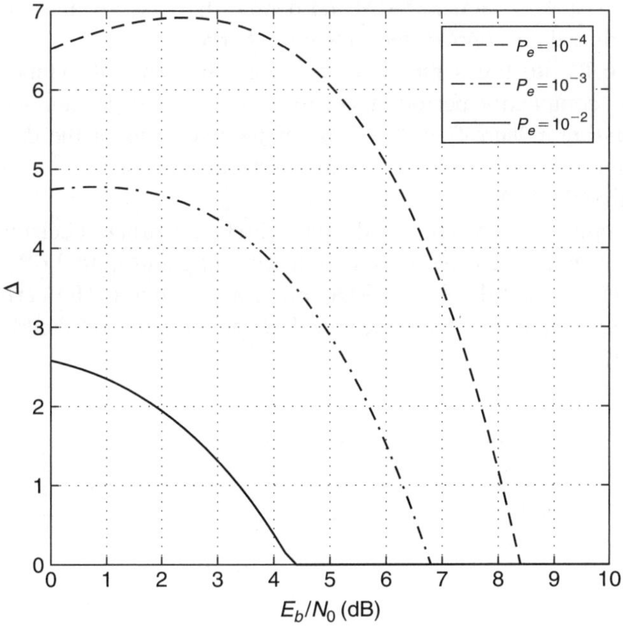

FIGURE 6.11 Evolution of Δ as a function of E_b/N_0 for several values of P_e.

with a complex FSM would be possible for the CABAC, this would no more be the case for the H.264/AVC encoder. Nevertheless, performing JSCD is still possible.

Comparison of Various MA and SA

Simulations are performed using the CABAC defined in H.264/AVC standard. Binarized source symbols belong to the first 9 binary code words of the zero-order Exp-Golomb scheme (EG0) (Marpe et al., 2003)

$$\mathcal{C}_{\text{EG0}} = \{1, 010, 011, 00100, 00101, 00110, 00111 \\ 0001000, 0001001\}.$$

A simplified context modeling with three contexts is considered. Sequences containing 100 source symbols are randomly generated. Each sequence is ended with a special end of sequence symbol, encoded using 0001111, the presence of which is tested at decoder side.

MAP decoders based on both SA and MA are considered, and the following abbreviations are adopted. Basic stack algorithm (BSA) and BMA(M) stand for the basic SA and MA, M being the number of paths kept at each MA iteration. GSA(M) is the generalized SA, extending M paths at each iteration.

Finally, FGSA(M, α) and FMA(M, α) denote the Fast SA and the Fast MA, i.e., embedding the test presented in Section 6.4.5.

Figure 6.12 illustrates the error resilience, in terms of symbol error rate (SER), and complexity performance, in terms of average number of visited branches during erasure-free decoding, of four versions of the decoder using the SA. Figure 6.13 presents the same performance for four versions of the decoder using the MA.

When compared to a standard decoder carrying out hard decisions on noisy bits, the sequential decoders present an important gain (up to 3 dB for the SA-based decoders and 4 dB for the MA) in error correction. The performance is improved as the number of simultaneously explored paths (M) increases and as α decreases.

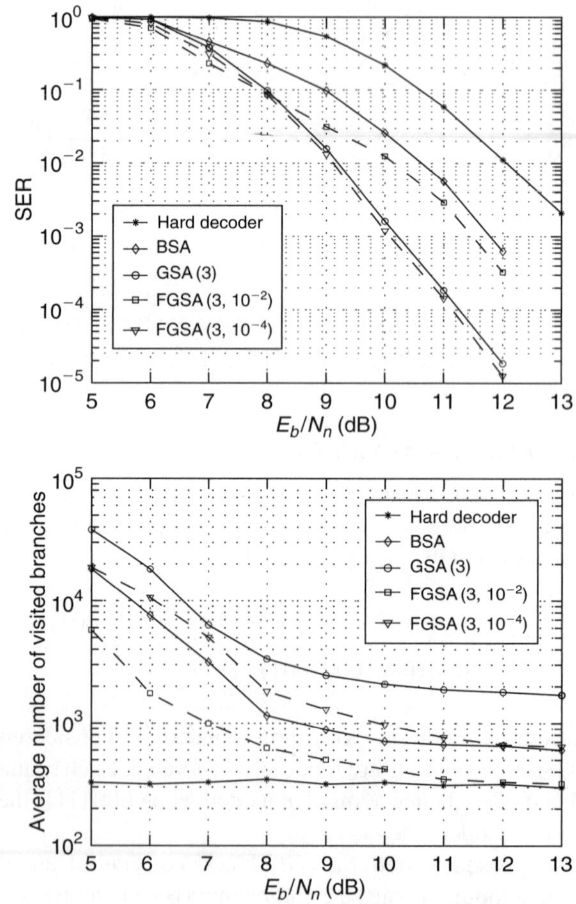

FIGURE 6.12 Performance in terms of error resilience and complexity of the decoder using the SA.

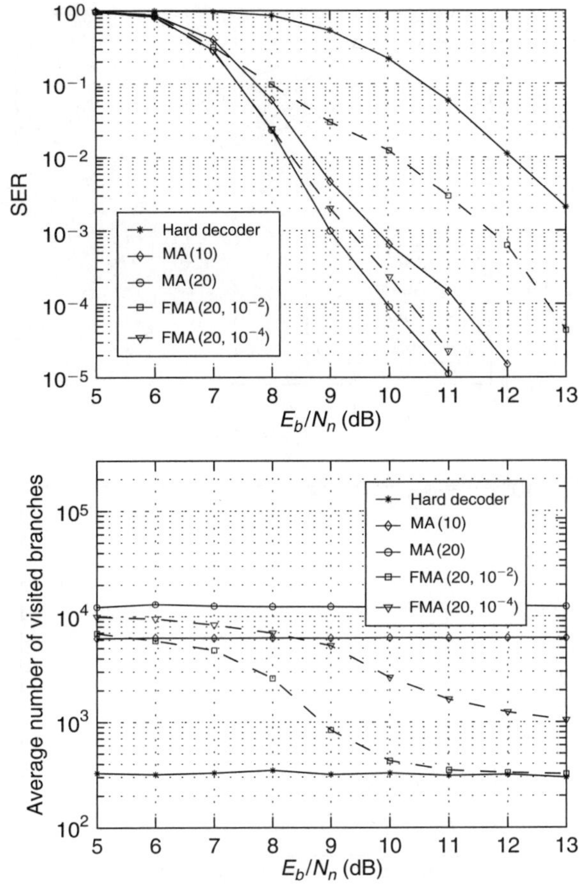

FIGURE 6.13 Performance in terms of error resilience and complexity of the decoder using the MA.

On the other hand, one may notice that the more efficient is the decoding in recovering errors, the higher is the complexity. Compared with the BSA, the FGSA(3, 10^{-4}) reaches a gain of 1.8 dB, at SER $= 10^{-3}$ for a doubled complexity at 12 dB. For the same SNR, FMA(20, 10^{-4}) presents a 20 times lower complexity compared with MA(10), with a gain of 2.5 dB at SER$= 10^{-3}$. This shows that the effects of α and M on the tradeoff between the complexity and the robustness may be combined to design JSCD schemes according to application needs.

Figure 6.14 depicts the packet erasure rate related to decoding failures for the same versions of the sequential algorithms. For the FGSA and FMA decoders, packet losses are mainly caused by the strategy introduced in Section 6.4.5. Indeed, for low values of the SNR, the number of stored paths often reaches zero

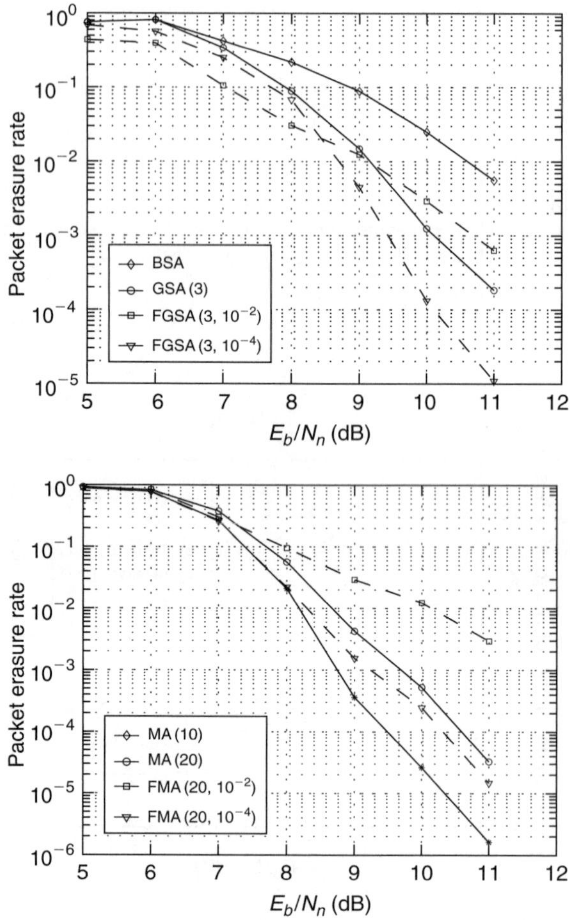

FIGURE 6.14 Packet erasure rate using SA- and MA-based decoders.

before any solution is obtained. Nevertheless, one can note, for example, that for a SNR of 10 dB, FGSA(3, 10^{-4}) presents a packet loss rate 200 times lower than the BSA. This avoids too frequent use of Automatic Repeat Quest (ARQ). For instance, supposing that the channel rate is equal to 64 kb/s, the effective rate available to the source, at SNR = 9 dB and taking the ARQ into account, is 57.6 kb/s using a BSA, 63.7 kb/s using a FGSA(3, 10^{-4}), 62.08 kb/s using a MA(10), and 63.93 kb/s using a FMA(20, 10^{-4}).

Decoding H.264/AVC-Encoded Sequences

For H.263, MPEG4, or H.264/AVC in the baseline profile, redundancy has been identified due to the structure of the VLCs used for entropy coding (Chen and Subbalakshmi, 2003; Kopansky and Bystrom, 1999; Lamy and Pothier, 2001), to the fact that video is encoded macroblock by macroblock (Nguyen and Duhamel, 2004) and to the packetization of the encoded macroblocks (Lee et al., 2005).

The redundancy in bitstreams generated by H.264/AVC in the main or high profile, when using CABAC, is much more difficult to identify. Nevertheless, the H.264/AVC decoder may be used as an error-detecting code. The encoder generates NALUs, which can be decoded independently. For a NALU consisting of n bits, not all 2^n possible combinations are consistent with the syntax of the H.264/AVC coder. Detecting successions of bits that are not syntax compliant helps to detect transmission errors. Figure 6.15 provides some examples corresponding to potential syntax errors.

To perform corrections, a syntax verification module (Sabeva et al., 2006) is combined with a sequential decoder, such as the SA or MA.

The simulations presented here involve a simplified transmission stack. The NALUs generated by H.264/AVC are protected by a cyclic redundancy check and transmitted over an AWGN channel. The 197 first frames of the sequence `foreman.cif` are encoded, modulated and transmitted to the decoder NALU by NALU through a noisy channel. Every altered NALU is fed to the MA in order to compute a set of M estimates. These estimates are then provided to the H.264/AVC syntax compliance checker, in decreasing metric order. If the current estimate of the NALU satisfies the H.264/AVC syntax, it is provided to

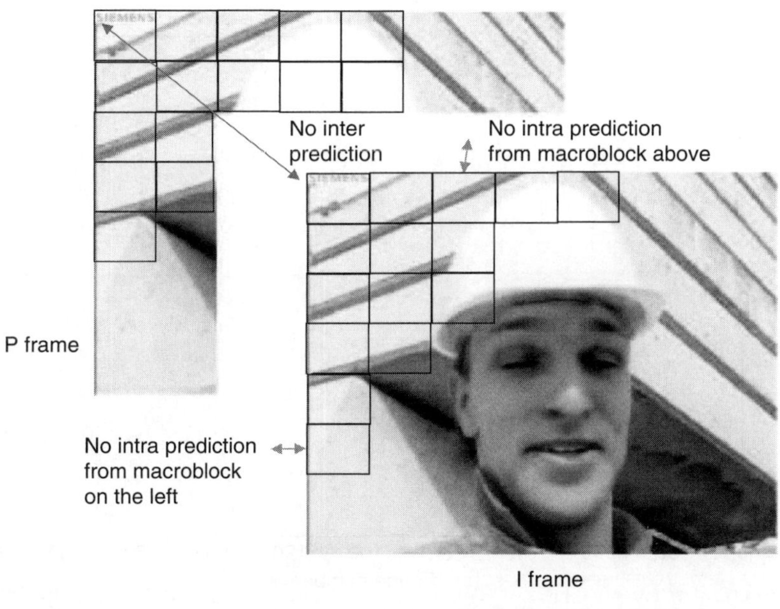

FIGURE 6.15 Example of errors that may be decoded by an H.264/AVC syntax checker.

the H.264/AVC decoder. If it is not the case, a syntax test is performed on the following estimate provided by the MA. When the whole set of M estimates for the current NALU has been tested without giving any acceptable solution, the H.264/AVC error concealment techniques are involved.

The JM 10.2 implementation of H.264/AVC (Wiegand et al., 2003) in its main profile has been used here. Some of the error resilience techniques supported by H.264/AVC (ITU-T and ISO/IEC JTC 1, 2003) have been put at work, namely slice-structured coding and insertion of a fully intra-coded frame every 16 encoded frames. The size of a slice (and thus of a NALU) has been limited to 120 bytes. This size limitation introduces some encoder performance reduction, equivalent to less than 10% redundancy introduction.

Figure 6.16 shows the PSNR for the luminance as a function of the frame number, for a channel SNR of 11 dB. The performance increase provided by the joint decoder is more than 5 dB compared with the hard decoder. Increasing M (the number of simultaneous paths processed by the MA) from 3 to 10 improves the performance of the decoder (with a decoding complexity that is also increased).

Syntax checks are time consuming, and were only performed and the end of each NALU in Figure 6.16. Figure 6.17 illustrates the performance increase when several syntax checks are performed for the same NALU during its decoding

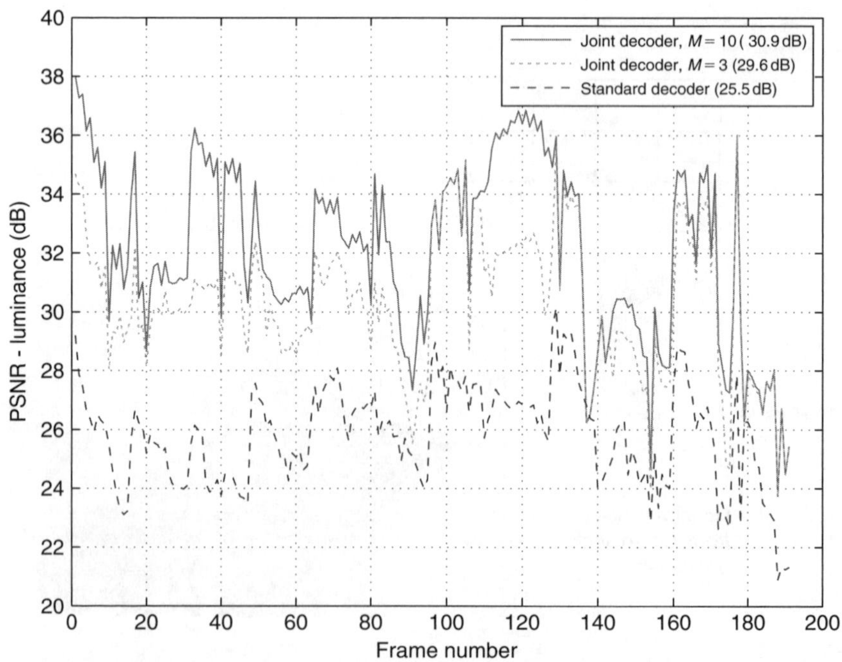

FIGURE 6.16 PSNR of the luminance, for a channel SNR of 11 dB.

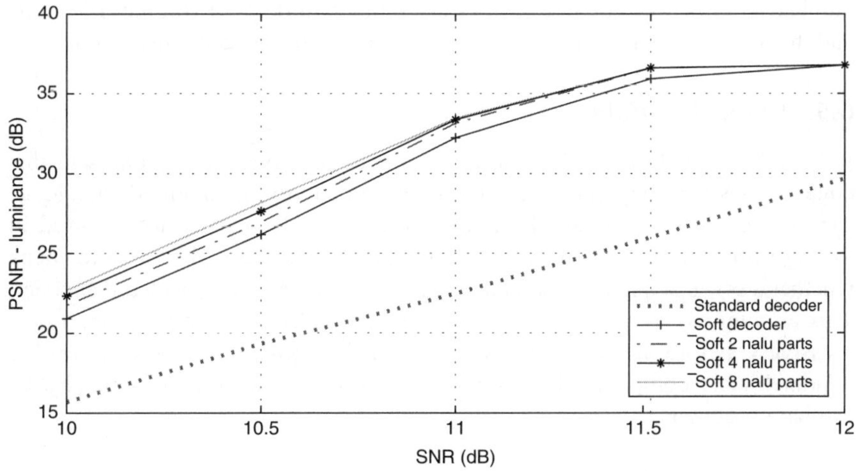

FIGURE 6.17 PSNR of the luminance as a function of the channel SNR, for a various amount of syntax checks and NALUs limited to 120 bytes.

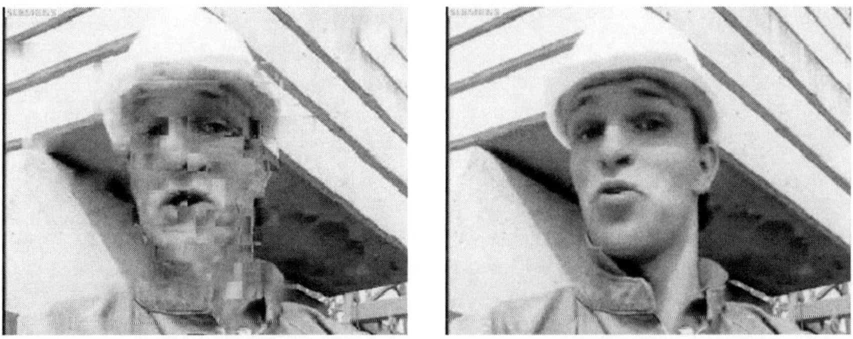

FIGURE 6.18 Picture 86 of `foreman.cif` sent over an AWGN channel with SNR = 11 dB; NALUs decoded with the standard hard decoder (left) and with a MA with $M = 10$ (right).

(1, 2, and 8 syntax checks per NALU). From 1 to 8 syntax checks, 1.5 dB is gained in average PSNR.

The computation complexity of the MA is negligible when compared to that of the H.264/AVC decoder. In optimized implementations, for packets without noise, the decoding complexity of the joint decoding scheme is of the same order of magnitude as that of the standard H.264/AVC decoder. For noisy packets, it is always less than M times the decoding complexity of a standard H.264/AVC decoder. M thus allows to tune some performance-complexity tradeoff.

Figure 6.18 shows the qualitative improvement brought by the proposed technique. An AWGN channel with an SNR of 11 dB is considered. Frame number 86 of the decoded sequences using the hard decoder (Figure 6.18, left) and the joint decoder with $M = 10$ (Figure 6.18, right) are both displayed.

6.5 CONCLUSION

It should be emphasized that the real strength of sequential decoders in this context goes well beyond the reduction of complexity. Its main advantage, as can be seen from the example, is that an explicit decoder is not required in the algorithm. In fact, the only requirement is that one should be able to check whether a given sequence could have been delivered by the considered encoder. This amounts to some kind of syntax check, which has to be performed progressively, before the end of the sequence. This seems to be the only feasible approach when complicated semantic constraints exist and have to be met by the estimated sequence.

Chapter 7

Protocol Layers

Traditionally, joint source-channel decoding (JSCD) studies are intended to make the best use of the redundancy found in the source bitstream in conjunction with channel decoding. One of the difficulties of this approach is that source coders improve quickly, and that the amount of redundancy decreases in each new standard. For example, it has been shown in Chapter 4 that exploiting the remaining redundancy in an H.263 bitstream could be done through some trellis, indicating that its structure can be made explicit. In contrast, the next video-coding standard H.264 had much less redundancy left, and moreover, its structure cannot be made explicit.

However, it has been already seen that packetization (introduced for networking purposes) was also useful in terms of available redundancy. It turns out that other sources of redundancy are present in the various protocol layers which, ultimately, may be exploited jointly with the redundancy left by the source coder to further improve the performance of a joint source-channel decoder.

Thus, like in the case of the source bitstream, one has to carefully examine the structure of the network layers to see where the redundancy is introduced and how it can be exploited for the benefit of a better decoding. Obviously, such detailed information about networking layers are seldom found in coding related works, but this is due to the nature of the JSCD topic, fully transversal in the whole communication process.

This chapter first presents the general architecture of networks and concentrates on the wireless network structure. The protocol involved in the WiFi receivers has been chosen as a representative of this category and is described with some details, to be explicit. The packet structure has even been recalled in Appendix A since some examples in the last chapters are linked to this information.

7.1 GENERAL ARCHITECTURE

7.1.1 Historical Notes

There are three main categories of networks, defined from the associated service, which in turn corresponded some years ago to various industries.

- Transfer services (*download*), which consist in transmitting data between two distant computers. This situation almost corresponds to an absence of delay constraint, but integrity of the data is essential.
- Conversational services (*real-time*) allowing users to communicate using speech, audio, or video signals. In this context, delay constraints are very strong, and the quality of the received signal must be sufficient (which does not directly translates in a constraint on the quality of the link).
- Broadcast services, in which a transmitter broadcasts (mostly) the same information to many users. Here, the communication is one way, there are strong delay constraints (even if less stringent than for conversational services), and the quality of the received signal must be sufficient.

Due to the variety of constraints, these networks have some specificities, and evolved separately for quite a long time. As a result, they were based on different technologies. However, they recently began to converge toward a unique network, coherent and efficient. The main driver in this trend, after the development of Internet in 1974, was the generalized use of digital technologies for processing all types of signal in all devices. The emergence of the second generation (2G) of mobile telephony (GSM was standardized in 1989) was a new step for speech communication: the signals turn digital and mobility is allowed. Their evolution toward the so-called 2.5G and 3G corresponds to a large increase of the bitrates and of allowed mobility [in Europe, this corresponds to the GPRS (Bates, 2002), UMTS (Lescuyer and Bott, 2004), and WiMAX (Andrews et al., 2007) standards]. At the same time, convergence between the Internet and mobile worlds is accelerating.

At the same time, broadcast networks were deeply impacted by the digital revolution after the definition of the DVB standard (Reimers, 2005) in 1993. The bitrate necessary for transmitting a TV program was heavily reduced (due to efficient video compression) and the standard was proposing packetized transmission schemes for various physical supports (terrestrial, satellite, cable, mobile).

To summarize, the emergence of the digital era allowed the networks to become more homogeneous and allowed a generalized mobility for the users. The general trend toward (1) the use of Internet to transport all types of digital communications (packetized, of IP type) and (2) universal wireless access to the network is already on its way.

7.1.2 Internet

The Internet is a worldwide computer network that interconnects millions of computing devices, from PCs, servers, up to personal digital assistants (PDAs), TVs, cell phones, and an increasing number of unexpected devices in the near future. End users are not attached to each other via a single communication link but are indirectly connected to each other by means of switches. The data are separated in packets, which are forwarded independently to their destination. Thus, each terminal has a unique address (the IP address), which depends on its localization (operator, local network, ...). Each packet traveling in the network contains the address of the source terminal as well as that of the destination; thus, allowing the routers to forward the data to the right direction.

When a router receives a packet, the final address is analyzed, and the packet is sent via the most appropriate path, which is not necessarily unique for each packet of a communication. In fact, for reliability purposes, the network is highly redundant, and numerous paths allow to reach the destination.

Because of the variability of traffic conditions, the packets may not arrive in the order they were sent by the source, and the IP address of the destination has to be complemented by a number corresponding to the packet, which must be available at the receiver, provided by the higher layers. Moreover, if the packet did not arrive after some time, the destination must be able to report this fact to the source, which also requires the knowledge of this address. The mechanisms implemented by the protocols at the various layers can perform this task.

Figure 7.1 illustrates a very simplified view of part of the Internet. Many different paths can connect the three terminals, and it is the task of the routers to decide which one a given packet will follow. Depending on the location and on the

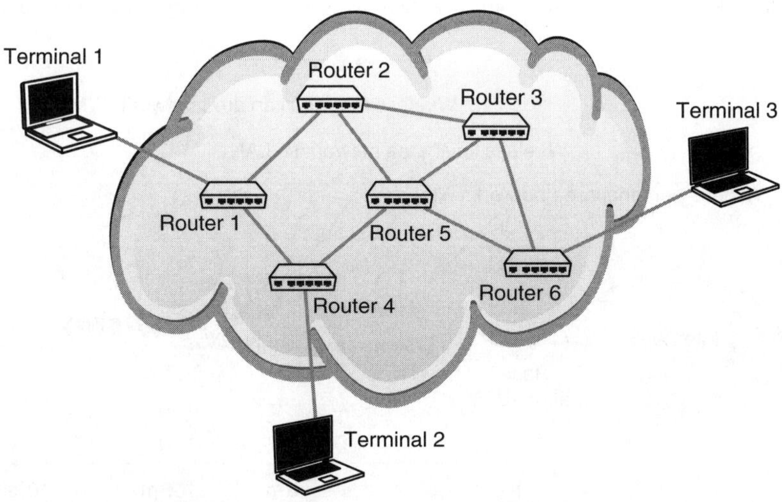

FIGURE 7.1 Schematic view of part of the Internet.

traffic in each part of the network, two types of physical links can be found: cables and optical fiber. Copper cables can reach bitrates in the order of 100 Mbps with negligible bit error rates (BERs) (in the order of 10^{-6}), whereas optical fiber has a very high capacity (up to 40 Gbps per fiber) with BERs three orders of magnitude smaller (lower than 10^{-9}). On another side, congestion due to the queues in the routers is quickly diminishing, due to the increase in processor power, storage capacity, and network capacity. As a result, the reliability of the transport allows the final users to have available bitrates in the order of 10 Mbps. With such bitrates, even very demanding applications (video transmission) become feasible.

7.1.3 Wireless Networks

There is currently an increasing trend toward mobile communications, which have a significant interest for terminal links, since they allow the users to remain connected within a given zone. Wireless communication standards are classified according to their coverage, i.e., the size of a cell. The most common ones are depicted in Figure 7.2. *Wireless personal area networks* (WPANs) have a coverage of a few meters, with bitrates ranging from 1 to 100 Mbps with the Bluetooth (Morrow, 2002) technology. A wider coverage (about 100 meters) is obtained with *wireless local area networks* (WLANs), e.g., with the 802.11*n* (802.11, 2005) standard, which can provide bitrates up to 250 Mbps. On a wider scale (about 10 km), the 802.16*e* (IEEE 802.16 Standard, 2004; IEEE 802.16e Amendment, 2005) allows bitrates in the order of 20 Mbps. This characterizes *wireless metropolitan area networks* (WMANs).

In contrast with wired connections that carry the signal to a given user, wireless links broadcast the signals to all users in a cell. This allows easily some mobility (within a cell, handover is another matter), but the corresponding drawback is that the radio resources have to be shared among all users in a cell.

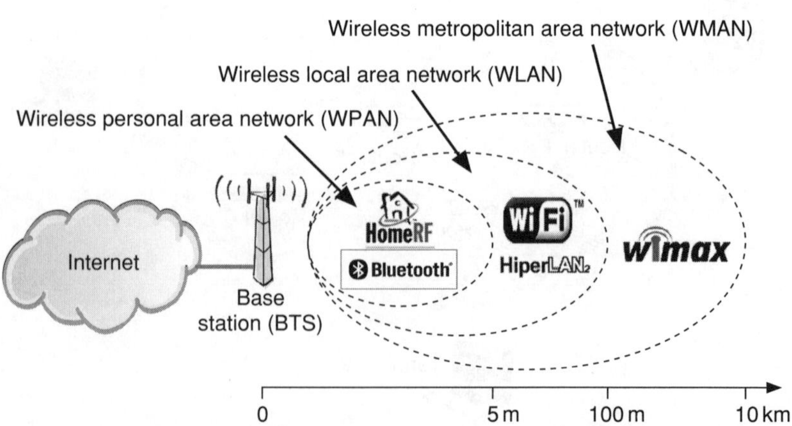

FIGURE 7.2 Some wireless communication standards.

Since the bandwidths are limited, the order of magnitude of the bitrates given in the previous section have to be divided by the number of users at a given time in a given cell. As a result, the available bitrate reduces quickly with the size of the cell. In the case of WiMAX, for example, it may happen that one user cannot obtain more than 100 kbps.

Moreover, wireless links are impaired by some phenomena (such as inter symbol interference, interference between cells, ...), which reduce the quality of the received signal, and result in a rather high BER. Therefore, there is a strong need for protecting the transmitted information, and the corresponding mechanisms are integrated in the mobile networks. A description of these methods, which obviously will have an impact on the JSCD issues, is provided in Section 7.1.4. They are based on three mechanisms.

- Packetization: even if the data are already in the form of packets when issued by the wired network, the packet size has to be adapted to the characteristics of the wireless link.
- Retransmission: when a given packet has been detected as erroneous by the receiver, the transmitter may be asked to resend the corrupted packet.
- Addition of redundancy to the useful data: used to correct most of the transmission errors and reduce the amount of retransmissions.

The combination of these techniques allows to drastically reduce the transmission errors, but this is obtained at the cost of a reduced throughput, useful bitrate, dedicated to the end user. As an example, for WiMAX, the useful bitrate seldom exceeds 50 kbps per user.

To summarize, the Internet carries the information by means of packets, oriented by the routers. These packets contain various identifiers, which characterize the sender, the receiver, and the packetization. This architecture allows to attain high transmission rates and small BER. At the termination link, mobility is ensured by wireless communication, but the available useful bitrate can be moderate to small, and may hardly convey multimedia signals, such as video, on large areas. As a consequence, the wireless link is the critical link of the transmission chain.

In this book, the influence of the signal packetization is the only impact of the wired part of the communication that is considered, and the Internet is assumed to be working perfectly otherwise. As a result, the corresponding situation is that of a real-time transmission of video from a server to a receiver connected to the Internet. The Internet provides without any error the packets to the base station. Finally, the base station delivers the signal to the end user through some wireless link.

7.1.4 Wireless Networks Architecture

Obviously, the tasks of transmitting, routing, segmenting the data according to some constraint, addressing the right user, etc. are highly complex tasks.

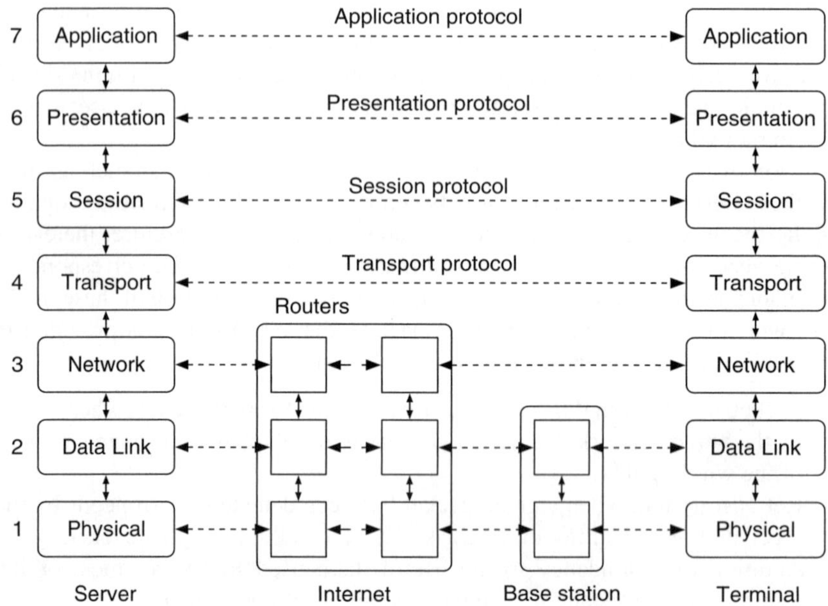

FIGURE 7.3 Mixed Internet and wireless transmission scheme according to the OSI model.

Moreover, they have to be compatible between very heterogeneous systems when connecting them. The classical solution in this case is usually to separate the work to be done into separate tasks, each one having a small number of interfaces so that its design is not too complex, and can be adapted to a variety of situations. This is the reason that motivated the layered architecture of communication systems.

The universal reference is the Open Systems Interconnection (OSI) model, depicted in Figure 7.3. This model is made of a stack of seven layers, which define the rules that each task must follow. Within a receiver, each task receives information from the lower layers and forwards the result of its own processing to the next one. Each processing is assumed perfect. Therefore, each layer in the transmission stack can be assumed being connected to its corresponding layer in the receiver through a protocol that, therefore, defines a virtual link between corresponding layers.

The task of each OSI layer is clearly defined and corresponds to a specific role. One may distinguish between the lower layers (1, 2, 3, and 4) that control the delivery of the data, and the upper layers (5, 6, and 7) that are doing some local processing involving only the server and the terminal. At all levels, the data is organized in blocks for which we use the generic name *packets*, which take different names depending on the layer. Each layer is described below.

1. The *physical layer* ensures the final shaping of the binary data so that they can correctly be received. This corresponds to the following tasks: (1) transformation of the binary bitstream into an analog waveform, adapted to the channel (cable, fiber, wireless channel), (2) channel coding, for a better protection of the data, (3) synchronization of the data between the transmitter and receiver.
2. The *link layer* takes care of communications between two adjacent devices, connected through the physical medium. Its main task is to ensure the packetization of the data into slices, the size of which is adapted to the channel. Upon reception, it controls the integrity of the corresponding packets by incorporating error-detection mechanisms and may ask for the retransmission of erroneous packets. When the resources are shared among several users, it performs (jointly with PHY layer) the corresponding multiplex, as well as the resource allocation. In this level, a packet is named *frame*.
3. The *network layer* establishes the communication link between the server and the terminal, and contains routing protocols that determine the routes the packets take. It is the only layer concerned with the network topology and is also the last layer supported by the central network equipments, whereas the upper layers are only supported by the terminal parts. At this level, the packet is named *datagram* when the protocol is *connectionless* (e.g., IP layer) or *segment* if it is connection-oriented (e.g., X.25 network layer).
4. The *transport layer* manages the transport of the application-layer messages between the client and the server sides of the application, by adapting the data transmission in the network and the service required by the user. It performs data multiplexing between the various processes, fragmentation of the packets coming from the upper layers into packets, the size of which is adapted to the network. There are two types of transport protocol, depending on the nature of the data to be transmitted. In the connection-oriented mode, the service includes guaranteed delivery of application-layer messages; there is no packet loss and the received packets are correct. This may generate high transmission delays. In the connectionless mode, packets are transmitted only once by the transmitter, and there is no acknowledgment procedure on the packets. This is a low delay, but risky transmission. At this level, the packet can take two names. It is named datagram when the protocol is *connectionless* (e.g., UDP) or segment if it is connection-oriented (e.g., TCP).
5. The *session layer* organizes and synchronizes the dialog between distant tasks. It establishes a link between two programs that have to cooperate and control their connection, e.g., by token management. If ever a connection has been broken, this layer allows to reconnect by using synchronization markers inserted in the data stream. At this level, the packet is named *Message*.
6. The *presentation layer* ensures compatibility of the data between communicating tasks. It is in charge of the application-layer messages coding and

ensures the format transition between user-related data format and the stream, which is transported by the network. Typical processing is data compression, encryption, or conversion. At this level, the packet is named *Message*.

7. The *application layer* is where network applications and application-layer protocol reside. It is the interface between the user and the network. At this level, the packet is named *Message*.

This model is fully generic, but when the Internet was developed, the choice was made of a simpler protocol. The number of layers was reduced to 4, and is now named after the names of the first two layers: the IP (Postel, 1981a) of the network layer and the TCP (Postel, 1981b) of the transport layer. TCP/IP model, simpler, was preferred. It is, thus, known under the acronym TCP/IP. It was first implemented, and then standardized. Both OSI and TCP/IP architectures are compared in Figure 7.4.

In contrast with the OSI model, the session and presentation layers are not independent anymore; their tasks are taken in charge by the application layer. Moreover, the physical and link layers are grouped in the Host-to-Network layer (Tanenbaum, 2002) (note that other authors consider two separate layers at this level). The implementation of this layer is not defined in the standard, only its role is defined: it must ensure access to the physical medium and to send packets to the network. Practically, the required standard depends on the technology used in the local network (Ethernet, Wi-Fi).

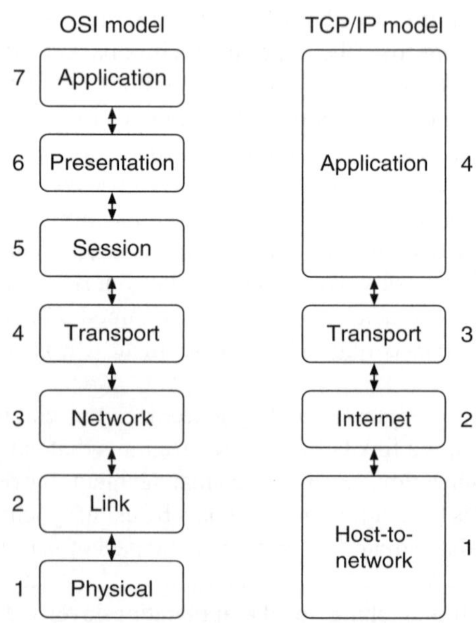

FIGURE 7.4 Comparison between the TCP/IP architecture and the OSI model.

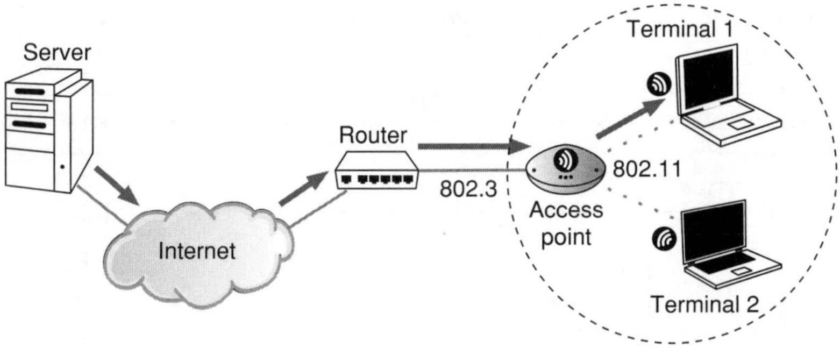

FIGURE 7.5 Transmission scheme considered in this book.

Since our main interest is in the real-time transmission (streaming) of video from a server to a mobile terminal, now concentrate on the corresponding protocols. In this context, the transfer between the transmitter and receiver must be very fast, a fact that prevents any dialog between both terminals. The TCP, being connection-oriented, is not suited to this situation. It is replaced by a set of two protocols that constitute the RTP/UDP (Postel, 1980; Schulzrinne et al., 1996) layer. Moreover, even if the principles of JSCD do not depend on a precise situation, its actual implementation is strongly dependent on the detailed structure of the Host-to-Network layer, which corresponds to a given wireless communication standard. In this context, most of our results will be illustrated on the Wi-Fi 802.11 (IEEE, 1999) standard that will, thus, be described with some details. This standard provides the specifications of the physical (PHY) and link (MAC) layers. The complete transmission chain is shown in Figure 7.5.

7.1.5 The Layers in IEEE 802.11

The various protocols involved in a WiFi connection carrying encoded video are represented in Figure 7.6. Upon reception, the situation corresponds to the successive application of the following layers: PHY 802.11/MAC 802.11/IP/UDP/RTP. Encapsulation and fragmentation mechanisms are found across all layers.

The Physical Layer in 802.11

The various tools used in the PHY layer aim at transmitting reliably the data between the access point and the mobile while minimizing the radio resource usage. The PHY layer allows carrier sense/clear channel assessment (CS/CCA) to communicate to the MAC layer whether the channel is busy or not through internal messages between the layers. When the channel is clear, the PHY layer sends a CCA to the link layer. The PHY layer of the IEEE 802.11 standard provides bitrates on the order of 1 or 2 Mbps in the 2.4-GHz frequency band.

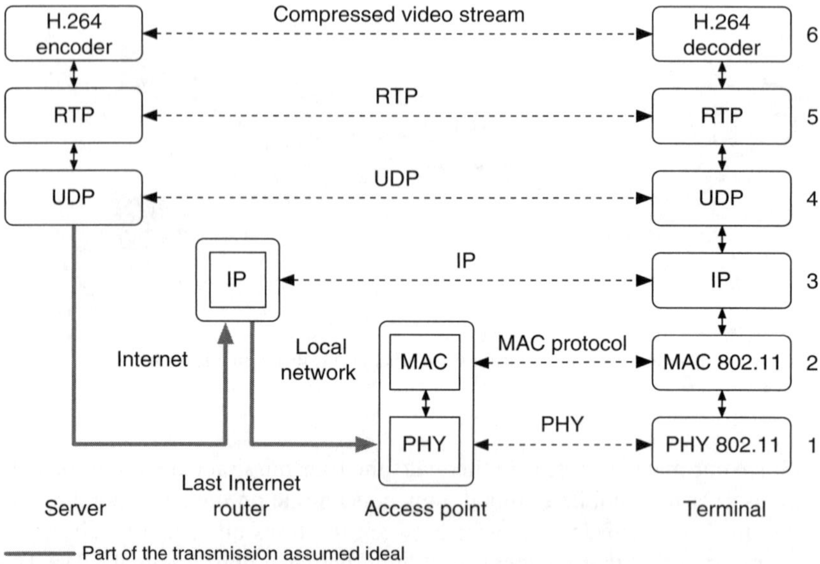

FIGURE 7.6 The various protocols between the terminal and the other equipments.

Two spread spectrum techniques are defined in the standard: FHSS (*frequency-hopping spread spectrum*) and DSSS (*direct-sequence spread spectrum*).

Frequency-Hopping Spread Spectrum

In this mode, the transmitter makes successive use of several carrier frequencies, according to a sequence known only from the transmitter and the receiver. These synchronized changes are defined through a code that defines the amplitude of the hops. To allow this process, the frequency band around 2.4 GHz is split into 79 channels of 1-MHz bandwidth. The transmission on each of these channels can last from tens of milliseconds to some hundreds of milliseconds (generally 300–400 ms). This allows to attain base bitrates of about 1 Mbps. One of the advantages of this mode is that the signal is not sensitive to narrowband interferers since it makes use of frequency diversity. Figure 7.7 illustrates this FHSS mode.

FHSS is used in combination with a GFSK modulation (Gaussian Frequency-Shift Keying), in which the information is carried by a shift of frequency around the carrier frequency. In this mode, bits are processed by a 2-GFSK modulator and successively transmitted. The two binary states (0 and 1) are defined by two symmetric frequencies around the carrier frequency, separated by 320 kHz. In the 2 Mbps mode, two bits are simultaneously transmitted via a 4-GFSK modulator and are represented by four frequencies shifted by 144 kHz, centered around the carrier frequency. Both modes are represented in Figure 7.8.

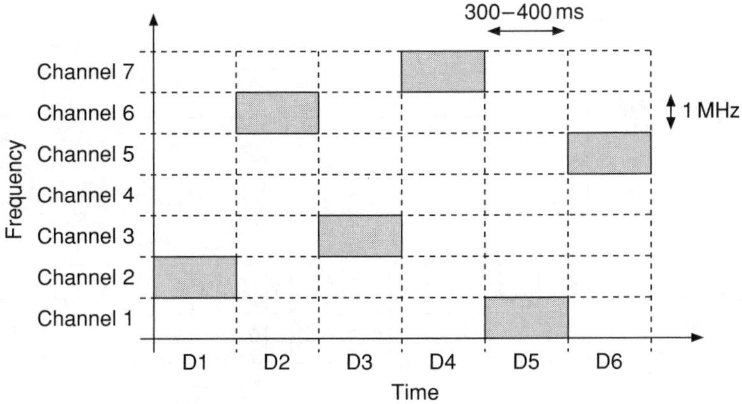

FIGURE 7.7 Time–frequency plane in frequency-hopping spread spectrum.

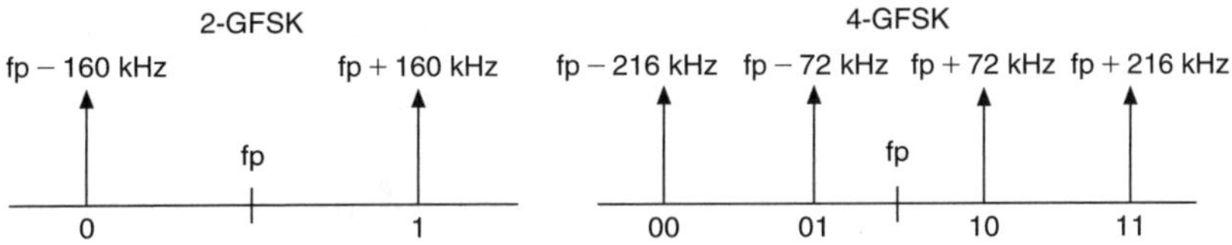

FIGURE 7.8 2-GFSK and 4-GFSK modulations.

Direct-Sequence Spread Spectrum

In this method, one bit is spread over several samples with very fast transitions, driven by a given binary sequence. Therefore, in contrast with the previous method, DSSS exploits a continuous spectrum. The longer the code, the higher the transition frequency, and the larger the spectrum is. In the 1 Mbps mode of the 802.11 standard, a Barker code with 11 bits (*chips*) is used to spread the spectrum. The logical state 0 is represented by the sequence 10110111000 and the state 1 by its complement 01001000111. Chip frequency is 11 MHz and the energy of the baseband signal is concentrated in a first lobe of 22 MHz. This method exhibits a strong noise immunity, due to the use of the code, which improves transmission efficiency. This encoding is illustrated in Figure 7.9.

In the DSSS mode of the 802.11 standard, the binary information is carried by the phase of the signal, using a *phase-shift keying* (PSK) modulation. 1 Mbps is provided by a *binary phase-shift keying* (BPSK) modulation (with two possible phases: 0 and π) while a *quadrature phase-shift keying* (QPSK) modulation (with possible phases 0, $\frac{\pi}{2}$, π and $\frac{3\pi}{2}$) doubles this bitrate since each symbol carries two bits. Figure 7.10 illustrates these two modulations.

In this book, most simulations illustrating the performance of JSCD in practical contexts are based on the DSSS mode because of its simplicity and noise robustness, due to the use of a Barker code.

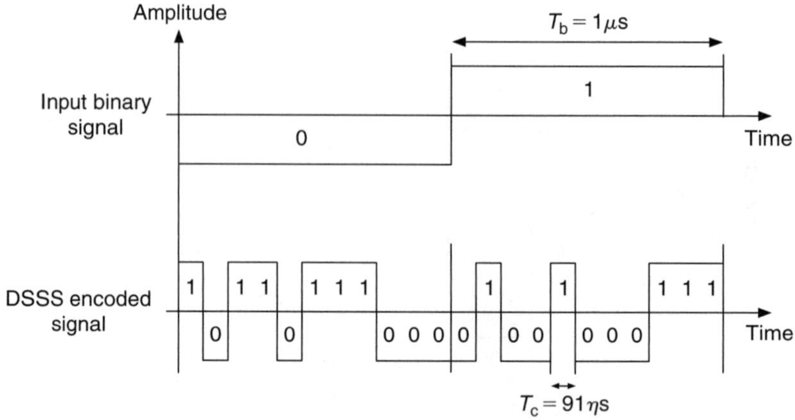

FIGURE 7.9 DSSS encoding based on a 11-bit Barker code.

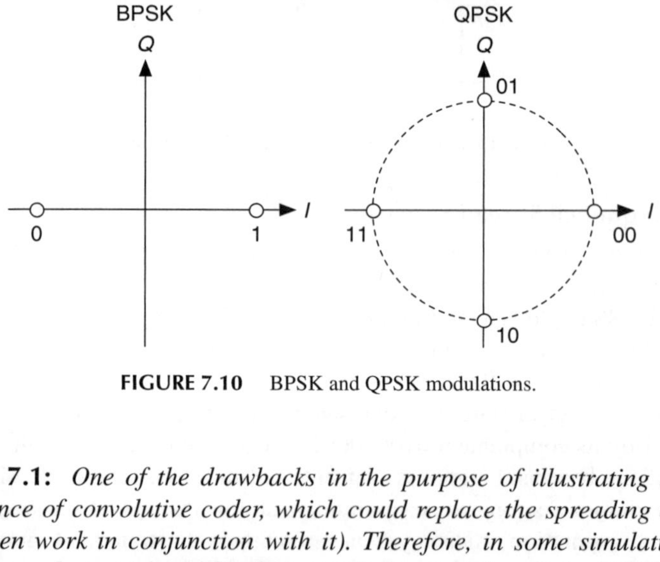

FIGURE 7.10 BPSK and QPSK modulations.

Remark 7.1: *One of the drawbacks in the purpose of illustrating JSCD is the absence of convolutive coder, which could replace the spreading code (or could even work in conjunction with it). Therefore, in some simulations, the spreading code is replaced by the convolutive encoder of the IEEE* 802.11*a standard. Although this combination does not correspond to any practical situation, it illustrates the ability of JSCD to be combined with the use of forward error correction (FEC) coders. The chosen FEC is depicted in Figure 7.11. Each new input bit results in two output bits (the code rate is* $R = 1/2$*), computed as an* exclusive OR *of the previous information bits stored in a shift register. The bits that are combined are selected according to generator polynomials, which are, in this case,* $g_0 = 133_8$ *and* $g_1 = 171_8$*, as illustrated in Figure 7.11.*

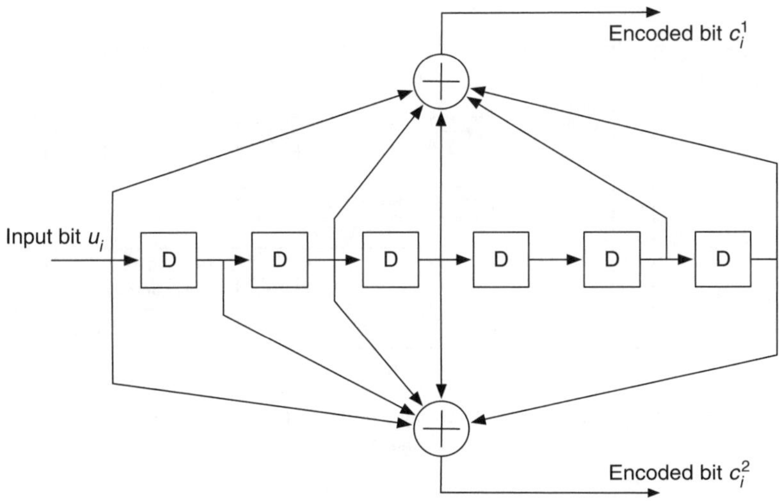

FIGURE 7.11 Convolutive coder of the IEEE 802.11*a* standard.

The Link Layer 802.11 (MAC)

This section describes the mechanisms of the link layer of the IEEE 802.11 standard. Their main task is to perform resource allocation between the various constituents of the network. Two access modes are defined in the specifications: the Distributed Coordination Function (DCF) protocol and the Point Coordination Function (PCF) protocol. The link layer also performs data encryption using the Wireless Encryption Protocol (WEP). This protocol ensures confidentiality and is obviously present in many practical situations. However, for simplicity, simulations will be performed without WEP.

The DCF Protocol

DCF is the default mode used to share the physical medium. It is used in both cases of ad hoc and infrastructure networks. Any Wi-Fi device must contain an implementation of this mode. This DCF corresponds to an improved version of Carrier Sense Multiple Access with Collision Avoidance (CSMA/CA).

Just like in the plain CSMA mechanism, a base station willing to transmit data first waits during a predefined period, denoted as DCF Interframe Space (DIFS) of 50 μs. If the physical layer did not detect any signal during this period of time, a CCA is sent to the MAC layer, and the base station still wait during a random duration (contention window). If no signal has been detected during this time, the base station sends to the receiver a Request To Send (RTS). This short control frame contains the MAC address of the transmitter and receiver. It also contains a *Duration* field, which contains the estimated duration of the first MAC fragment. This allows the other base stations to determine their Network Allocation Vector, NAV(RTS). During this period, the adjacent base stations are

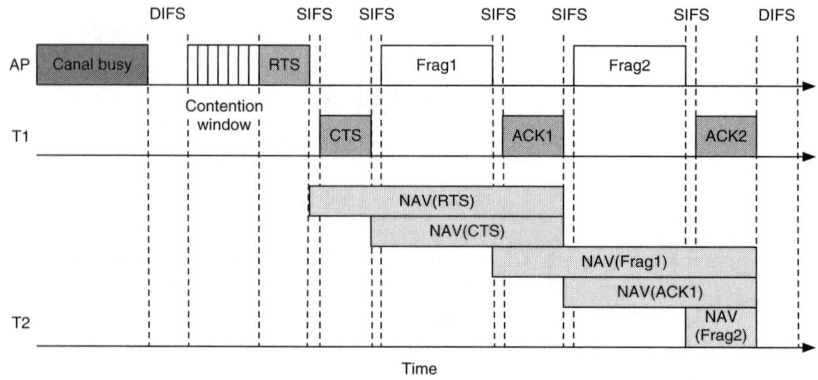

FIGURE 7.12 DCF protocol of the MAC layer in IEEE 802.11 involving an access point (AP) that communicates with a terminal (T1), a second terminal (T2) is idle.

not allowed to transmit data to avoid collisions. After a Short Interframe Space (SIFS) of 10 μs, the receiver sends a Clear To Send (CTS) control frame that indicates to the transmitter that it is ready for receiving the signal. This frame also contains an important information that allows to adjust NAV(CTS). The reservation procedure being terminated, the transmitters sends the data corresponding to an IP packet. If this packet is too large, it can be cut into several MAC fragments that are successively sent onto the channel. The receiver acknowledges each fragment by sending an Acknowledgment (ACK) frame. When the transmitter has not received the ACK of some fragment, it sends this fragment again. All frames contain a field indicating the estimated transmission duration of the next fragment, which allows to update the NAV of all adjacent base stations. An SIFS separates each frame to avoid any overlap. At the end of the transmission, the reservation procedure begins again, and another user can transmit. Figure 7.12 illustrates the DCF procedure in the case where the transmission goes from the access point and terminal 1, the data of the IP packet being cut into two MAC fragments.

The PCF Mode

Since the previous DCF mode is nondeterministic, a terminal cannot predict when it will have access to the channel. The PCF mode intends to solve this problem by providing a synchronous transmission mode. This is obtained by a centralization of the information at the access point (infrastructure mode), which allows the access point to supervise the access to the channel of each terminal. To avoid collisions, the coordinator sends a special request (CF-Poll) to each terminal. If the access point is willing to send data to the current terminal, it inserts date in the reservation frame (CF-Poll + D). When the current terminal is willing to send data to the network, the answer to the access point is an ACK frame that also contains data (CF-ACK + U). During this time, the other terminal remain

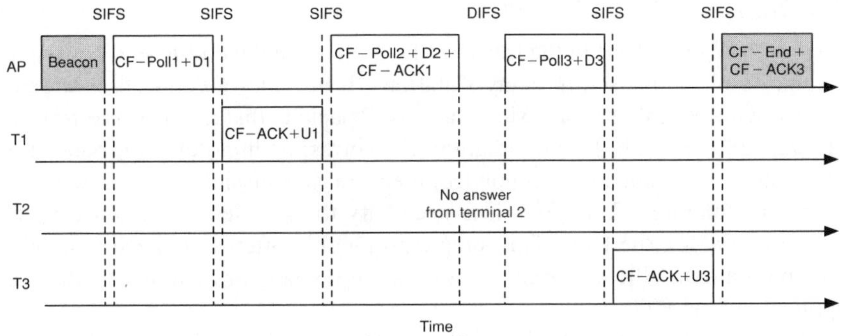

FIGURE 7.13 PCF protocol of the MAC layer in IEEE 802.11, where an Access Point (AP) supervises the access of three terminals (T1–T3).

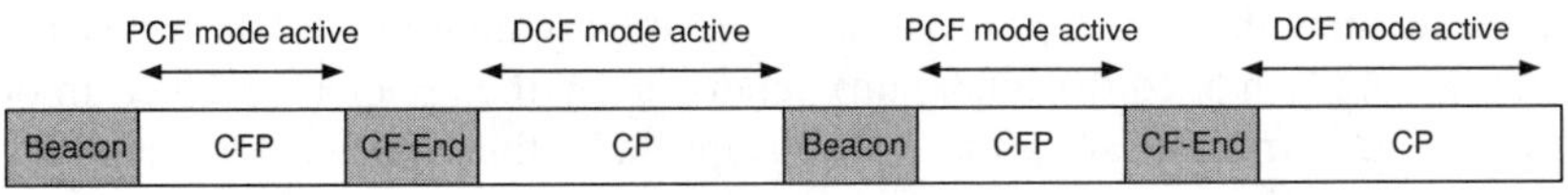

FIGURE 7.14 Access point: alternance between PCF and DCF modes.

silent. Once the data is transmitted, the access point turns to the next terminal while acknowledging the data received from the previous terminal (CF-Poll + CF-ACK or CF-Poll + CF-ACK + D). If some terminal is not answering during the PCF Interframe Space (PIFs) of duration 30 μs, the access point assumes that it is not willing to transmit and goes to the next terminal. This is summarized in Figure 7.13 where the access point takes care of three terminals. A communication in the PCF mode is initialized by a beacon frame and is terminated by a CF-End or CF-End + CF-ACK frame.

The PCF mode is optional in the IEEE 802.11 standard and is always used in conjunction with the main access mechanism. Therefore, some coexistence procedure between DCF and PCF is necessary. It is outlined in Figure 7.14 and will not be detailed here.

In practice, the PCF mode does not seem to be used in any device. All provided simulations, therefore, assume that only the DCF mode is used.

The Network Layer (IP)

The main task of the IP layer is to guide the data from the transmitter to the receiver by providing the routers with the appropriate information. These routers make use of tables to send the data in the right direction. Another task is fragmentation of IP packets if necessary. This option is not activated in our simulations. Furthermore, even if the IP is not fully reliable, our simulations assume that all packets arrive safely at the access point (no problem due to congestion or packet losses in the core network). However, the packets may arrive in any order upon reception. Finally, in our simulations, we assume that IPV4 is used.

The Transport Layer (UDP/RTP)

The protocols involved in the transport layer manage the end-to-end communication. The TCP is used in many situations. It provides a connection-oriented service with reliable transmission since TCP packets that are lost or erroneous are transmitted again. In our situation, the corresponding delay introduced by the protocol mechanisms (acknowledgment, retransmissions, sliding window) cannot be tolerated. Therefore, it is necessary to use a less constrained protocol (nonreliable), the UDP. This simple protocol is often associated to another one to obtain an efficient connection between the network and the multimedia application: the RTP.

The RTP takes care of IP data with real-time constraints. It is directly associated with the current application, and some of its data are used at the upper layer. It allows to rearrange the received packets in sequential order, to control the safe reception of the packets, to synchronize the decoded bitstreams (time markers), and to identify the data types. All these are mostly additional data (other than what is usually named the bitstream) generated by the application layer. In our case of video transmission, the RTP encapsulates the NAL packets generated by the H.264/AVC encoder (see Chapter 3). The data are placed in the *payload* field, and a header containing the additional data, useful at RTP level, is added in front of each packet. The RTP format is detailed in Appendix A.

Finally, the UDP packets encapsulate the data from the RTP packets. The corresponding data are placed in a payload field, and a *header* contains the data useful for the UDP, such as *Server Port*, *Terminal Port*, and the size of the UDP packet, as detailed in Appendix A.

7.2 IDENTIFYING THE REDUNDANCY

As can be seen from the description of the various layers in the above section, and from the detailed description of the header syntax in Appendix A, each layer adds a lot of information on top of the other ones. This can take several forms which are as follows:

1. The header of each layer is considered as part of the payload of lower layers.
2. The headers may contain data that do not change from packet to packet.
3. The headers may contain data that can be computed from quantities available in the previous packets at the same layer.
4. The headers may contain data that can be deduced from quantities available at other layers.
5. The length of some packets is sometimes provided in the headers.
6. Some fields may not take all possible values.
7. CRCs are pure redundancy, used for checking data or headers integrity.
8. The channel outputs corresponding to the various CRCs are available at the receiver.

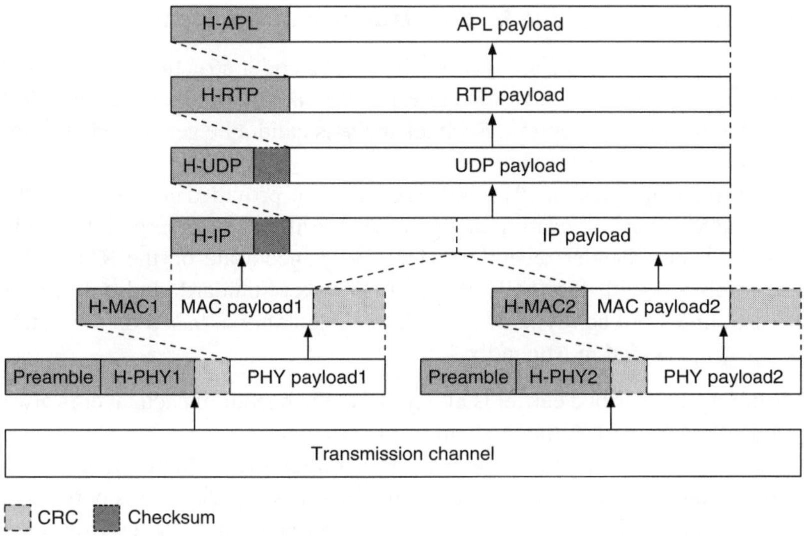

FIGURE 7.15 Schematic view of the protocol stack for video transmission over Wi-Fi.

As a result, plain transmission of the resulting bitstream would result in a situation where the true payload (corresponding to video data) can represent only as little as half to total bitstream circulating onto the channel.

This can be schematically represented in Figure 7.15 for a real-time video transmission based on a RTP/UDP/IP stack (Kurose and Ross, 2005). This figure provides an example of the segmentation and encapsulation mechanisms implemented at each protocol layer in the case of video transmitted with a 802.11 radio interface. It is easily realized that the form of redundancy corresponding to Property 7 is of a different nature as the previous ones. Properties 1–6 are due to the syntax and meaning of the information that has to circulate within the protocol. It is of the same nature as that described in Chapter 4 concerning the video bitstream: the semantics of the network layers are such that not all combinations of bits are feasible. In contrast, Properties 7 and 8 come from an explicit redundancy: a certain number of bits the cyclic redundancy check (CRC) are computed from the protected data, appended to the packet, and the corresponding measurements are available at the receiver. Upon reception, the classical use of the CRC is as follows: one estimates the useful data and CRC from the received sequence, and the match between the estimated data and estimated CRC is done. If they do not match, an erroneous information is detected. It is designed and used as an error-detection code.

As a result, it may well happen that the redundancy introduced by the protocol, as outlined previously, is much higher than that due to the structure of the source.

7.2.1 Redundancy due to the Structure of the Headers

Obviously, just like in the analysis of the source coder structure, only a careful analysis of the protocol stack of interest will make this redundancy usable for improving the video reception. Such an analysis cannot be generic and has to be performed in a given situation.

A careful analysis of the PHY and MAC layers is provided in Chapter 8, where they are taken as an application of a general algorithm able to exploit this redundancy for robust header reception. Here, we concentrate on the RTP/UDP/IP stack combined with the IEEE 802.11 PHY layer standard, and only provide some examples of each type of redundancy cited above, based on the detailed description provided in Appendix A:

- Property 1 described earlier is always true: the header of each layer is always considered as part of the payload of lower layers.
- The MAC address of the receiver and transmitter do not change during a given communication and is hence found in each MAC packet. This corresponds to Property 2: the headers may contain data that do not change from packet to packet.
- In the same MAC layer, MAC addresses of receiver and transmitter are specified in RTS/CTS packets and are also used in other packets. This is an example of Property 3.
- The length of the IP packets is sometimes provided in the corresponding headers. This is an example of Property 5.
- Property 6 is characterized by the situation in which some fields may not take all possible values. One such example is the signal field at PHY layer, which is 1 byte but can take only two values.

Careful examination of the headers can provide many more examples, as is easily seen from the detailed description of the headers in Appendix A. A general framework allowing to exploit these various redundancies is explained in Chapter 8.

7.2.2 Redundancy due to the CRCs and Checksums

The various CRCs in IEEE 802.11

CRCs or checksums are found at most layers, and their size depends on the precise layer and on the type of data that are protected. The details are given in Appendix A, but the important characteristics are recalled below.

- At PHY layer: CRC on part of the header (2 bytes)
- At MAC layer: CRC on both header and data (4 bytes)
- At IP layer: checksum on the header (2 bytes)
- At UDP layer: checksum on the header and data (2 bytes)

This clearly corresponds to a rather significant amount of redundancy.

The CRC as an Error-Detection Device

As described in Appendix A, the classical use of CRCs and checksums is error detection; they are used to check the integrity of some data, and if some error is detected, some appropriate action can be taken (usually, ask for retransmission or discard the corresponding packet).

Parity Checks

The most simple form of error detection is certainly based on a single parity bit. Assume that the packet to be protected has B bits. The parity bit, appended to the packet, can be computed, for example, as the "exclusive OR" of all the bits, which corresponds to choosing a parity bit equal to 1 if the number of ones in the packet is odd. This is named as "even parity scheme" because in this case, the total number of 1s in the sequence of length $B+1$ is even. Upon reception, the consistency check amounts to compute an exclusive OR of all the $B+1$ bits, and if the result is zero, the received sequence follows the same rule as the transmitted sequence, and it will pass the check.

Obviously, this does not ensure that the received sequence is correct, it means that the number of errors was even. However, if the probability of error is small and if errors are assumed to arise independently from one bit to the next, the probability of multiple errors (even 2) will be much smaller than the probability of one error.

Obviously, also, the error-detection capacity of this scheme is very poor and must be reserved for very small bit-error probabilities. Moreover, in practical situations, the errors are known to be clustered in bursts, and such a scheme will not be very efficient.

Example 7.2: Let the initial packet take the following binary value: 0110 0001 (which is represented by 61 in hexadecimal.)

The number of 1s in the word being odd, the parity bit is set to 1, and the word plus the parity bit reads (in binary form) 0110 0001 1. Upon reception, if no transmission error occurred, the exclusive OR of the received bits is zero, where any odd number of errors will have as a consequence that the exclusive OR of all bits will be 1, thus allowing error detection.

Obviously, any even number of errors will not be detected.

Checksums

This is a generalization of the previous technique; rather than working bit by bit, the packet is treated as a sequence of integers, each one of l bits. These words are added with 1s complement arithmetic. This means that any carry getting out the l bits should be added to the least significant bit (LSB) of the word. The checksum that is added to the packet is then the 1s complement of this 1s complement sum. By construction, the consistency check is obtained as follows: perform the

TABLE 7.1 The 1s Complement Representation

Binary	Decimal	Hex
0000 0000	0	00
0000 0001	1	01
0000 0010	2	02
0000 0011	3	03
...	...	...
1111 1111	−0	*FF*
1111 1110	−1	*FE*
1111 1101	−2	*FD*
1111 1100	−3	*FC*

1s complement addition of all l bits' sections (including the checksum), which should be an "all 1s" word if no error occurred.

This should be clear with the following example:

The 1s complement representation of fixed-point integers (8-bit) is recalled in Table 7.1 and is characterized by the fact the zero has two representations, in contrast with more familiar 2s complement representations.

The 1s complement addition, as recalled above, requires some feedback of the carry (carries), as illustrated in Example 7.3.

Example 7.3 (1s complement addition): This example illustrates the addition of −3 and +7 when 1s complement representation is used.

First, the numbers must be represented in the appropriate manner: −3 is represented by *FC* and +7 is represented by 07. Then, addition may take place.

First, a classical binary addition (here represented in hexadecimal) is performed: $FC+07=01\,03$ in which 01 is the carry that has to be added to the LSB of the result (03) to give the correct result:

$01+03=04$, which is the 1s complement representation of the correct result: +4

So, the 1s complement sum is done by summing the numbers and adding the carry (or carries) to the result. We are now ready to process a checksum example.

Example 7.4 (Simple Internet checksum example): Suppose we have an 8-bit machine and that we want to send out the packet *FE* 05. Let's calculate and verify the Internet checksum. This requires the following:

1. Perform 1s complement addition of the words. This is done in two steps:
   - Binary addition of the words (here represented in hexadecimal form)

     $$FE+05=01\,03.$$

   - The 1s complement sum requires the addition of the carry to the 8-bit word as seen in the example above

     $$03+01=04.$$

   The 1s complement sum of $FE+05$ is thus 04.
2. The 1s complement of the 1s complement sum defines the Internet checksum, which turns out to be $-04=FB$.

As a result, the packet will be sent as

$$FE\ 05\ FB.$$

Now, at the receiving end, we add all the received bytes, including the checksum (again using classical binary addition)

$$FE+05+FB=01\,FE.$$

The 1s complement sum is

$$FE+01=FF=-0,$$

which checks that the transmission was OK.

Checksums require relatively little overhead but offer a relatively weak protection against multiple errors, compared with CRCs that are described below. In fact, even if the checksum appears good on a message that has been received, the message may still contain an undetected error. The probability of this is bounded by 2^{-C}, where C is the number of checksum bits.

Cyclic Redundancy Check

An error-detection technique that is widely used in networks is based on Cyclic Redundancy Check codes. In this framework, the digits of the bit strings are thought of as the coefficients of a polynomial in some variable x, coefficients that are elements of the finite field GF(2) instead of more familiar numbers. Therefore, the operations on the bit string are interpreted as polynomial arithmetic, with operations performed modulo 2.

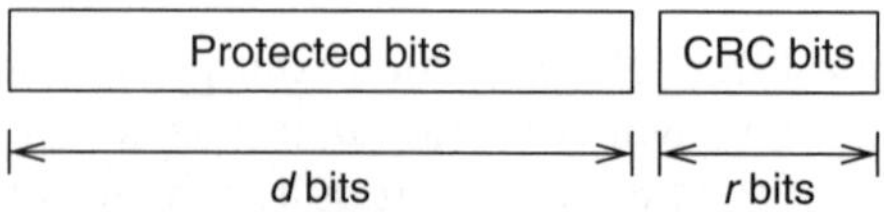

FIGURE 7.16 The data and the CRC in a block.

For example, let D be the data to be protected (d bits) and R be the redundancy bits (there are r such bits) to be appended to form the block to be sent (see Figure 7.16. Since all computations are to be expressed modulo 2, the preferred tool is exclusive OR (denoted as $\oplus$), which corresponds to addition as well as subtraction. The fact of appending the CRC on the right-hand side of the protected data and assuming the LSBs to be located on the right can be written mathematically as the following operations: shift the data D to the right by r bits (i.e., multiply D by 2^r) and add the CRC.

In terms of words, this reads

$$D2^r \oplus R \tag{7.1}$$

or, in terms of polynomials, with operations mod 2 (i.e., in GF(2)),

$$D(x)x^r \oplus R(x), \tag{7.2}$$

with the degree of $R(x)$ equal to $r-1$. Let n be the total number of coordinates (information to be protected, k bits plus redundancy, r bits). We have $n = k + r$.

Once stated as above, one can always think of $R(x)$ as being the reminder of some division by a polynomial of degree r. Let $G(x)$ be this polynomial and consider the division of $D(x)x^r$ by $G(x)$,

$$D(x)x^r = Q(x)G(x) \oplus R(x). \tag{7.3}$$

Since addition and subtraction mod 2 are similar, adding $R(x)$ to both sides of the equality results in

$$D(x)x^r \oplus R(x) = Q(x)G(x), \tag{7.4}$$

which provides a precise interpretation to (7.2): the CRC, expressed as a polynomial, is the reminder of the division of the polynomial corresponding to the protected bits, appropriately shifted ($D(x)x^r$) by some generator polynomial $G(x)$ of degree $r-1$. This is precisely the structure of some systematic cyclic code (Blahut, 1983).

Such codes can also be described as being the multiplication of the (row) data vector D multiplied by some generator matrix, $\underline{G} = [\underline{P}|\underline{I_k}]$. The representation as a generator matrix will be useful in Chapter 9 for the derivation of decoding metrics taking the CRC into account.

Example 7.5 (Simple systematic cyclic code example): Consider the $(n=7, k=3)$ binary cyclic code with generator polynomial $G(x) = x^4 \oplus x^3 \oplus x^2 \oplus 1$, and encode 101, represented by the polynomial $D(x) = 1 \oplus x^2$.

- Step 1: $x^r D(x) = x^4(x^2 \oplus 1) = x^6 \oplus x^4$.
- Step 2: $x^6 \oplus x^4 = (x^2 \oplus x \oplus 1)(x^4 \oplus x^3 \oplus x^2 \oplus 1) \oplus (x \oplus 1)$ so that $R(x) = (x \oplus 1)$.
- Step 3: The resulting code word is thus $x^6 \oplus x^4 \oplus (x \oplus 1) = 1 \oplus x \oplus x^4 \oplus x^6$ and the transmitted code word is 1100101.

An alternative representation is obtained by the generator matrix G, by selecting as rows the code words associated with the messages 100, 010, and 001 (due to the linearity of the code). This yields

$$\underline{G} = \left[\begin{array}{cccc|ccc} 1 & 0 & 1 & 1 & 1 & 0 & 0 \\ 1 & 1 & 1 & 0 & 0 & 1 & 0 \\ 0 & 1 & 1 & 1 & 0 & 0 & 1 \end{array}\right],$$

and it is easily checked that the transmitted code word is obtained by adding (modulo 2) the first and last row.

Standards have been defined for 8-, 16-, and 32-bit generators, and the 32-bit standard that has been adopted in a number of link-level IEEE protocols uses the following generator:

$$100000100110000010001110110110111.$$

Each of the CRC standards can detect error bursts smaller than or equal to r bits. Moreover, under appropriate assumptions, a burst of length greater than $r+1$ bits is detected with probability $1-0.5^r$, and each of the CRC standards can detect any odd number of bit errors. A deeper development of this topic is out of scope of this book, but more explanations can be found in the books by Blahut (1983) and Schwartz (1980).

7.2.3 Possible Extensions to Other Standards (WiMAX)

The above description is strongly linked to the Wi-Fi standard, and one may wonder to what extent the mentioned properties still hold in other contexts. The redundancy properties of the MAC and PHY layer are outlined below in comparison with IEEE 802.11 (For a more detailed presentation, see the appendices X, Y and Z at website: http://www.elsevierdirect.com/companions/9780123744494).

Implicit Redundancy

- Property 5 is also found in the WiMAX standard at MAC layer. In contrast with the Duration field in MAC header of Wi-Fi, the length in bytes of the MAC packet data unit (PDU) is explicitly provided as a *Length (LEN)* field (11 bits long) in its own header. The length includes the MAC header and the CRC, if present.
- Some redundancy properties are lost: the transmission to subscriber stations (SSs) in WiMAX is connection based (All the SSs are connected to a single base station); therefore, in the downlink situation, only some *Connection*

Identifier (CID) field is required to identify the destination of the packet. This CID cannot be treated as a redundancy as it is required to identify the SS for which the MAC packet is meant, and this is in contrast to MAC address fields (Terminal Address, Access Point Address, Router Address) of WiFi.
- Each SS can read all MAC packets in the *burst* [burst decoded by SS is specified by burst profile and location information present in downlink frame prefix (DLFP)] and CID field helps the SS to identify MAC packets meant for it. Therefore, even if the CID field cannot be considered as constant, Property 6 holds for this field as it does not take all possible values.
- Several other fields of the MAC packet do not change from packet to packet; thus, Property 2 holds. One such example is the Header Type (HT) field, which is set to 0 for downlink MAC packets. Similar is the case for *Encryption Key Sequence* (EKS) and *Reserved* (Rsv) fields.

Explicit redundancy: The various CRCs in IEEE 802.16

Several CRCs or checksums are found in both the MAC and PHY layers of WiMAX. The important characteristics for each of them are as follows.

At MAC layer:

- Header Check Sequence (HCS) on the header (1 byte)
- CRC (optional) on both header and data (4 bytes)

At PHY layer:

- HCS (1 byte) is applied on the DLFP of the Downlink PHY frame.

7.3 GENERAL PROPERTIES

7.3.1 Lower Layers (PHY and MAC)

Beyond the details obtained by scrutinizing the standards, these two examples are clear illustrations that some mechanisms are very general since it is likely that some quantities will have to be transmitted along with the data to secure the reception. Clearly, some information about the packet lengths at various levels is such a quantity. The information may not be fully accurate (may be in bytes, which requires some zero padding), but it is still usable.

CRCs and checksums are also found in almost every standards since they allow in the classical processing to reject (and retransmit) packets that could cause dramatic performance loss.

7.3.2 Higher Layers

It should also be clear that the involved mechanisms apply at all network layers, including the RTP/UDP/IP ones, which are common to all these standards. Any redundancy analysis and use of these layers will be valid for all current wireless communication standards.

7.4 CONCLUSION

Network layers are usually discarded when studying the robust transmission of multimedia, which comes in contrast with the classical way of working since these layers play an important role in the reliable delivery of the multimedia stream.

Therefore, we described the role of each protocol layer and gave more details on the structure of the Wi-Fi (IEEE 802.11) to provide examples of the various types of redundancy that can be found in the headers of these layers. Special emphasis was given on the error-detection capabilities of the checksums and CRCs that are also found in many layers.

A short section was also devoted to the description of another standard (WiMAX: IEEE 802.16), demonstrating that even if the details were different, similar redundancy properties held.

Note also that the redundancy properties of headers in the network layers are well known. In fact, it may happen that a large percentage of the bitstream that would circulate on the channel is due to the packet structuration, and not to the actual payload. Therefore, Robust Header Compression (RoHC) has been proposed at several places and is even available in many standards. In this case, the situation is similar to that of a payload: the actual data to be transmitted (here the headers) has some properties (syntax, semantics) that can be used to obtain robust transmission. Moreover, it seems that in the actual state of the art, such techniques are reserved for higher layers and are not (yet) applied to PHY and MAC layers.

Chapter 8

Joint Protocol-Channel Decoding

As seen in Chapters 4–6, joint source-channel decoding (JSCD) techniques can be used to correct the content of damaged packets using robust source decoders at application (APL) layer. Several sources of redundancy have been identified in Chapter 4: constraints in the syntax of variable-length codes, redundancy due to the semantic properties of the source, or redundancy due to the packetization of compressed data. Altogether, the various redundancies can attain an unexpected amount. Furthermore, redundancy introduced by channel codes at physical (PHY) layer can also be used in combination with residual redundancy to build iterative decoders, as will be introduced in Chapter 9. As already seen, these joint decoding schemes provide improved performance when compared to classical schemes and could be of great use in many applications. They are, however, not compliant with the standard protocol stacks in several ways: (1) they require exchange of soft information (e.g., a posteriori probabilities or likelihood ratios) between the channel decoder at PHY layer and the robust source decoder at APL layer, (2) they are not compatible with the use of acknowledgment procedures: in a JSCD setting, a packet received in error needs not be retransmitted unless the robust receiver cannot recover the error, (3) the headers of packets at a given layer must absolutely be available without error since they contain information necessary for driving the layer in question (at the receiver).

Problem (1) above can be circumvented in some circumstances: a mobile receiver contains all the layers and can choose to forward soft values between layers. In what follows, we assume that it is the case. The second aspect is addressed in Chapter 11. The main compatibility problem seems to be the third one: standard protocol stacks do not allow damaged packets to reach the APL layer, the main reason being that the errors may impact some essential information contained in the headers, which is necessary even for the robust APL decoders.

This chapter proposes tools allowing to receive the various headers with an inherent robustness (even more than the robustness brought by JSCD to the payload) by using techniques widely used in JSCD and applied here to the whole protocol layers. Using joint protocol-channel decoding (JPCD), more headers are, thus, correctly interpreted at each layer, increasing the number of packets reaching the APL layer. We show that robustness of the header is much higher than that of the corresponding payload, which is a prerequisite for implementing a fully permeable protocol layer mechanism (Jenkac et al., 2005).

Two aspects of JPCD are considered in this chapter, aiming at solving two problems that are usually overcome when dealing with JSCD: packet header error recovery and burst segmentation. The first concerns the recovery of headers at various protocol layers, which is a necessity for the protocols to work properly. The second is about robust segmentation of packets aggregated in large bursts, which aims at estimating the location and the content of the headers of the packets forming the burst. The chapter starts with the introduction of an improved permeable layer mechanism in Section 8.1. A maximum a posteriori estimation (MAP) of packet headers is described in Section 8.2, with application to the estimation of the headers at PHY and Media Access Control (MAC) layers of Wi-Fi 802.11. Section 8.3 presents some tools to perform a robust segmentation of packets aggregated within bursts of several packets, with application to the segmentation of MAC packets aggregated in WiMAX bursts. These two JPCD techniques rely on the capability of exploiting the error-correcting capabilities of cyclic redundancy checks (CRCs) and checksums, as detailed in Section 8.4. Some open questions are presented in Section 8.5.

8.1 PERMEABLE LAYER MECHANISM

Packetized multimedia transmission is usually based on an RTP/UDP/IP protocol stack (Kurose and Ross, 2005). Figure 7.15, page 207, illustrates an example of segmentation and encapsulation mechanisms implemented at each protocol layer in case of a multimedia packet transmission with the Wi-Fi 802.11 (IEEE, 1999) standard. Error-detection mechanisms implemented at each layer, presented in Chapter 7, are briefly summarized below.

At PHY layer, a CRC protects the header fields (the preamble and the payload are not protected). Received packets with damaged headers are discarded. At MAC layer, a CRC protects the corresponding header and payload. When an error occurs, the packet is retransmitted. At IPv4 layer, the header fields are protected by a checksum. Received packets with damaged headers are discarded. At UDP layer, a checksum protects the header and the payload. When an error occurs, the packet is discarded. We assume here that packet fragmentation only occurs at MAC layer, which is a reasonable assumption for a wireless transmission.

The error-detection mechanisms provided by CRCs and checksums, combined with the retransmission mechanism at MAC layer, allow the APL layer to receive only error-free packets. The price to be paid is a reduced throughput due to MAC level retransmissions which increase when the channel conditions worsen or frequent use of error concealment when errors are detected at IPv4 or UDP layers (generally due to time-out constraints: the limit on the number of retransmissions at MAC level has been reached).

The JSCD methods presented in Chapters 5 and 6 allow many errors to be corrected at APL layer based on soft information provided by lower protocol layers. The recently introduced UDP-Lite (Larzon et al., 2004) mechanism, combined with lower permeable protocol layers (Jenkac et al., 2005), allow damaged APL packets to reach the APL layer. Martini et al. (2007) and Panza et al. (2005), propose an efficient video transmission scheme using this type of protocol to achieve a wireless network transparency. With UDP-Lite, a checksum protects a limited number of bytes (generally including the UDP-Lite, RTP, and APL header fields). Thus, packets with erroneous headers are still discarded. Considering the order of magnitude of the length of the packets and that of the various headers in actual wireless communications when tuned for difficult situations (Van der Schaar and Shankar, 2005), this may happen more than expected. The bottleneck of such permeable transmission schemes is the fact that packets are discarded due to erroneous headers. This section aims at reducing this problem.

The proposed robust header recovery technique detailed in Section 8.2 involves two main ideas. First, intralayer and interlayer redundancies are present in the protocol stack, see Section 7.2, page 206. This redundancy has been exploited in the *RObust Header Compression* (ROHC) mechanism (Bormann et al., 2001), at upper protocol layers, by replacing the headers introduced by the RTP, UDP, and IP layers by a compressed version. Here, redundancy is used to build some a priori information on the erroneous headers, improving their estimation. Second, CRCs and checksums are used as error-correcting codes, as proposed by Marin et al. (2007) and Nieto and Furman (2007).

Figure 8.1 illustrates the principle of the proposed decoding technique to build a permeable layer L. Assume that soft information associated to the payload of the τth packet has been transmitted by layer $L-1$ to layer L. The payload of the τth packet at layer $L-1$ contains the header, the payload, and the CRC related to layer L. At layer L, the header recovery block may combine the soft information provided by layer $L-1$, the properties of the CRC (or the checksum), and the a priori information obtained by additional sources of redundancy (intralayer and interlayer redundancies) to recover the header of layer L. The processing details of the header recovery are discussed in Section 8.2. Intralayer and interlayer redundancies are determined by a careful examination of the network protocols and is mainly due to the correlations between the various headers. It may take

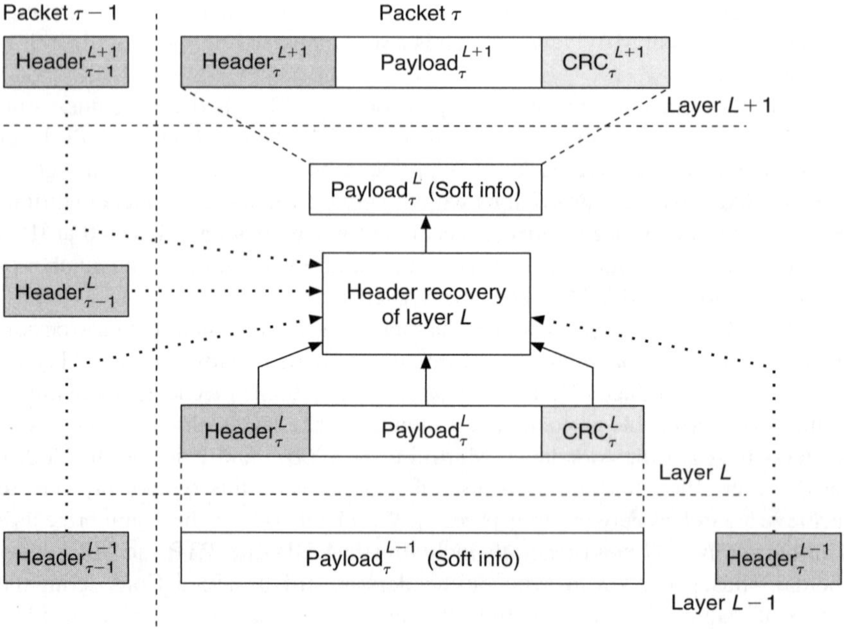

FIGURE 8.1 Proposed permeable layer mechanism.

the form of information issued by previous packets already available, may be at other layers. The a priori information exchange due to this type of redundancy is represented in Figure 8.1 by dotted lines. Finally, layer L uses the decoded header to feed the soft information associated to the payload from layer L to layer $L+1$. Repeating this operation in other protocol layers makes it possible to get a complete permeable protocol stack.

The various examples provided in this chapter consider the PHY and MAC layers of Wi-Fi as generic examples. In fact, the proposed permeable layer mechanism may be applied to any protocol layer. Section 8.2 focuses on the robust recovery of a single packet. These results are extended to the recovery of the header of several packets aggregated in a burst in Section 8.3.

8.2 MAP ESTIMATOR FOR ROBUST HEADER RECOVERY

8.2.1 Notations

As a general situation, at a given layer L, the τth incoming packet may include three items: a header, a payload, and a CRC. Information protected by the CRC $\mathbf{c}_{\tau}^{L}$ may have various properties, as far as the corresponding redundancy is concerned. All vectors here are row vectors.

- The constant fields, represented by vector $\mathbf{k}_\tau^L$, are assumed to be known.
- The predictable fields are embedded in the vector $\mathbf{p}_\tau^L$. In contrast with the known fields, the predictable fields are estimated by exploiting the intralayer and interlayer redundancies represented by $\mathcal{R}_\tau^L$, which will be defined formally in what follows. They are predicted from information contained in the previously received packets. The predictable fields are assumed to be entirely determined if the previous packets have been correctly received.
- The important unknown fields are collected in the vector $\mathbf{u}_\tau^L$. These parameters are either completely unknown or limited to a configuration set $\Omega_u^L(\mathbf{k}_\tau^L, \mathbf{p}_\tau^L, \mathcal{R}_\tau^L)$ the content of which is determined by the values of $\mathbf{k}_\tau^L$, $\mathbf{p}_\tau^L$, and $\mathcal{R}_\tau^L$. This set contains the actual information on the data that the receiver must estimate.
- Finally, the vector $\mathbf{o}_\tau^L$ contains the other fields covered by the CRC. This last part contains unknown data, which are not required for the processing of the packet at layer L, but may be important at layer $L+1$.

$\mathcal{R}_\tau^L$ contains all the header information of the $\tau-1$th packet (at layers $L-1$, L, and $L+1$) and that of the τth packet at layer $L-1$

$$\mathcal{R}_\tau^L = \Big\{\mathbf{k}_{\tau-1}^{L-1}, \mathbf{k}_{\tau-1}^{L}, \mathbf{k}_{\tau-1}^{L+1}, \mathbf{k}_{\tau}^{L-1}, \mathbf{p}_{\tau-1}^{L-1}, \mathbf{p}_{\tau-1}^{L}, \\ \mathbf{p}_{\tau-1}^{L+1}, \mathbf{p}_{\tau}^{L-1}, \mathbf{u}_{\tau-1}^{L-1}, \mathbf{u}_{\tau-1}^{L}, \mathbf{u}_{\tau-1}^{L+1}, \mathbf{u}_{\tau}^{L-1}\Big\}.$$

In addition, data not covered by the CRC at layer L are denoted by $\mathbf{x}_\tau^L$.

All bits protected by the CRC are collected in the vector $\mathbf{r}_\tau^L = (\mathbf{k}_\tau^L, \mathbf{p}_\tau^L, \mathbf{u}_\tau^L, \mathbf{o}_\tau^L)$, which contains the above-defined fields. Note that the order of the bits in $\mathbf{r}_\tau^L$ does not correspond to the order in which data are actually transmitted in the τth packet, but we use this notation for mathematical convenience. The CRC $\mathbf{c}_\tau^L$ associated to $\mathbf{r}_\tau^L$ is evaluated as $\mathbf{c}_\tau^L = \mathcal{F}^L(\mathbf{r}_\tau^L)$, where $\mathcal{F}^L$ is a generic encoding function.

When there is no ambiguity, the indices τ and L are omitted in what follows.

The evaluation of $\mathbf{c}$ depends on a generator polynomial $g(x) = \sum_{i=0}^{\ell(\mathbf{c})} g_i x^i$ characterizing the CRC (Blahut, 1983). A systematic generator matrix $\mathbf{G} = [\mathbf{I}, \mathbf{\Pi}]$ can be associated to $g(x)$, taking into account the reordering of the bits in $\mathbf{r}$. Using $\mathbf{G}$, $\mathbf{c}$ can be obtained iteratively as follows

$$\begin{cases} \mathbf{c}^0 = \mathbf{0}, \\ \mathbf{c}^{j+1} = \mathcal{F}(\mathbf{r}^{j+1}) = \mathbf{c}^j \oplus \big(r_{j+1} \cdot \boldsymbol{\pi}(j+1)\big). \end{cases} \tag{8.1}$$

In (8.1), $\mathbf{r}^j = (r_1 \ldots r_j, 0 \ldots 0)$, $\oplus$ is the XOR operator, and $\boldsymbol{\pi}(j)$ represents the parity vector associated to bit r_j, which corresponds to the jth line of $\mathbf{\Pi}$. After $\ell(\mathbf{r})$ iterations, $\mathbf{c}^{\ell(\mathbf{r})} = \mathcal{F}(\mathbf{r}) = \mathbf{c}$.

Assume that the data have been transmitted over a memoryless channel and that soft values are forwarded inside the receiver from each layer to the next one. Noisy data and CRC coming from layer $L-1$ are denoted as

$\mathbf{y} = [\mathbf{y}_k, \mathbf{y}_p, \mathbf{y}_u, \mathbf{y}_o, \mathbf{y}_c]$, which includes observations (at PHY layer) or estimations (at other layers) of **k**, **p**, **u**, **o**, and **c**.

8.2.2 MAP Header Estimator

Since **k** and **p** are known or may be exactly predicted from the already received data, only **u** remains to be estimated. A MAP estimator

$$\widehat{\mathbf{u}}_{\text{MAP}} = \arg\max_{\mathbf{u}} P(\mathbf{u}|\mathbf{k}, \mathbf{p}, R, \mathbf{y}_u, \mathbf{y}_o, \mathbf{y}_c) \tag{8.2}$$

is thus developed, taking into account the observations **y**, the knowledge of **k**, **p**, and $\mathcal{R}$, as well as the CRC properties. After some derivations, one obtains

$$\widehat{\mathbf{u}}_{\text{MAP}} = \arg\max_{\mathbf{u}} P(\mathbf{u}, \mathbf{y}_u, \mathbf{y}_o, \mathbf{y}_c|\mathbf{k}, \mathbf{p}, \mathcal{R}). \tag{8.3}$$

Since the channel is memoryless and assuming that **o** is independent of R, one gets

$$P(\mathbf{u}, \mathbf{o}, \mathbf{y}_u, \mathbf{y}_o, \mathbf{y}_c|\mathbf{k}, \mathbf{p}, \mathcal{R}) = P(\mathbf{u}|\mathbf{k}, \mathbf{p}, \mathcal{R})P(\mathbf{y}_u|\mathbf{u})P(\mathbf{o}, \mathbf{y}_o, \mathbf{y}_c|\mathbf{k}, \mathbf{p}, \mathbf{u}). \tag{8.4}$$

For the sake of generality, assume that **u** does not necessarily take all the $2^{\ell(\mathbf{u})}$ values and that a study of the protocol allows to define $\Omega_u = \Omega_u(\mathbf{k}, \mathbf{p}, \mathcal{R})$, the set of possible values of **u**. Further, assume that these values are equally likely. Thus, one may write

$$P(\mathbf{u}|\mathbf{k}, \mathbf{p}, \mathcal{R}) = P(\mathbf{u}|\Omega_u) = 1/|\Omega_u|, \tag{8.5}$$

where $|\Omega_u|$ denotes the cardinal number of Ω_u.

Marginalizing (8.4) over the $2^{\ell(\mathbf{o})}$ combinations of **o**, one obtains

$$P(\mathbf{u}, \mathbf{y}_u, \mathbf{y}_o, \mathbf{y}_c|\mathbf{k}, \mathbf{p}, \mathcal{R}) = P(\mathbf{u}|\Omega_u)P(\mathbf{y}_u|\mathbf{u}) \sum_{\mathbf{o}} P(\mathbf{o}, \mathbf{y}_o, \mathbf{y}_c|\mathbf{k}, \mathbf{p}, \mathbf{u}), \tag{8.6}$$

where $\sum_{\mathbf{o}} P(\mathbf{o}, \mathbf{y}_o, \mathbf{y}_c|\mathbf{k}, \mathbf{p}, \mathbf{u})$ obviously involves the properties of the CRC. Finally, substituting (8.6) in (8.3), the MAP estimator is expressed as

$$\widehat{\mathbf{u}}_{\text{MAP}} = \arg\max_{\mathbf{u}\in\Omega_u} P(\mathbf{y}_u|\mathbf{u})\Psi(\mathbf{k}, \mathbf{p}, \mathbf{u}, \mathbf{y}_o, \mathbf{y}_c), \tag{8.7}$$

with

$$\Psi(\mathbf{k}, \mathbf{p}, \mathbf{u}, \mathbf{y}_o, \mathbf{y}_c) = \sum_{\mathbf{o}} P(\mathbf{o}, \mathbf{y}_o, \mathbf{y}_c|\mathbf{k}, \mathbf{p}, \mathbf{u}). \tag{8.8}$$

$\Psi(\mathbf{k}, \mathbf{p}, \mathbf{u}, \mathbf{y}_o, \mathbf{y}_c)$ is very important, since this term allows to take into account the presence of the CRC protecting the important parts of the header (**k**, **p**, and **u**) and parts of the header that are less important and parts of the payload (**o**). All possible values of **o** are taken into account in (8.8), through the marginalization with respect to **o**.

Being very general, the above equations encompass many different situations. For the sake of clarity, the following section details the evaluation of $\widehat{\mathbf{u}}_{\text{MAP}}$ in several practical situations.

8.2.3 Practical Evaluation of the MAP Estimator

The Set $\mathbf{o}$ *is Empty*

There are many circumstances in which all bits covered by the CRC belong only to the sets $\mathbf{k}$, $\mathbf{p}$, or $\mathbf{u}$. In these cases, there is no $\mathbf{o}$ and (8.3) simplifies to

$$\widehat{\mathbf{u}}_{\mathrm{MAP}} = \arg\max_{\mathbf{u}\in\Omega_u} P(\mathbf{y}_u|\mathbf{u})P\big(\mathbf{y}_c|\mathcal{F}([\mathbf{k},\mathbf{p},\mathbf{u}])\big), \tag{8.9}$$

where $\mathcal{F}([\mathbf{k},\mathbf{p},\mathbf{u}])$ is directly evaluated by (8.1). Hence, an elementary CRC computation replaces the sum over all the possible values of $\mathbf{o}$, and the computational complexity is heavily reduced.

The Set $\mathbf{o}$ *is Not Empty*

When $\mathbf{o}$ is present, the sum in (8.6) becomes

$$\Psi(\mathbf{k},\mathbf{p},\mathbf{u},\mathbf{y}_o,\mathbf{y}_c) = \sum_{\mathbf{o}} P(\mathbf{o})P(\mathbf{y}_o|\mathbf{o})P\big(\mathbf{y}_c|\mathcal{F}([\mathbf{k},\mathbf{p},\mathbf{u},\mathbf{o}])\big). \tag{8.10}$$

Evaluating (8.10) requires the computation of the sum of probabilities related to the $2^{\ell(\mathbf{o})}$ combinations of $\mathbf{o}$ and to their corresponding CRCs. A direct evaluation has obviously a complexity exponential in $\ell(\mathbf{o})$. Section 8.4 proposes two methods with reduced complexity: the first one is an exact computation while the second one provides an approximate solution.

8.2.4 Application to the 802.11 Standard

We focus on the downlink multimedia transmission over the 802.11 radio interface (IEEE, 1999). First, the format of packets at PHY and MAC layers are briefly recalled, see Chapter 7 and Appendix A for more details. Intralayer and interlayer redundancies, evidenced in Section 7.2, are then described before introducing the resulting processing details for the enhanced permeable layer mechanism.

DSSS PHY Layer Description

At PHY layer, the 802.11 standard provides 1 or 2 Mbps transmission rate in the 2.4 GHz band using either frequency-hopping spread spectrum (FHSS) or direct-sequence spread spectrum (DSSS). In DSSS, an 11-chip Barker code sequence is used for spreading the 1 Mbps bitstream. The coded flow thus represents an 11 MHz baseband signal. A DBPSK or DQPSK modulation is applied depending on the required bitrate, see Section 7.1.5.

The DSSS PHY packet format is illustrated in Figure 8.2, see also Appendix A. The preamble and the header are transmitted by using the 1 Mbps DBPSK modulation while the payload is modulated either in 1 Mbps DBPSK or 2 Mbps DQPSK. In such PHY packets, the SYNC and SFD fields consist of 144 known bits, which are not protected by the CRC. These fields are used to estimate the variance of the channel noise, see Section 8.2.4.

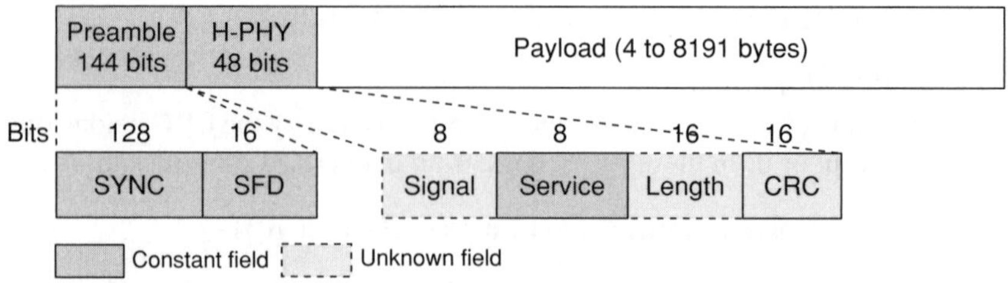

FIGURE 8.2 PHY packet format in 802.11 standard.

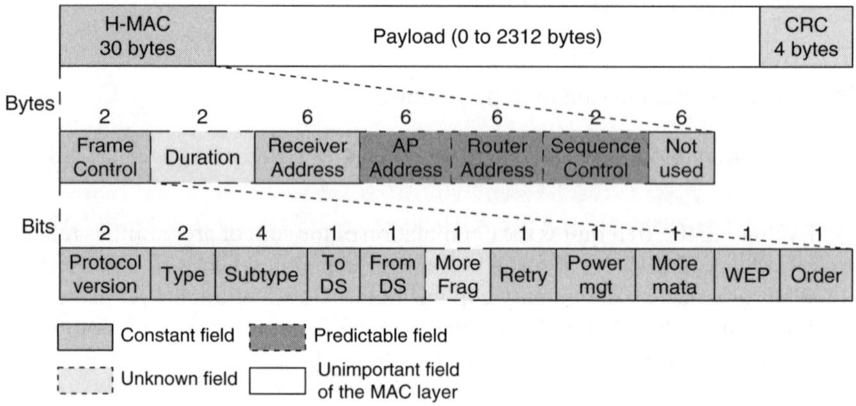

FIGURE 8.3 MAC packet format in 802.11 standard.

The CCITT CRC-16 $\mathbf{c}^{\mathrm{PHY}}$ of 2 bytes protects the *Signal*, *Service*, and *Length* fields; its associated encoding function is denoted by $\mathcal{F}^{\mathrm{PHY}}$. The payload, assumed to contain only one MAC packet, is not protected at this layer. *Service* is reserved for future recommendation. It is set to 00_{16}, and included in $\mathbf{k}^{\mathrm{PHY}}$, according to the notations of Section 8.2. *Signal* indicates the payload modulation and is equal to $0A_{16}$ or 14_{16} for 1 or 2 Mbps bitrate, respectively. *Length* indicates on 2 bytes the number of microseconds required to transmit the payload. It depends on both the bitrate and the payload size. It ranges from 16 to $2^{16}-1$. *Signal* and *Length* form, thus, $\mathbf{u}^{\mathrm{PHY}}$. At this layer, $\mathbf{p}^{\mathrm{PHY}} = \mathbf{o}^{\mathrm{PHY}} = \emptyset$ and $\mathbf{x}^{\mathrm{PHY}}$ contains the $\ell(\mathbf{x}^{\mathrm{PHY}})$ bits of payload.

MAC Layer Description

The MAC packet format is depicted in Figure 8.3. The content of each field is detailed in Appendix A. In this packet, the CRC $\mathbf{c}^{\mathrm{MAC}}$ of 4 bytes protects both the header fields and the payload; its encoding function is $\mathcal{F}^{\mathrm{MAC}}$.

Considering a nonencrypted downlink transmission of ordered MAC data packets with deactivated retransmission and power-save mode, the 2-byte *Frame Control* field except the *More Frag* flag are assumed to be known. The 6-byte

Receiver Address field contains the MAC address of the receiver and is thus known. The last field of the MAC header is reserved for local wireless networks and is composed of 6 bytes of zeros in this study. Using the notations of Section 8.2, all previously mentioned fields are embedded in $\mathbf{k}^{\mathrm{MAC}}$.

The 6-byte *AP Address* field contains the MAC address of the access point AP. This address is transmitted during the medium reservation procedure Request To Send-Clear To Send (RTS-CTS) and can be totally deduced by the receiver. The 6-byte *Router Address* field corresponds to the MAC address of the router. Assuming that the AP is connected to a single router and that the router address has been already received in other information packets, *Router Address* can also be predicted by the receiver. The 2-byte *Sequence Control* field contains two parameters: a sequence number and a fragment number. The sequence number represents the value of the current IP packet counter. The fragment number indicates the value of the current MAC data packet counter. In this study, packets are transmitted in order and these parameters are easily determined: the sequence number is incremented by one for each RTS-CTS and the fragment number is incremented by one for each received MAC data packet. *Sequence Control* can be estimated by the receiver. All these predictable fields are represented by $\mathbf{p}^{\mathrm{MAC}}$.

The *More Frag* flag specifies if the current MAC data packet is the last fragment composing an IP packet. The 2 bytes of *Duration* indicate the number of microseconds required to transmit the next MAC fragment and some control packets. Its value depends on the current modulation and the size of the coming MAC data packet. These two fields are embedded in $\mathbf{u}^{\mathrm{MAC}}$. Finally, the payload contains the data to be transmitted and its size is between 0 and 2312 bytes. It is represented by $\mathbf{o}^{\mathrm{MAC}}$.

Identifying Intralayer and Interlayer Correlations

To evidence these correlations, the transactions at MAC layer, assumed to use the DCF protocol (see Section 7.1.5, page 203), have to be described.

In the 802.11 standard, transmission of each IP packet is initialized by a medium reservation procedure at MAC layer consisting of an RTS-CTS exchange. MAC fragments composing the IP packet are then transmitted to the receiver, which acknowledges them (ACK). In this work, control packets such as RTS, CTS, and ACK are assumed to be correctly received. This assumption is reasonable since these packets are small and DBPSK-modulated. Only errors in data packets (or fragments) will be considered. A *Short Interframe Space* (SIFS) of 10 μs separates two packets successively transmitted over the channel. A *Duration* field is included in each packet and its value indicates the number of microseconds required to transmit the next fragment and some specific packets (CTS and ACK). *Duration* allows to adjust the *Network Allocation Vector* (NAV) for the other terminals. The other stations cannot communicate during the NAV period to avoid interferences.

Assume that D_τ^{MAC} and B_τ^{PHY} represent the value of *Duration* and the transmission bitrate (coded in *Signal*) associated to the τth packet transmitted by the AP (either RTS or data packets). Following the MAC layer specifications of 802.11 standard, D_τ^{MAC} is defined as

$$D_\tau^{\mathrm{MAC}} = 3T_{\mathrm{SIFS}} + 3T_{\mathrm{OVH}} + 2\ell_{\mathrm{C\text{-}A}}/B_\tau^{\mathrm{PHY}} + \ell\left(\mathbf{x}_{\tau+1}^{\mathrm{PHY}}\right)/B_\tau^{\mathrm{PHY}}, \tag{8.11}$$

except for the last fragment of an IP packet, i.e., when the value of *More Frag* $M_\tau^{\mathrm{MAC}} = 0$. In this case, one has

$$D_\tau^{\mathrm{MAC}} = T_{\mathrm{SIFS}} + T_{\mathrm{OVH}} + \ell_{\mathrm{C\text{-}A}}/B_\tau^{\mathrm{PHY}}. \tag{8.12}$$

In (8.11) and (8.12), T_{SIFS} denotes the duration of a SIFS and T_{OVH} represents the duration for transmitting at 1 Mbps the PHY overhead (composed of the preamble and the header of constant size). The other terms depend on the current bitrate B_τ^{PHY}. CTS and ACK have the same constant size $\ell_{\mathrm{C\text{-}A}}$, and $\ell_{\mathrm{C\text{-}A}}/B_\tau^{\mathrm{PHY}}$ thus corresponds to the duration for sending one of these packets. Finally, $\ell\left(\mathbf{x}_{\tau+1}^{\mathrm{PHY}}\right)/B_\tau^{\mathrm{PHY}}$ refers to the transmission duration of the next PHY payload of $\ell\left(\mathbf{x}_{\tau+1}^{\mathrm{PHY}}\right)$ bits.

PHY Header Recovery

For the τth packet at PHY layer, the observations associated to $\mathbf{k}_\tau^{\mathrm{PHY}}$, $\mathbf{u}_\tau^{\mathrm{PHY}}$, and $\mathbf{c}_\tau^{\mathrm{PHY}}$ defined in Section 8.2.4 are collected in $\mathbf{y}_\tau^{\mathrm{PHY}} = \left[\mathbf{y}_{k,\tau}^{\mathrm{PHY}}, \mathbf{y}_{u,\tau}^{\mathrm{PHY}}, \mathbf{y}_{c,\tau}^{\mathrm{PHY}}\right]$. In addition, $\mathbf{y}_{x,\tau}^{\mathrm{PHY}}$ denotes the observations associated to the $\ell\left(\mathbf{x}_\tau^{\mathrm{PHY}}\right)$ bits of the payload $\mathbf{x}_\tau^{\mathrm{PHY}}$, which is not protected by the CRC.

The number of possible values taken by $\mathbf{u}^{\mathrm{PHY}}$ is significantly reduced when exploiting the *Duration* field contained in the previously received MAC packet (either an RTS or a data packet). Using $B_{\tau-1}^{\mathrm{PHY}}$ and $D_{\tau-1}^{\mathrm{MAC}}$, one may deduce $\ell(\mathbf{x}_\tau^{\mathrm{PHY}})$ from (8.11) as

$$\ell\left(\mathbf{x}_\tau^{\mathrm{PHY}}\right) = \left(D_{\tau-1}^{\mathrm{MAC}} - 3T_{\mathrm{SIFS}} - 3T_{\mathrm{OVH}} - 2\ell_{\mathrm{C\text{-}A}}/B_{\tau-1}^{\mathrm{PHY}}\right) B_{\tau-1}^{\mathrm{PHY}}. \tag{8.13}$$

Then, the duration L_τ^{PHY} coded in the *Length* field of the current PHY packet is computed as

$$L_\tau^{\mathrm{PHY}} = \ell\left(\mathbf{x}_\tau^{\mathrm{PHY}}\right)/B_\tau^{\mathrm{PHY}}. \tag{8.14}$$

In (8.13), $\ell\left(\mathbf{x}_\tau^{\mathrm{PHY}}\right)$ is totally determined assuming correct estimation of the header of the previous packet. Then, according to (8.14), L_τ^{PHY} may only take two values depending on B_τ^{PHY}, which are stored in $\Omega_{u,\tau}^{\mathrm{PHY}}$. Integrating these properties in (8.9), one obtains

$$\widehat{\mathbf{u}}_\tau^{\mathrm{PHY}} = \arg\max_{\mathbf{u}_\tau^{\mathrm{PHY}} \in \Omega_{u,\tau}^{\mathrm{PHY}}} P\left(\mathbf{y}_{u,\tau}^{\mathrm{PHY}} | \mathbf{u}_\tau^{\mathrm{PHY}}\right) P\left(\mathbf{y}_{c,\tau}^{\mathrm{PHY}} | \mathbf{c}_\tau^{\mathrm{PHY}}\right), \tag{8.15}$$

with $\mathbf{c}_\tau^{\mathrm{PHY}} = \mathcal{F}^{\mathrm{PHY}}\left(\left[\mathbf{k}_\tau^{\mathrm{PHY}}, \mathbf{u}_\tau^{\mathrm{PHY}}\right]\right)$.

MAC Header Recovery

The PHY layer provides $\mathbf{y}_{\tau}^{\text{MAC}} = \mathbf{y}_{x,\tau}^{\text{PHY}} = \left[\mathbf{y}_{k,\tau}^{\text{MAC}}, \mathbf{y}_{p,\tau}^{\text{MAC}}, \mathbf{y}_{u,\tau}^{\text{MAC}}, \mathbf{y}_{o,\tau}^{\text{MAC}}, \mathbf{y}_{c,\tau}^{\text{MAC}}\right]$ at the input of MAC layer.[1] It contains the observations associated to $\mathbf{k}_{\tau}^{\text{MAC}}$, $\mathbf{p}_{\tau}^{\text{MAC}}$, $\mathbf{u}_{\tau}^{\text{MAC}}$, $\mathbf{o}_{\tau}^{\text{MAC}}$, and $\mathbf{c}_{\tau}^{\text{MAC}}$ specified in Section 8.2.4.

The number of possible combinations for $\mathbf{u}_{\tau}^{\text{MAC}}$ may be significantly reduced when exploiting (8.11) and (8.12). Note that D_{τ}^{MAC} is fully determined when $M_{\tau}^{\text{MAC}} = 0$. When $M_{\tau}^{\text{MAC}} = 1$, the value of *Duration* depends on the next PHY payload size. The number of combinations is associated to the range of MAC payload size. Considering that the payload contains an entire number of bytes, the possible values of $\ell(\mathbf{x}_{\tau+1}^{\text{PHY}})$ in (8.11) are given by

$$\ell\left(\mathbf{x}_{\tau+1}^{\text{PHY}}\right) = \ell_{\text{HDR}} + 8i, \tag{8.16}$$

where $i = 1, 2 \ldots 2312$. In (8.16), ℓ_{HDR} specifies the known size of the header in a MAC data packet. Then, using (8.11), (8.12), and (8.16), one may show that $\mathbf{u}_{\tau}^{\text{MAC}}$ is limited to 2313 combinations, which are inserted in $\Omega_{u,\tau}^{\text{MAC}}$. Combining these properties in (8.7), one obtains

$$\widehat{\mathbf{u}}_{\tau}^{\text{MAC}} = \arg \max_{\mathbf{u}_{\tau}^{\text{MAC}} \in \Omega_{u,\tau}^{\text{MAC}}} P\left(\mathbf{y}_{u,\tau}^{\text{MAC}} | \mathbf{u}_{\tau}^{\text{MAC}}\right) \Psi\left(\mathbf{k}_{\tau}^{\text{MAC}}, \mathbf{p}_{\tau}^{\text{MAC}}, \mathbf{u}_{\tau}^{\text{MAC}}, \mathbf{y}_{o,\tau}^{\text{MAC}}, \mathbf{y}_{c,\tau}^{\text{MAC}}\right), \tag{8.17}$$

where the second term can be computed with methods presented in Sections 8.4.1.

Global Scheme

Figure 8.4 illustrates the improved permeable layer mechanism applied to the PHY and MAC layers at the receiver, emphasizing on the exchange of information between layers and between consecutive packets.

At PHY layer, the header recovery block first exploits the knowledge of the previous decoded fields *Duration* and *Signal* (contained in the previous packet) to construct a reduced set of combination Ω_{u}^{PHY} for $\mathbf{u}_{\tau}^{\text{PHY}}$. It then chooses the best header combination by taking into account the observations of the PHY header and the properties of the CRC included in the header. The recovered header is then used as reference to decode the header of the next PHY packet. At MAC layer, the header recovery block combines the soft information, the knowledge of the decoded field *Signal*, together with the CRC of the MAC layer to find an estimate of the MAC header. The estimated field *Duration* is exploited at the PHY layer to decode the header of the next PHY packet.

---

1. When encryption is activated, the *WEP* flag in the MAC header is set to 1. In addition, $\mathbf{y}_{o,n}^{\text{MAC}}$ and $\mathbf{y}_{c,n}^{\text{MAC}}$ are the observations of the encrypted bits (plainstream XORed with a pseudo-random keystream). Decryption may easily be performed at receiver side by inverting some LLRs in $\mathbf{y}_{o,n}^{\text{MAC}}$ and $\mathbf{y}_{c,n}^{\text{MAC}}$ according to the known keystream.

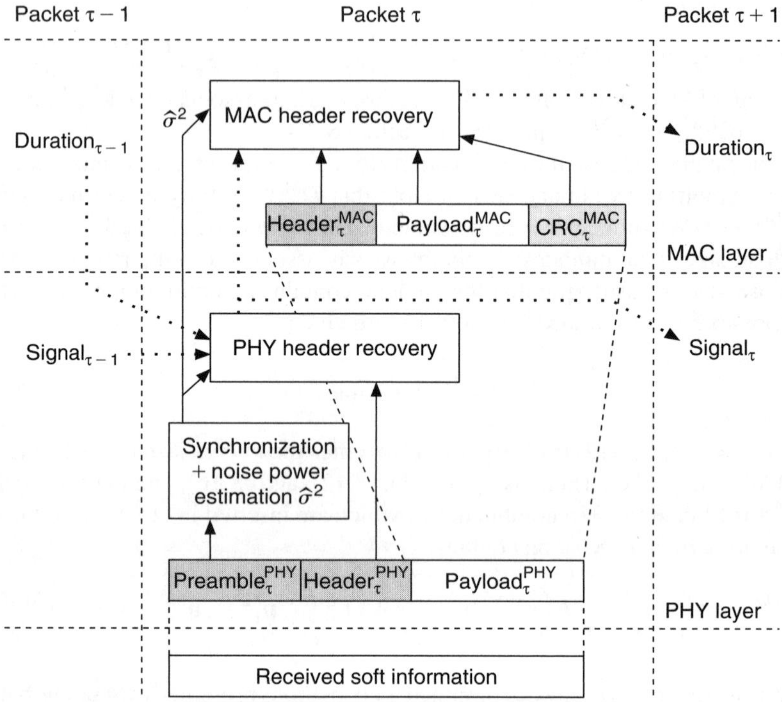

FIGURE 8.4 Proposed scheme for PHY and MAC layers.

In addition, we consider that $\mathbf{y}_{s,\tau}^{\text{PHY}}$ represents the observations of the known preamble $\mathbf{s}^{\text{PHY}}$. As explained previously, the receiver synchronization is performed with $\mathbf{s}^{\text{PHY}}$. We simultaneously estimate σ^2 from $\mathbf{s}^{\text{PHY}}$ and $\mathbf{y}_{s,\tau}^{\text{PHY}}$. This measure is essential for working with soft information, as it allows the evaluation of all the likelihoods. The estimator $\widehat{\sigma}^2$ is given by

$$\widehat{\sigma}^2 = \|\mathbf{y}_{s,\tau}^{\text{PHY}} - \mathbf{s}^{\text{PHY}}\|^2 / \ell\left(\mathbf{s}^{\text{PHY}}\right). \tag{8.18}$$

Computational complexity is minimized by deactivating the robust header recovery processing when the normal CRC check is successful. It should also be deactivated when the quality of the soft information provided by the lower layer is too poor, i.e., when the signal power is lower than a predefined threshold. In such a case, the packet may be retransmitted.

Simulation Results

The improved permeable scheme for 802.11 PHY and MAC layers has been implemented (see Figure 8.4). A transmission device consisting of a transmitter (AP), an AWGN channel, and a receiver has been simulated using a C program. The AP generates PHY and MAC packets following the format defined in

Figures 8.2 and 8.3. The MAC payload consists of a variable amount of randomly generated bytes, which are transmitted using DBPSK modulation. The channel coding relative to the 802.11 DSSS standard is deactivated in these simulations. This operation just consists in spreading the data stream by using an 11-chip Barker code sequence. The consequences are thus limited: when the channel coding is activated, curves are shifted toward low signal-to-noise ratios (SNRs) but relative gains are preserved.

Three types of header recovery methods are considered at each layer of the receiver. The *standard* decoder performs hard decisions on the data at the channel output. The *robust* decoder exploits only the intralayer and interlayer redundancies through a soft-decoding algorithm, neglecting the information provided by the CRC. Finally, the *CRC-robust* decoder combines the intralayer and interlayer redundancies together with the information provided by the CRC through the soft decoding algorithm implementing (8.9) and (8.10). Performance analysis is performed in terms of *header error rate* (HER) versus SNR.

In Figure 8.5, the standard, robust, and CRC-robust PHY decoders are compared under the assumption that the *Duration* field of the previous MAC packet has been correctly received. Obviously, robust decoders outperform the standard one. An HER of less than 10^{-5} is obtained with the robust decoder for an SNR of 4 dB and with the CRC-robust decoder for an SNR of 2 dB. With the standard decoder, an SNR of at least 15 dB is required to get a comparable HER. At PHY

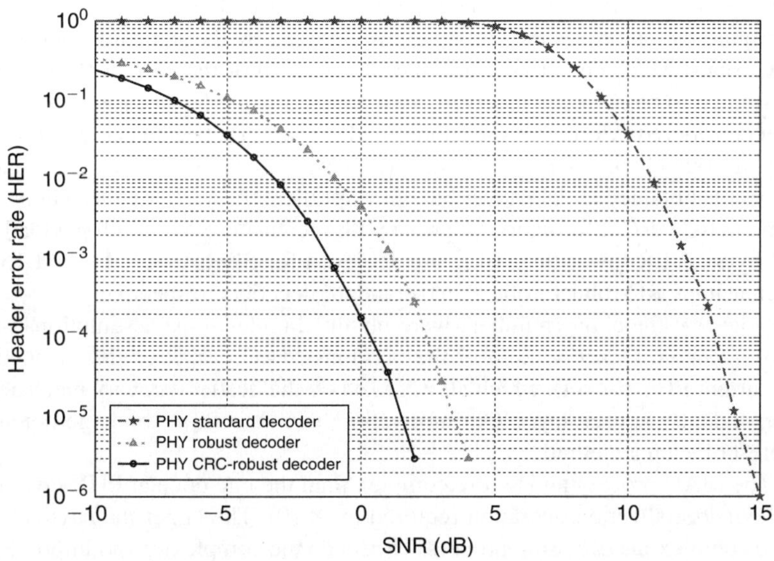

FIGURE 8.5 Header error rate (HER) versus SNR for the standard, robust, and CRC-robust decoders applied to the PHY layer. In this simulation, we have studied the PHY header-correction capacity according to the three types of decoders.

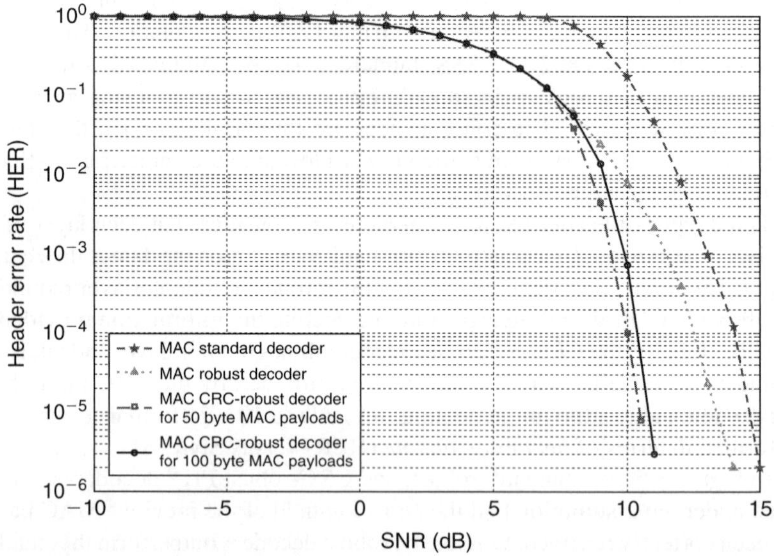

FIGURE 8.6 Header error rate (HER) versus SNR for the standard, robust, and CRC-robust decoders applied to the MAC layer. In this simulation, we have considered two payload sizes: 50 and 100 bytes.

layer, considerable coding gains for a relatively low additional complexity are, thus, observed, since (8.9) is used to perform the decoding.

Figure 8.6 compares the coding gains obtained by the standard, robust, and CRC-robust MAC decoders. Here, the *Bitrate* field of the current PHY packet is assumed to be correctly received. Two payload sizes (50 and 100 bytes) have been considered. Moreover, the suboptimal method has been used, dividing the CRC in four blocks of 1 byte each. The shape of the curves is very similar to the results obtained at PHY layer but with significantly smaller gains. Gains due to the MAC CRC information improve with increasing SNR. With payloads of 100 bytes, HER lower than 10^{-5} are achieved for SNRs of 11, 14, and 15 dB when using CRC-robust, robust, and standard decoders, respectively.

Note that the above numbers were obtained under some assumptions (correctly received *Duration* field of the previous MAC packet or *Bitrate* field of the current PHY packet), which allows to study the header recovery mechanism independently at each layer. Our motivation here is to show the large potential interest of such a method.

The MAC processing is more complex than the one done at PHY layer due to the marginalization operation required in (8.10). The larger the payload, the more complex the decoding process. To reduce the complexity and improve the MAC header recovery performance, the principle of UDP-Lite has been applied at the MAC layer, resulting in a permeable MAC layer (called *MAC-Lite*) where the CRC protects the MAC header field only. In this case, $\mathbf{o}_{\tau}^{\text{MAC-L}} = \emptyset$ and (8.17) becomes

$$\widehat{\mathbf{u}}_{\tau}^{\text{MAC-L}} = \arg \max_{\mathbf{u}_{\tau}^{\text{MAC-L}} \in \Omega_{u,\tau}^{\text{MAC-L}}} P\left(\mathbf{y}_{u,\tau}^{\text{MAC-L}} | \mathbf{u}_{\tau}^{\text{MAC-L}}\right) P\left(\mathbf{y}_{c,\tau}^{\text{MAC-L}} | \mathbf{c}_{\tau}^{\text{MAC-L}}\right), \quad (8.19)$$

where $\mathbf{c}_{\tau}^{\text{MAC-L}} = \mathcal{F}^{\text{MAC-L}}\left(\left[\mathbf{k}_{\tau}^{\text{MAC-L}}, \mathbf{p}_{\tau}^{\text{MAC-L}}, \mathbf{u}_{\tau}^{\text{MAC-L}}\right]\right)$.

Standard, robust, and CRC-robust MAC-Lite decoders are depicted in Figure 8.7. Comparison with Figure 8.6 does not show any difference between MAC and MAC-Lite situations for the standard and robust decoders. This is normal since the information provided by the CRC is not used by these decoders. However, Figure 8.7 demonstrates that the CRC-robust decoder is now significantly more efficient for decoding MAC-Lite headers than for decoding classical MAC headers. HER is lower than 10^{-5} for SNRs larger than 3 dB when exploiting the CRC redundancy, whereas the two other methods need at least 14 dB. Additionally, the CRC-robust decoding is significantly less complex when processing MAC-Lite headers instead of classical MAC headers since (8.19) does not require any marginalization.

Consequently, the combination of the proposed permeable PHY and MAC-Lite layer mechanisms recovers eventually all the PHY and MAC headers from 3 dB SNR onward. The combination of the proposed permeable PHY and MAC layers reaches this result when the SNR is about 11 dB for an increased complexity. This result demonstrates the potential of replacing the classical MAC layer by the proposed MAC-Lite layer if appropriate tools are used at other layers.

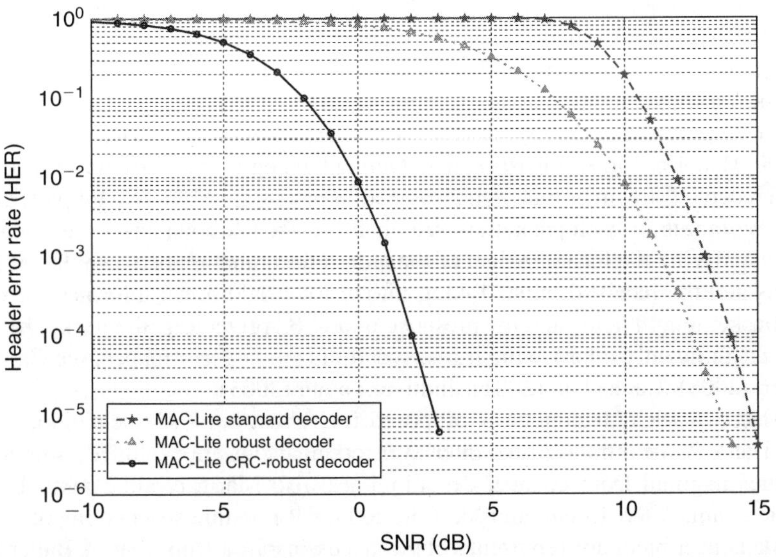

FIGURE 8.7 Header error rate (HER) versus SNR for the standard, robust, and CRC-robust decoders of the MAC-Lite layer. In this case, CRC covers only the MAC header fields.

8.2.5 Summary

The proposed method for recovering headers is based on the various sources of redundancy in the protocol stack. It increases the amount of packets that can be used for robust decoding at APL layer. As a result, the efficiency of the decoding at APL layer is improved in all the cases: (1) when retransmissions are allowed, only packets that were not corrected at APL layer are retransmitted, decreasing the number of retransmitted packets, (2) when higher layer redundancy has been introduced to circumvent the problem (e.g., the so-called MPE-FEC of the MAC layer in DVB-H [ETSI]), our strategy at least allows to reduce the amount of redundancy, and finally, (3) when no retransmission is allowed, it improves the quality of the multimedia content because error concealment is used less frequently.

8.3 ROBUST BURST SEGMENTATION

In many communication systems, at given layers of the protocol stack, small packets are aggregated into larger packets (called burst in what follows), which are then forwarded to lower protocol layers or directly sent over the channel. This is, for example, the case in WiMAX where several small MAC packets may be aggregated to form a MAC burst, which is then forwarded to the PHY layer (IEEE, 2004; Wang et al., 2007). The main advantage of packet aggregation is the reduction of header overhead: aggregating packets improves the payload-to-header ratio and thus the throughput when the channel is clear. The price to be paid is a higher sensitivity of the burst to transmission errors: isolating each aggregated packet may become difficult when the received burst is corrupted by noise. The presence of length fields and synchronization markers may be helpful but not sufficient.

This problem is closely connected to that of *frame synchronization* (Choi, 1990; Haberle, 1969) consisting in detecting the beginning of frames or packets in a received stream of bits. Synchronization markers, i.e., known bit patterns at the beginning of each packet are searched for. Packet boundaries are obtained by identifying the local maxima of the correlation function between the bit patterns and the received burst. Taking into account additional information, such as packet length fields and the presence of a CRC protecting, the packet header may significantly improve the frame synchronization performance, see Cantillo et al. (2008), Ueda et al. (2001), and Ueda et al. (2003).

Here, a JPCD technique is proposed for robust burst segmentation. As in Section 8.2, the intralayer and interlayer redundancies are exploited using techniques inspired from channel decoding, see also Marin et al., 2008; Martini and Chiani, 2009; Chiani and Martini, 2007. All possible successions of MAC packets in a burst are represented by a trellis inspired from that of Bauer and Hagenauer (2000c), and a modified BCJR algorithm (Bahl et al., 1974) is used on this trellis to obtain the packet boundaries. The technique has been applied to the WiMAX MAC layer, but it easily extends to other protocols where packet aggregation is performed.

Section 8.3.2 introduces the problem of robust segmentation of aggregated packets using channel information and redundancy present in the protocol stack. The principle of the robust decoder is provided in Section 8.3.3. WiMAX MAC layer is briefly introduced as an application of the proposed segmentation algorithm in Section 8.3.4. Simulation results are shown for the AWGN channel in Section 8.3.5 before drawing some conclusions.

8.3.1 Structure of a Burst

Consider a burst $\mathbf{q}$ of ℓ_{q} bits formed by $t-1$ *data* packets and an additional *padding* packet containing only padding bits to fill the burst, see Figure 8.8 (the padding packet may be empty). Assume that each of these packets, except the padding packet, contains a header and a payload and follows the same syntax. The header $\mathbf{h}_\tau$ of the τth packet consists of the following:

- A constant field $\mathbf{k}$, corresponding, e.g., to a synchronization marker indicating the beginning of the packet
- A length field $\mathbf{u}_\tau$, corresponding to the size in bits λ_τ of the packet, including the header
- Other fields $\mathbf{o}_\tau$, which are not useful to perform the burst segmentation
- A CRC or check-sum $\mathbf{c}_\tau$, covering only the header

The payload of the τth packet is denoted by $\mathbf{b}_\tau$. The bits of $\mathbf{b}_\tau$ are assumed to be generated by a binary symmetric source. The CRC for all packets is assumed to be obtained from the systematic generator matrix $\mathbf{G}=[\mathbf{I}, \mathbf{\Pi}]$.

Before being put in a burst, the length λ of a packet may be described as the realization of a stationary memoryless process Λ characterized by

$$\pi_\lambda = \Pr(\Lambda = \lambda) \neq 0 \text{ for } \ell_{\min} \leqslant \lambda \leqslant \ell_{\max}, \tag{8.20}$$

where $\ell_{\min}$ and $\ell_{\max}$ are the minimum and maximum size of a packet. Assuming that the length ℓ_{q} of the burst is fixed before packet aggregation and that J is not determined a priori, the accumulated lengths ℓ of the τ first aggregated packets

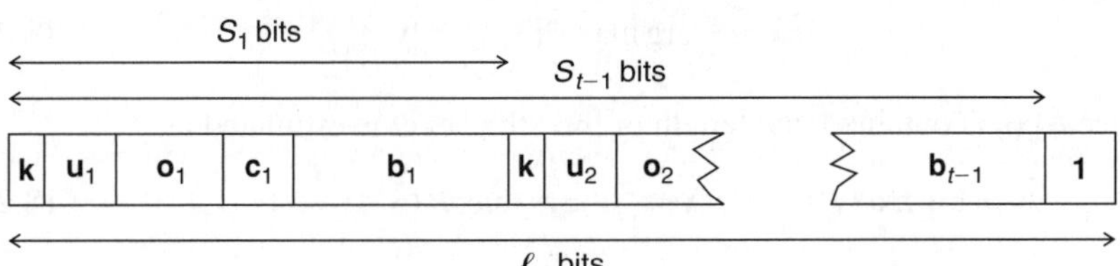

FIGURE 8.8 A burst at PHY layer containing several MAC packets.

are now realizations of a Markov process, which state is denoted by S_τ. If $\ell < \ell_q$, then

$$P(S_\tau = \ell \mid S_{\tau-1} = \ell') = \begin{cases} \pi_{\ell-\ell'} & \text{if } \ell_{\min} \leqslant \ell - \ell' \leqslant \ell_{\max} \\ 0 & \text{else,} \end{cases}$$

and if $\ell = \ell_q$, then

$$P(S_\tau = \ell_q \mid S_{\tau-1} = \ell') = \begin{cases} 0, \text{ if } \ell_q - \ell' > \ell_{\max} \\ 1, \text{ if } 0 < \ell_q - \ell' < \ell_{\min} \\ 1 - \sum\limits_{\ell=\ell'+1}^{\ell_q-1} P\left(S_\tau = \ell \mid S_{\tau-1} = \ell'\right), \text{ else.} \end{cases}$$

If $0 < \ell_q - \ell' < \ell_{\min}$, there is not enough space in the burst to put a MAC packet; the τth packet is thus necessarily a padding packet. If $\ell_q - \ell' > \ell_{\max}$, there is enough space to put a MAC packet of any allowed length; thus, the τth packet cannot end the burst. In the other cases, only MAC and padding packets of $\ell_q - \ell'$ bits are allowed to finish the burst.

8.3.2 Estimating the Number of Packets and Their Boundaries

Assume that a burst $\mathbf{q}$ is sent over a noisy channel and that a vector $\mathbf{y}_{1:\ell_q}$ is obtained at channel output. Our aim is to obtain joint MAP estimates $\widehat{t}$ of t and $\widehat{\lambda}_\tau$ of λ_τ, $\tau = 1, \ldots, \widehat{t}$, using the knowledge of $\mathbf{y}_{1:\ell_q}$ and all identifiable sources of redundancy in the packet aggregation process

$$\left(\widehat{t}, \widehat{\lambda}_1, \ldots, \widehat{\lambda}_{\widehat{t}}\right) = \arg \max_{t, \lambda_1, \ldots, \lambda_t} p\left(t, \lambda_1, \ldots, \lambda_t \mid \mathbf{y}_{1:\ell_q}\right). \tag{8.21}$$

The main difficulty comes from the fact that t is only known to satisfy

$$t_{\min} \leqslant t \leqslant t_{\max} \tag{8.22}$$

with $t_{\min} = \lfloor \ell_q / \ell_{\max} \rfloor$ and $t_{\max} = \lceil \ell_q / \ell_{\min} \rceil$ and that the number of padding bits is unknown at receiver side.

Obtaining directly the solution to (8.21) is quite difficult. This is why, we resort to a suboptimal estimator. The MAP estimate $\widehat{t}_{\mathrm{MAP}}$ of t is

$$\widehat{t}_{\mathrm{MAP}} = \arg\max_\tau P\left(S_\tau = \ell_q | \mathbf{y}_{1:\ell_q}\right). \tag{8.23}$$

Once $\widehat{t}_{\mathrm{MAP}}$ is obtained, the length of the τth packet is estimated as

$$\widehat{\lambda}_\tau = \arg\max_\ell P\left(S_\tau = \ell | \mathbf{y}_{1:\ell_q}\right) - \arg\max_\ell P\left(S_{\tau-1} = \ell | \mathbf{y}_{1:\ell_q}\right). \tag{8.24}$$

This estimator is suboptimal when compared to the joint MAP estimator (8.21) since the estimated length of the τth data packet may not necessarily satisfy the length constraint (8.20).

Both (8.23) and (8.24) require the evaluation of $P\left(S_\tau = \ell | \mathbf{y}_{1:\ell_q}\right)$, which is done in Section 8.3.3.

8.3.3 BCJR Algorithm for Robust Burst Segmentation

The successive values that may be taken by S_τ, $\tau = 0, \ldots, t_{\max}$ are described by a *packet-clock* trellis in Figure 8.9, where $\ell_{\min} = 4$, $\ell_{\max} = 7$, and $\ell_q = 16$. Dashed transitions correspond to padding packets and plain transitions correspond to data packets. For the last packet of the burst (when $S_\tau = \ell_q$), dotted and plain transitions may be parallel. This trellis is directly inspired from the symbol-clock trellis presented in Section 5.2.2, page 132. Each packet may be seen as a symbol encoded with a variable-length code. The main difference comes from the fact that only the packet boundaries are of interest here.

As seen in Section 8.3.2, the segmentation of aggregated packets requires the evaluation of the a posteriori probabilities $P(S_\tau = \ell | \mathbf{y}_{1:\ell_q})$ for all possible values of τ and ℓ_q. This may be performed with a variant of the BCJR algorithm (Bahl et al., 1974) by writing the joint probability $P(S_\tau = \ell, \mathbf{y}_{1:\ell_q})$ as

$$P(S_\tau = \ell, \mathbf{y}_{1:\ell_q}) = \alpha_\tau(\ell)\, \beta_\tau(\ell), \tag{8.25}$$

where $\alpha_\tau(\ell) = P(S_\tau = \ell, \mathbf{y}_{1:\ell})$ and $\beta_\tau(\ell) = P(\mathbf{y}_{\ell+1:\ell_q} | S_\tau = \ell)$.

Evaluation of α_τ and β_τ

Now for $\tau = 1, \ldots, t_{\max}$, classical BCJR forward

$$\alpha_\tau(\ell) = \sum_{\ell'} \alpha_{\tau-1}(\ell')\gamma_\tau(\ell', \ell)$$

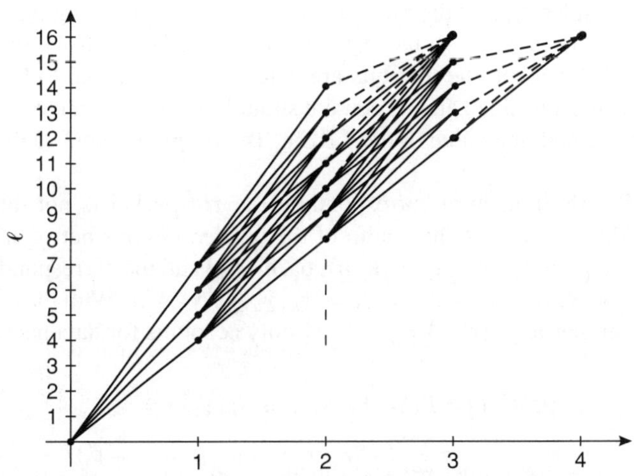

FIGURE 8.9 Trellis representing all MAC frame combinations within a MAC burst.

and backward recursions

$$\beta_\tau(\ell) = \sum_{\ell'} \beta_{\tau+1}(\ell')\gamma_{\tau+1}(\ell',\ell),$$

with

$$\gamma_\tau(\ell',\ell) = P(S_\tau = \ell, \mathbf{y}_{\ell'+1:\ell} | S_{\tau-1} = \ell') \qquad (8.26)$$

may be performed.

For the forward recursion, the initial state is perfectly known and $\alpha_0(0) = 1$. For the backward recursion, the number of packets is only known to satisfy (8.22); thus there are $t_{\max} - t_{\min} + 1$ allowed final states. Assuming that all these final states are equally likely (which is a quite coarse approximation), one gets

$$\beta_\tau(\ell_q) = \frac{1}{t_{\max} - t_{\min} + 1}, \; t_{\min} \leqslant \tau \leqslant t_{\max}.$$

Evaluation of γ_τ

When performing the evaluation of $\gamma_\tau(\ell',\ell)$, one implicitly assumes that the τth packet starts at the $\ell'+1$th bit and ends at the ℓth bit of the burst. When $\ell < \ell_q$ and provided that $\ell_{\min} \leqslant \ell - \ell' \leqslant \ell_{\max}$, the τth packet is not the last one; thus, only data packets have to be considered. When $\ell = \ell_q$, depending on the value of ℓ', data and padding packets have to be considered simultaneously (parallel transitions in Figure 8.9) or only padding packets have to be taken into account (dashed transitions in Figure 8.9).

Many sources of redundancy may be taken into account for the evaluation of $\gamma_\tau(\ell',\ell)$. Considering that the τth packet contains only padding bits, these bits are perfectly determined (assume, e.g., that they are all equal to 1). When the τth packet is a data packet, the first bits are determined and equal to $\mathbf{k}$, the length field $\mathbf{u}_\tau$ is also determined and its content should be $\ell - \ell'$. Albeit other fields, $\mathbf{o}_\tau$ are not determined; the value of the CRC $\mathbf{c}_\tau$ is strongly influenced by $\mathbf{k}$ and $\mathbf{u}_\tau$.

Case $\boldsymbol{\ell} < \boldsymbol{\ell}_\mathbf{q}$: the transition corresponding the τth packet is not the last one. Assuming that $\ell - \ell' \geqslant \ell_\mathrm{h}$, the length of the header, the bits between $\ell'+1$ and ℓ may be interpreted as $\mathbf{q}_{\ell'+1:\ell} = [\mathbf{k}, \mathbf{u}_\tau, \mathbf{o}_\tau, \mathbf{c}_\tau, \mathbf{b}_\tau]$ and the corresponding observation may be written as $\mathbf{y}_{\ell'+1:\ell} = \mathbf{y}_\tau = [\mathbf{y}_k, \mathbf{y}_u, \mathbf{y}_o, \mathbf{y}_c, \mathbf{y}_b]$. With these notations, the transition metric $\gamma_\tau(\ell',\ell) = \gamma_\tau^\mathrm{d}(\ell',\ell)$ only accounts for data packets

$$\begin{aligned}\gamma_\tau^\mathrm{d}(\ell',\ell) &= P(S_\tau = \ell, \mathbf{y}_{\ell'+1:\ell} | S_{\tau-1} = \ell') \\ &= \sum_{\mathbf{p}\neq\mathbf{1}} P(S_\tau = \ell, \mathbf{y}_{\ell'+1:\ell}, \mathbf{p} | S_{\tau-1} = \ell') \\ &= p(S_\tau = \ell | S_{\tau-1} = \ell')\,\varphi(\mathbf{y}_{\ell'+1:\ell}),\end{aligned} \qquad (8.27)$$

where

$$\begin{aligned}\varphi(\mathbf{y}_\tau) &= \sum_{\mathbf{p}\neq\mathbf{1}} P(\mathbf{y}_\tau, \mathbf{p}|S_{\tau-1}=\ell', S_\tau=\ell)\\ &= \sum_{\mathbf{p}\neq\mathbf{1}} P(\mathbf{y}_\tau|\mathbf{p}, S_{\tau-1}=\ell', S_\tau=\ell)\, P(\mathbf{p}|S_{\tau-1}=\ell', S_\tau=\ell). \end{aligned} \tag{8.28}$$

The sum in (8.28) is over all possible packets $\mathbf{p}$ of $\ell-\ell'$ bits, except $\mathbf{p}=\mathbf{1}$ corresponding to the content of a padding packet. Nevertheless, only the $\mathbf{p}$s starting with $\mathbf{k}$ and $\mathbf{u}_\tau=\mathbf{u}(\ell-\ell')$ (the binary representation of $\ell-\ell'$) have to be considered since these fields are fully determined. Moreover, assuming that the channel is memoryless, taking into account the fact that $\mathbf{k}, \mathbf{u}_\tau, \mathbf{o}_\tau$, and $\mathbf{c}_\tau$ does not depend on $\mathbf{b}_\tau$, and the fact that CRC $\mathbf{c}_\tau$ is fully determined by $\mathbf{k}$, $\mathbf{u}_\tau$, and $\mathbf{o}_\tau$, (8.28) simplifies to

$$\begin{aligned}\varphi(\mathbf{y}_\tau) = P(\mathbf{y}_k|\mathbf{k})\, P(\mathbf{y}_u|\mathbf{u}(\ell-\ell')) \sum_{\mathbf{b}} P(\mathbf{y}_b|\mathbf{b})\, P(\mathbf{b})\\ \sum_{\mathbf{o}} P(\mathbf{o})\, P(\mathbf{y}_o|\mathbf{o})\, P(\mathbf{y}_c|\mathbf{c}=\mathbf{f}(\mathbf{k}, \mathbf{u}_\tau, \mathbf{o})). \end{aligned} \tag{8.29}$$

Assuming that all values that may be taken by $\mathbf{b}$ are equally likely, one gets $P(\mathbf{b})=2^{-\ell(\mathbf{b})}$. In (8.29), the sum over all possible $\mathbf{o}$ may be quite complex to evaluate for long vectors $\mathbf{o}$. A reduced-complexity algorithm for evaluating this sum may be found in Section 8.4.

Case $\boldsymbol{\ell=\ell_q}$: the τth packet is the last one, and $\mathbf{p}=\mathbf{1}$ has also to be considered in the sum of (8.27). Then,

$$\gamma_\tau(\ell', \ell_q) = \gamma_\tau^{\mathrm{d}}(\ell', \ell_q) + \gamma_\tau^{\mathrm{p}}(\ell', \ell_q), \tag{8.30}$$

where

$$\begin{aligned}\gamma_\tau^{\mathrm{p}}(\ell', \ell_q) = p(S_\tau=\ell_q|S_{\tau-1}=\ell')\, P(\mathbf{y}_\tau|\mathbf{p}=\mathbf{1}, S_{\tau-1}=\ell', S_\tau=\ell)\\ P(\mathbf{p}=\mathbf{1}|S_{\tau-1}=\ell', S_\tau=\ell_q), \end{aligned} \tag{8.31}$$

accounts for the padding bits. In (8.31), $P(\mathbf{p}=\mathbf{1}|S_{\tau-1}=\ell', S_\tau=\ell_q)$ corresponds to the probability that the last packet contains only padding bits, i.e., the τth data packet is not of $\ell_q-\ell'$ bits. Thus,

$$P(\mathbf{p}=\mathbf{1}|S_{\tau-1}=\ell', S_\tau=\ell_q) = 1-\pi_{\ell_q-\ell'}. \tag{8.32}$$

Using (8.32), for a memoryless channel, (8.31) becomes

$$\gamma_\tau^{\mathrm{p}}(\ell', \ell_q) = p(S_\tau=\ell_q|S_{\tau-1}=\ell')\, P(\mathbf{y}_\tau|\mathbf{p}=\mathbf{1})\, (1-\pi_{\ell_q-\ell'}). \tag{8.33}$$

8.3.4 WiMAX MAC Burst Segmentation

In the WiMAX standard, the scheduler manages the downlink (DL) and uplink (UL) subframes and controls how much resources are allocated in each direction.

DL subframes begin by a frame control section that contains the downlink map (DL-MAP) for the current downlink frame and the uplink map (UL-MAP) for a frame in future. The DL-MAP informs all subscribers stations (SSs) of which part of the current frame they should listen to. The UL-MAP informs SSs of their transmission opportunities as a response to their dynamic bandwidth requests or on the basis of prior service agreements. These are then followed by the transmission of the DL subframe and the UL subframe. The DL subframe is divided into a number of time division multiplex (TDM) portions, which are so-called burst profiles: all traffic associated to a particular burst profile is transmitted sequentially. Each burst profile is characterized by the use of a particular modulation and coding scheme (there are as many burst profiles as there are modulation and coding schemes), which are negotiated between Base Station (BS) and SSs while establishing connection and are identified by the Downlink Interval Usage Code (DIUC).

Each burst can contain multiple concatenated fixed-size or variable-size packets or fragments of packets received from the higher layers. Each unused byte would be set to stuff byte value (0xFF) (IEEE, 2004). The packets are queued at the BS, the scheduler assigns the MAC packets to the MAC burst, and the burst is filled with MAC packets in order, until there is not enough space left and finally padding bytes (0xFF) are added. Lengths of burst profiles are assigned by the scheduler and are communicated to users in the DL-MAP. Then, the DL subframe is transmitted. The UL subframe is quite different as it requires coordination between the various SSs transmitting upward. In what follows, we consider only the downlink scenario and thus the DL subframe only.

First, we explain the structure of WIMAX MAC PDU, followed by the description of MAC PDU header. MAC PDUs are illustrated in Figure 8.10. Each MAC PDU begins with a fixed-length header, followed by a variable-length payload and an optional CRC.

The header first begins by the generic header containing either MAC management messages or Convergence Sublayer (CS) data. Then, one finds the bandwidth request header used to request additional bandwidth. As we are considering only the downlink case, where the connection is already established, MAC PDUs inside a burst contains only CS data, so only the Generic MAC header is possible inside a burst. Generic MAC header as specified in IEEE 802.16-2004 (IEEE, 2004) is shown in Figure 8.10.

The following assumptions are made for the sake of simplicity.

- No CRC is used (the CRC is mandatory only for MAC management messages in OFDM and OFDMA).
- No ARQ is used.
- No packing and fragmentation is used.
- No Encryption is used for payload.
- No extended subheader is present.

Some fields are already fixed in a MAC header, but with the considered assumptions, fields such as Header Type, Encryption Control, Sub Headers and

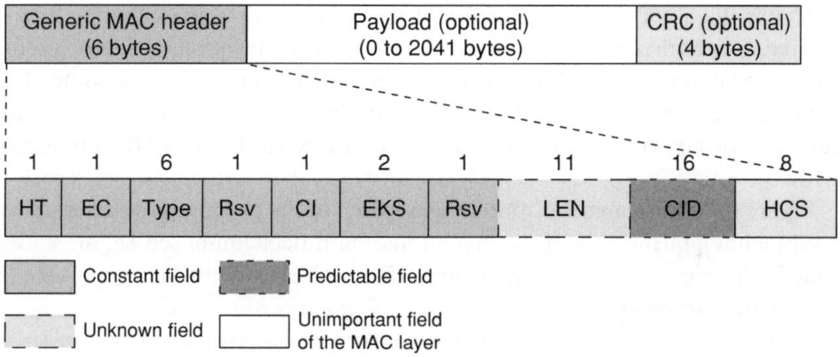

FIGURE 8.10 Generic MAC Header.

Special Payload Types, Reserved, CRC Indicator, and Encryption Key Sequence remain constant. The LEN field, representing the length in bytes of the MAC PDU including the MAC header and the CRC, if present, is an unknown field. Connection Identifier represents identifier of the receiver's connection and is thus predictable. Header Check Sequence, an 8-bit field used to detect errors in the header, is a variable field and is a function of the contents of all the header fields.

As indicated in Section 8.3.2, we assume that, before aggregation, the size of each MAC packet is independent from the previous one. To determine the set of possible lengths of each MAC packet, assume that H.264 video NALUs are RTP, UDP, and IP packetized before they reach the WiMAX MAC layer. The headers of each of these layers have following sizes: 12 bytes for RTP, 8 bytes for UDP, and 20 bytes for IP. Hence, the total length of RTP/UDP/IP headers is 40 bytes. Thus, the minimum size of the MAC packet is $\ell_{\min} = 47$ bytes, which includes the MAC header of 6 bytes and at least 1 data byte. Maximum size of MAC packet can be $\ell_{\max} = 2048$ bytes (due to the 11 bits LEN field), but in practice, for wireless transmission, much smaller values are considered.

8.3.5 Simulation Results

A simulator consisting of a MAC burst generator, an AWGN channel, and a receiver has been implemented. Random sized data packets with $\ell_{\min} = 47$ bytes and $\ell_{\max} = 250$ bytes are aggregated in bursts of $\ell_q = 1000$ bytes. A burst can, thus, contain between 4 and 21 data packets. MAC headers are constructed with the format defined in Section 8.3.4, and the payloads of MAC packets consist of a variable amount of randomly generated bytes. If the generated packet cannot be inserted in the remaining space of a burst, a padding packet is inserted to fill the burst before passing it to a BPSK modulator and sending it over an AWGN channel. For performance analysis, erroneous packet location rate (EPLR) is evaluated as a function of the channel SNR. It should be noted that to recover MAC packets correctly, both the starting and ending frontiers of the packet should be correct.

To use the algorithm introduced in Section 8.3.3, the size of the burst ℓ_q is required. It is transmitted in DL-MAP for WiMAX and is protected with a more robust modulation and coding scheme. In our simulation, we have assumed that the burst size ℓ_q is received without any error. We have compared our proposed segmentation algorithm with conventional hard decision for an SNR from −5 dB to 10 dB.

Figure 8.11 compares the simulation results for the proposed robust segmentation technique (plain) and a conventional hard decision-based segmentation (dotted). A gain of more than 8 dB in SNR is achieved with the proposed soft segmentation method.

These segmented packets may contain errors. They may then be forwarded to upper layers using the techniques presented in Section 8.3 and robustly decoded at APL layer using the JSCD techniques of Chapters 4–6.

8.3.6 Summary

To perform JPCD applied to the robust segmentation of aggregated packet, soft information provided at the output of the channel (or channel decoder) are exploited in conjunction with a priori information due to the redundancy present at various layers of the protocol stack (known fields, presence of CRC

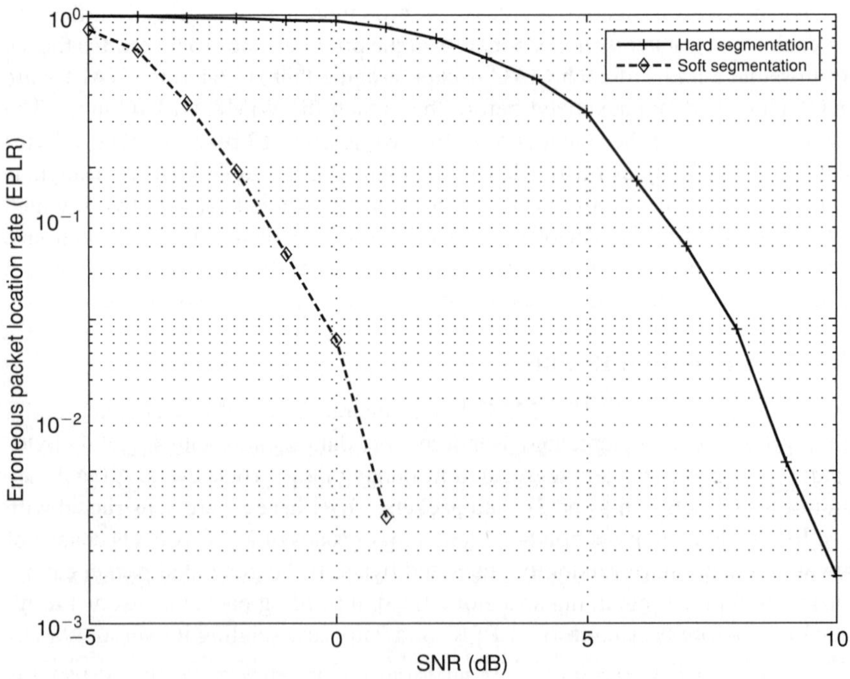

FIGURE 8.11 Erroneous packet location rate as a function of the PSNR on the channel for Wi-Fi MAC packets between 47 and 250 bytes aggregated in PHY bursts of 1000 bytes.

or checksums, etc). For the case of WiMAX, all successions of MAC packets within a MAC burst may be represented by a trellis on which a variant of the BCJR algorithm has been put at work to robustly identify the boundaries of each MAC packet. For similar MAC packet localization error, on an AWGN channel, a gain of about 8 dB in SNR compared to a standard segmentation technique is observed.

In WiMAX/IEEE 802.16, header suppression/compression (Nuaymi et al., 2007) have been defined. Headers are then less subject to errors, as they are shorter, but much more sensitive. It would be interesting to adapt the proposed segmentation technique to compressed headers.

8.4 COMPUTING APPs OF INPUTS OF BLOCK CODES

In Sections 8.2 and 8.3, the presence of CRCs or checksums protecting the headers and parts of the payloads is taken into account with

$$\Psi(\mathbf{k},\mathbf{u},\mathbf{y}_o,\mathbf{y}_c)=\sum_{\mathbf{o}} P(\mathbf{o})P(\mathbf{y}_o|\mathbf{o})P(\mathbf{y}_c|\mathcal{F}([\mathbf{k},\mathbf{u},\mathbf{o}])), \tag{8.34}$$

see (8.8) or (8.29). Since **k** and **p** play the same role in (8.8), **p** has been omitted in (8.34).

The evaluation of (8.34) is clearly similar to computing the APP of the inputs (here parts of inputs) from the measured outputs of block codes. Therefore, our method has many similarities with that of Wolf (1978), which deals with soft decoding of block codes. In his paper, Wolf proposes a method based on a trellis, built from the parity check matrix, for the decoding of linear block codes. Here, the computation is different since the code is divided in three parts: a known portion (vectors **k**), a candidate value (**u**), and an unknown part (**o** and **c**). We want to find the best combination of **u** by taking into account the redundancy of the code (given by **c**). The trellis is thus applied to the portions **o** and **c** for given **k** and **u**. Additionally, the technique does not estimate **o** and **c** but evaluates the probability associated to the $2^{\ell(\mathbf{o})}$ combinations of $[\mathbf{o},\mathbf{c}]$.

8.4.1 Exact Computation

Assuming that the CRC is obtained from the systematic generator matrix $\mathbf{G}=[\mathbf{I},\mathbf{\Pi}]$, it can be evaluated iteratively over the data **r**, as shown by (8.1). More precisely, the value of the CRC associated to the $j+1$ first bits of **r** (shortly, at time $j+1$) only depends on the value of the CRC at time j and on the $j+1$th bit of **r**. Each value of the CRC at time j leads to two different values of the CRC at time $j+1$. Consequently, the evolution of the CRC values according to the bits of **r** can be described by a trellis. In this trellis, states correspond to the $2^{\ell(\mathbf{c})}$ possible values of the CRC. Transitions are determined by the bits of **r**. At each time $j=1,\ldots,\ell(\mathbf{r})$, we study the contribution of r_j (the jth bit of **r**) over the global CRC, see Example 8.1.

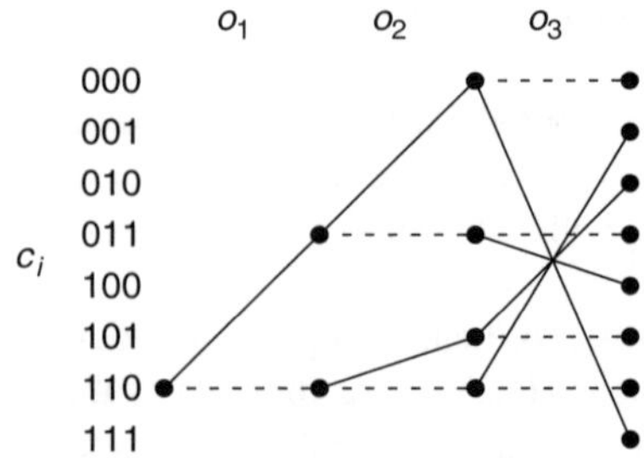

FIGURE 8.12 Trellis obtained by forward construction, the dotted lines correspond to transitions associated to $c_i=0$, the plain lines to $c_i=1$.

Example 8.1: The trellis that may be obtained for a systematic binary Hamming (7,4) code with

$$\Pi=\begin{bmatrix}1 & 1 & 0\\ 1 & 0 & 1\\ 0 & 1 & 1\\ 1 & 1 & 1\end{bmatrix}$$

is represented in Figure 8.12. In the example, it is assumed that $[\mathbf{k},\mathbf{u}]=[1]$ and that $\ell(\mathbf{o})=3$.

In our case, $\mathbf{r}=[\mathbf{k},\mathbf{u},\mathbf{o}]$. Data contained in $\mathbf{k}$ are assumed to be known, thus have a fixed contribution to the estimate of $\mathbf{u}$. Consequently, for a given value of $\mathbf{u}\in\Omega_u$, we want to determine its probability according to the observations $\mathbf{y}$ and to the CRC properties (depending on the bits belonging to $\mathbf{o}$). Each value of $\mathbf{u}$, in conjunction with $\mathbf{k}$ defines the initial state in the trellis (there is no contribution from $\mathbf{o}$). Each new bit in $\mathbf{o}$ may provide two new possible states, thus defining a trellis. For any value of $\mathbf{o}$, one gets a path starting from the same state associated to $\mathcal{F}([\mathbf{k},\mathbf{u},\mathbf{0}])$ and ending in the state associated to $\mathcal{F}([\mathbf{k},\mathbf{u},\mathbf{o}])$.

In the following, we propose a solution based on a backward construction of the trellis to directly evaluate (8.10) for all the values of $\mathbf{u}\in\Omega_u$. For this purpose, we consider that $V_{\mathbf{s}_i}(j)$ denotes the probability associated to state i at time j in the trellis. We also assume that $\bar{\mathbf{o}}^j$ represents the $\ell(\mathbf{o})-j$ last bits of $\mathbf{o}$ ($\bar{\mathbf{o}}^j=\left[o_{j+1}\ldots o_{\ell(\mathbf{o})}\right]$) and that $\mathbf{y}_{\bar{o}}^j=\left[y_{o_{j+1}}\ldots y_{o_{\ell(\mathbf{o})}}\right]$ corresponds to the respective observations. More precisely, $V_{\mathbf{s}_i}(j)$ represents the sum of probabilities associated to each combination of $\bar{\mathbf{o}}^j$ and its corresponding CRC when starting from $\mathbf{s}_i$ at time j such as

$$V_{\mathbf{s}_i}(j)=\sum_{\bar{\mathbf{o}}^j}P(\bar{\mathbf{o}}^j)P(\mathbf{y}_{\bar{o}}^j|\bar{\mathbf{o}}^j)P\left(\mathbf{y}_c|\mathbf{s}_i\oplus\mathcal{F}([\mathbf{0},\bar{\mathbf{o}}^j])\right), \tag{8.35}$$

for $i=0,1\ldots 2^{\ell(\mathbf{c})}-1$.

Applying (8.35) to state i at time $j-1$ results in

$$
\begin{aligned}
V_{\mathbf{s}_i}(j-1) &= \sum_{\bar{\mathbf{o}}^{j-1}} P(\bar{\mathbf{o}}^{j-1}) P\left(\mathbf{y}_{\bar{o}}^{j-1} | \bar{\mathbf{o}}^{j-1}\right) P\left(\mathbf{y}_c | \mathbf{s}_i \oplus \mathcal{F}\left(\left[\mathbf{0}, \bar{\mathbf{o}}^{j-1}\right]\right)\right) \\
&= P(o_j=0) P(y_{o_j}|o_j=0) \sum_{\bar{\mathbf{o}}^j} P(\bar{\mathbf{o}}^j) P(\mathbf{y}_{\bar{o}}^j | \bar{\mathbf{o}}^j) P\left(\mathbf{y}_c | \mathbf{s}_i \oplus \mathcal{F}([\mathbf{0}, \bar{\mathbf{o}}^j])\right) \\
&\quad + P(o_j=1) P(y_{o_j}|o_j=1) \sum_{\bar{\mathbf{o}}^j} P(\bar{\mathbf{o}}^j) P(\mathbf{y}_{\bar{o}}^j | \bar{\mathbf{o}}^j) P\left(\mathbf{y}_c | \mathbf{s}_q \oplus \mathcal{F}([\mathbf{0}, \bar{\mathbf{o}}^j])\right) \\
&= P(o_j=0) P(y_{o_j}|o_j=0) V_{\mathbf{s}_i}(j) + P(o_j=1) P(y_{o_j}|o_j=1) V_{\mathbf{s}_q}(j),
\end{aligned}
\tag{8.36}
$$

where $\mathbf{s}_q = \mathbf{s}_i \oplus \boldsymbol{\pi}(j)$. (8.36) above is the key for computing $V_{\mathbf{s}_i}(j)$ through a backward iteration over the bits of $\mathbf{o}$.

After $\ell(\mathbf{o})$ iterations, $V_{\mathbf{s}_i}(0)$ may be expressed as

$$
V_{\mathbf{s}_i}(0) = \sum_{\mathbf{o}} P(\mathbf{o}) P(\mathbf{y}_o | \mathbf{o}) P\left(\mathbf{y}_c | \mathbf{s}_i \oplus \mathcal{F}([\mathbf{0}, \mathbf{o}])\right), \tag{8.37}
$$

for $i = 0, 1 \ldots 2^{\ell(\mathbf{c})} - 1$. This method allows to simultaneously compute (8.10) for all values of $\mathbf{u} \in \Omega_u$. It is no more necessary to construct a new trellis for each value of $\mathbf{u}$. For a given $\mathbf{u}$, the corresponding probability is given by

$$
\begin{aligned}
V_{\mathcal{F}([\mathbf{k},\mathbf{p},\mathbf{u},\mathbf{0}])}(0) &= \sum_{\mathbf{o}} P(\mathbf{o}) P(\mathbf{y}_o | \mathbf{o}) P\left(\mathbf{y}_c | \mathcal{F}([\mathbf{k}, \mathbf{p}, \mathbf{u}, \mathbf{0}]) \oplus \mathcal{F}([\mathbf{0}, \mathbf{o}])\right) \\
&= \Psi(\mathbf{k}, \mathbf{p}, \mathbf{u}, \mathbf{y}_o, \mathbf{y}_c).
\end{aligned}
$$

The steps for evaluating (8.10) backward are summarized below. The trellis is constructed by starting from $j = \ell(\mathbf{o})$ and going backward to $j = 0$.
Step 1: At time $j = \ell(\mathbf{o})$, $V_{\mathbf{s}_i}(\ell(\mathbf{o})) = P(\mathbf{y}_c | \mathbf{s}_i)$ for $i = 0, 1 \ldots 2^{\ell(\mathbf{c})} - 1$.
Step 2: For $j = \ell(\mathbf{o}) - 1 \ldots 1, 0$, $V_{\mathbf{s}_i}(j)$ is updated according to (8.36) as

$$
\begin{aligned}
V_{\mathbf{s}_i}(j) &= P(o_{j+1}=0) P(y_{o_{j+1}} | o_{j+1}=0) V_{\mathbf{s}_i}(j+1) \\
&\quad + P(o_{j+1}=1) P(y_{o_{j+1}} | o_{j+1}=1) V_{\mathbf{s}_q}(j+1),
\end{aligned}
\tag{8.36'}
$$

where $i = 0, 1 \ldots 2^{\ell(\mathbf{c})} - 1$ and $\mathbf{s}_q = \mathbf{s}_i \oplus \boldsymbol{\pi}(j+1)$.
Step 3: After $\ell(\mathbf{o})$ iterations, for any value $\mathbf{u} \in \Omega_u$,

$$
\Psi(\mathbf{k}, \mathbf{p}, \mathbf{u}, \mathbf{y}_o, \mathbf{y}_c) = V_{\mathcal{F}([\mathbf{k},\mathbf{p},\mathbf{u},\mathbf{0}])}(0). \tag{8.37'}
$$

With this method, we can directly evaluate (8.10) for each state i such as $\mathbf{s}_i = \mathcal{F}([\mathbf{k}, \mathbf{p}, \mathbf{u}, \mathbf{0}])$ with $\mathbf{u} \in \Omega_u$. The global complexity of the process is thus $O\left(\ell(\mathbf{o}) 2^{\ell(\mathbf{c})}\right)$. Example 8.2 illustrates the trellis constructed for a backward evaluation of $\Psi(\mathbf{k}, \mathbf{p}, \mathbf{u}, \mathbf{y}_o, \mathbf{y}_c)$.

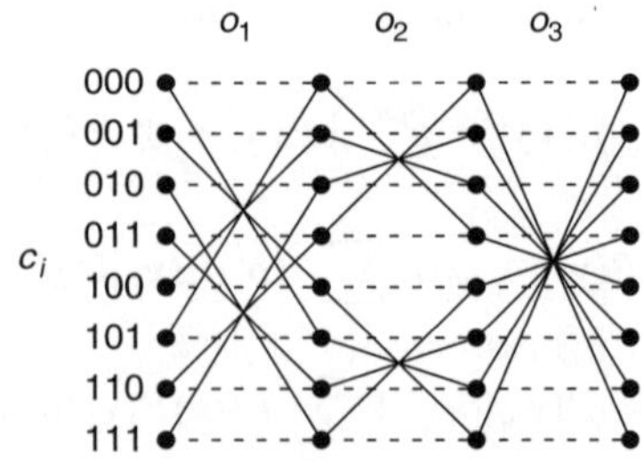

FIGURE 8.13 Trellis obtained with the backward construction.

Example 8.2: The trellis obtained for a systematic binary Hamming $(7,4)$ code with

$$\Pi = \begin{bmatrix} 1 & 1 & 0 \\ 1 & 0 & 1 \\ 0 & 1 & 1 \\ 1 & 1 & 1 \end{bmatrix}$$

is represented in Figure 8.13. In this example, we assume that $\ell([\mathbf{k}, \mathbf{p}, \mathbf{u}]) = 1$ bit, $[\mathbf{k}, \mathbf{p}, \mathbf{u}] = [1]$, and $\ell(\mathbf{o}) = 3$ bits. We may notice that the sum in (8.10) is simultaneously computed for the two possible initial states $\mathcal{F}([\mathbf{k}, \mathbf{p}, \mathbf{u}, \mathbf{0}]) = [0, 0, 0]$ and $\mathcal{F}([\mathbf{k}, \mathbf{p}, \mathbf{u}, \mathbf{0}]) = [1, 1, 0]$ obtained when $[\mathbf{k}, \mathbf{p}, \mathbf{u}] = [0]$ and $[\mathbf{k}, \mathbf{p}, \mathbf{u}] = [1]$, respectively.

This method allows to compute directly the sum (8.10) for all values taken by $\mathbf{u}$ and by the resulting CRC $\mathcal{F}\left([\mathbf{k}, \mathbf{p}, \mathbf{u}, \mathbf{o}]\right)$ with a complexity $O(\ell(\mathbf{o})\, 2^{\ell(\mathbf{c})})$. When $\ell(\mathbf{o})$ and $|\Omega_u|$ are small (typically $\ell(\mathbf{o}) < \ell(\mathbf{u})$), it may be more interesting to keep on the forward iteration, since fewer computations are performed to obtain the forward trellis.

8.4.2 Approximate Computation

Most CRCs are larger than 16 bits, and the complexity $\mathcal{O}\left(\ell(\mathbf{o}) 2^{\ell(\mathbf{c})}\right)$ is therefore too large to allow a real-time implementation of the method presented in Section 8.4.1. An approximate computation consists in splitting the CRC into M_b blocks of $\ell(\mathbf{c})/M_\mathrm{b}$ bits, each block being assumed statistically independent from the others. Thus, $\mathbf{y}_c$ may be written as $\mathbf{y}_c = \left[\mathbf{y}_{c_1}, \mathbf{y}_{c_2} \cdots \mathbf{y}_{c_{M_\mathrm{b}}}\right]$. Using the independence approximation, the sum in (8.10) becomes

$$\Psi(\mathbf{k}, \mathbf{u}, \mathbf{y}_o, \mathbf{y}_c) \approx \prod_{m=1}^{M_\mathrm{b}} \Psi_m\left(\mathbf{k}, \mathbf{u}, \mathbf{y}_o, \mathbf{y}_{c_m}\right), \tag{8.38}$$

with

$$\Psi_m\left(\mathbf{k}, \mathbf{u}, \mathbf{y}_o, \mathbf{y}_{c_m}\right) = \sum_{\mathbf{o}} P(\mathbf{o}) P(\mathbf{y}_o|\mathbf{o}) P\left(\mathbf{y}_{c_m}|\mathcal{F}_m([\mathbf{k}, \mathbf{u}, \mathbf{o}])\right), \tag{8.39}$$

where $\mathcal{F}_m$ is the encoding function associated to the columns $(m-1)\frac{\ell(\mathbf{c})}{M_\mathrm{b}}+1$ to $m\frac{\ell(\mathbf{c})}{M_\mathrm{b}}$ of Π, corresponding to a partial CRC of $\ell(\mathbf{c})/M_\mathrm{b}$ bits.

The evaluation of (8.39) is similar to that of Ψ described in Section 8.4.1. The only difference lies in the size of the trellis: $2^{\ell(\mathbf{c})/M_\mathrm{b}}$ states have to be considered at any depth (instead of $2^{\ell(\mathbf{c})}$ states without splitting the CRC). The total complexity for evaluating (8.38) is now $O\left(M_\mathrm{b}\ell(\mathbf{o})2^{\ell(\mathbf{c})/M_\mathrm{b}}\right)$ at the cost of a slightly suboptimal performance.

8.5 DISCUSSION

Robust segmentation of aggregated packets and robust header recovery are key elements to build an efficient permeable layer, allowing packets corrupted by transmission errors at the PHY layer to reach the APL layer. The main tool of these mechanisms consists of an MAP estimators exploiting jointly the structural properties of the protocol stack along with the redundancy provided by CRCs or checksums through a soft decoding algorithm. Note that the MAP estimator, which can be obtained via a procedure very similar to the BCJR algorithm, is not used to perform a sequence estimate, as usual. Instead, it is used for computing the a posteriori probabilities of the states from which the unknown parts of the fields are estimated.

These techniques have been illustrated on the Wi-Fi 802.11 Protocol but may readily be applied to other layers for various transmission protocols. As a result, an enhanced permeable layer mechanism (compared, e.g., to UDP-Lite) may be defined. This mechanism is particularly well suited when combined with JSCD techniques at APL layer, as detailed in Chapter 9. Simulations with PHY and MAC layers of WiFi illustrate the significant performance gains achieved with the proposed decoding technique. As a result, such techniques allow the headers to be much morc robust to channel impairments than the payload, thus avoiding the necessity of packet retransmission in most cases.

When considering IP, UDP-Lite, and RTP layers, the CRC is replaced by a checksum, and it involves modifications in the metric derivation. Nevertheless, decoding still employs a trellis, such as that used in Section 8.4.1, but the trellis transitions of which are determined by the way checksums are evaluated.

Among the open problems, it would be interesting to compare and/or combine the proposed solutions to solutions based on the ROHC protocols. Such compression protocol allows to get smaller headers, which are thus less subject to transmission error, but much more affected by errors.

Another aspect concerns the fact that since CRC and checksums have been used for error correction, it is no more possible to use them to determine whether the corrected header has a good chance to be the actual one. Hypotheses tests have been used to determine, up to some confidence level, if the correction has been performed in a satisfying way or if the packet has to be retransmitted, when this is possible. See Bouchireb et al. (2008) for a first approach.

Chapter 9

Joint Cross-Layer Decoding

Obviously, joint source-channel decoding (JSCD) is intrinsically of a cross-layer nature since it involves source decoding at the application (APL) layer based on some information coming from the physical (PHY) layer. However, the level of cross-layering can vary a lot, depending on the building blocks that are considered jointly with the decoding process. Some of these building blocks, such as the modulation, are very often considered, whereas some other building blocks, such as forward error correction (FEC), are seldom taken into account. The redundancy coming from the headers and cyclic redundancy check (CRC), which takes the form of side information, is usually omitted as well.

The fact that many early papers on JSCD do not consider the existence of FEC at PHY layer is a real drawback. In fact, FEC improves all situations so much so that procedures which seem to add a lot of robustness if FEC is not taken into account, can turn out to be almost equivalent to separate decoding once the error-correction capacity of FECs is activated. This is a crucial point for evaluating JSCD efficiency, which is recalled in Section 9.1. Section 9.2 then describes a classical way to recover some of the improvements by using iterations between source and channel decoders.

On another side, the process of using the available redundancy remaining in the source bitstream by scrutinizing the structure of the encoder may look tedious and of reduced usefulness since a large portion of the bitstream sent on the wireless channel comes from the intermediate layers. Section 9.1 provides some insight about the relative usefulness of both approaches.

Two aspects of joint cross-layer decoding are considered in this chapter. Section 9.1 shows the way redundancy at intermediate protocol layers may be combined with redundancy at APL layer to improve decoding performance. Section 9.2 focuses on iterative decoding. When taking all sources of redundancy simultaneously into account is not possible for complexity reasons, a solution is to use a fraction of the redundancy to perform a first decoding and to feed the results to a second decoder, exploiting the remaining part of the redundancy. The results of this second decoder may be fed back to the first decoder to correct residual errors. Iterations may be performed between both decoders until convergence. This "turbo-decoding" scheme, introduced for channel codes, is also interesting when one of the constituent codes is a source code presenting some redundancy.

9.1 NETWORK AND PHY LAYERS MAY JOINTLY HELP THE APPLICATION LAYER

In wireless transmission of multimedia contents, the received signal may be heavily corrupted and may not directly be usable by the source decoder. A first solution to alleviate this problem consists in grouping data into packets protected by an error-detection code (CRC or checksum) (Blahut, 1983; Kurose and Ross, 2005), as explained with some details in Chapter 7. Packets whose integrity is not ensured at receiver side may then be retransmitted. Nevertheless, retransmissions may become difficult in scenarios with strong delay constraints (e.g., for visiophony), or even impossible when broadcasting data, e.g., in satellite television.

9.1.1 Motivation: The Interplay between CRCs and Redundancy at APL Layer

In such situations, the standard solution is to make use of very strong error-correction codes, e.g., turbo codes or LDPC, at PHY layer combined with packet erasure codes, e.g., Reed–Solomon codes, at intermediate protocol layers (MacKay, 2003; Richardson and Urbanke, 2008). However, due to the high channel variability, redundancy is rarely optimally dimensioned. The codes may be oversized when the channel is clear, reducing the bandwidth allocated for the data. In contrary, some corrupted packets cannot be recovered in bad channel conditions and are lost. Error-concealment techniques (Hong et al., 1999; Kung et al., 2006) may then be used by the source decoders at APL layer. They exploit the redundancy (temporal or spatial) in the decoded multimedia stream for estimating the missing information. However, even if very efficient for providing decoded contents of acceptable quality, error concealment cannot always replace a clean reception in terms of quality.

We have seen in the previous chapters that CRCs belong to the received stream and that they could be considered as additional redundancy for the payload, on top of the one left by the source coder, and the side information provided by the packet lengths. This section provides the algorithmic tools that allow a full exploitation of these additional sources of redundancy for robust decoding of video data in a downlink situation.

More specifically, we derive a sequential decoding algorithm jointly exploiting the syntax and semantic properties of the encoded video stream together with the redundancy at Media Access Control (MAC) layer provided by the CRC. Here, the CRC is not only used to detect errors but is also considered as an error-correcting code. This CRC-based decoding approach has been introduced by Lamy and Merigeault (2002) and Nieto and Furman (2007) for correcting erroneous packets. We propose here to make a simultaneous usage of the CRC and the source redundancy to improve the video decoding performance. This technique is based on a variety of techniques already described in this book, such as source decoding depending on syntax and semantic of bitstream (see Chapters 5 and 6), combined with soft decoding of block codes outlined in Chapter 8, and with sequential decoding, see Chapter 6, to attain our objective.

Note that all robust techniques introduced above require soft information to be delivered from the PHY layer to the APL layer. Obviously, this does not correspond to a classical structure of the decoder and requires the use of some additional tools, some of which have been presented in Chapter 8, such as robust segmentation of aggregated packets or header recovery techniques exploiting the intralayer and interlayer redundancies along with the CRCs or checksums, see also Marin et al. (2007). With these techniques, the header is very likely to be correctly decoded even for poor signal-to-noise (SNRs), and the payload may be forwarded to the upper layers, resulting in a *permeable* protocol stack. A complementary work, introducing a *transparent* network architecture, may be found in a paper by Huusko et al. (2007). In this chapter, we assume that, thanks to the use of such techniques, the headers are correctly received, and we concentrate on the evaluation of the payload (i.e., the reception of the video).

The remainder of this section is organized as follows. After a brief description of the model for the permeable protocol stack in Section 9.1.2, Section 9.1.3 describes the derivation of the decoding metric and proposes a sequential decoding method. Simulation results are described in Section 9.1.4.

9.1.2 Model of Permeable Protocol Stack

Multimedia packetized transmission usually relies on a multilayer architecture (Kurose and Ross, 2005) based on the RTP/UDP/IP stack introduced in Chapter 7. Figure 7.15, page 207, illustrates an example of the segmentation and encapsulation mechanisms implemented at each protocol layer in the case of video transmitted with a Wi-Fi radio interface (802.11 standard) (IEEE, 1999). The data processed by the PHY layer are forwarded to the MAC layer that checks their integrity with the help of some CRC. For corrupted packets, a retransmission is requested. Correctly received data are assembled to form the binary stream that is then fed to the video decoder (at APL layer) after removal of IP, UDP, and RTP protocol headers.

A protocol stack design where the PHY, MAC, and APL layers of the receiver work very closely together is considered here. Three changes are required to implement the proposed solution:

- The PHY layer includes a *soft-input soft-output* (SISO) channel decoder for processing the incoming protected data. The soft information are transmitted to the next layer.
- In the MAC layer, the CRC check is deactivated and no retransmission is allowed. Complete MAC packets (composed of header, payload, and CRC) are transferred to the upper layer for being integrated in the payload of IP packets.
- The MAC header and MAC CRC, which are usually not transmitted by the IP, UDP, and RTP layers, are now assumed to be available at the APL layer in the form of soft values.

These changes require some information to be available everywhere inside the receiver and are compatible with the usual transmission structure: only the receiver has to be modified, and both the transmitter operations and the signal sent are unchanged. As outlined above, they are facilitated by using the robust header recovery and permeable layer mechanisms discussed in Chapter 8. Here, the headers are assumed to be available without errors at all layers.

With these modifications, the APL layer receives a succession of MAC packets, containing soft information (provided by the PHY layer). The format of data received by the APL layer is depicted in Figure 9.1.

In the proposed architecture, the CRC still plays some error-detection role. When the normal CRC check is successful, the robust decoding process (which is computationally expensive) is deactivated, and the packet is decoded in a classical way. When some test, possibly also involving the CRC, see (Bouchireb et al., 2008), shows that the quality of the soft information provided by the lower layers is too poor, i.e., when the signal power is smaller than a predefined threshold, JSCD techniques have very few chances to recover the transmitted packet. The packet is discarded and, if possible, a retransmission request is sent, or its content is concealed at APL layer.

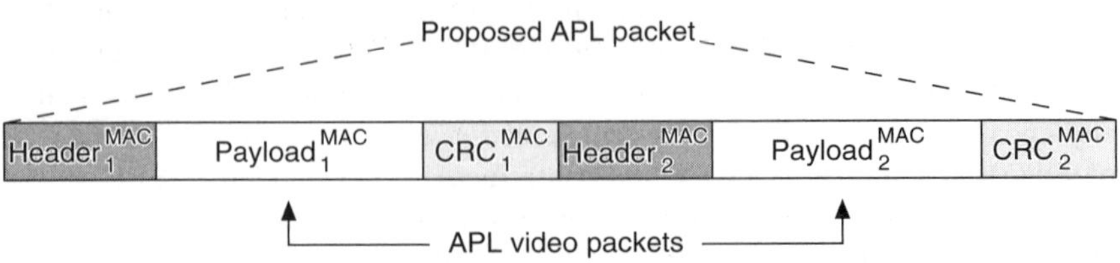

FIGURE 9.1 New format of the APL packet at the input of the decoder. In this example, the original APL packet has been fragmented in two MAC packets.

The next section presents the analytical derivation of the decoding metric, which may be used for robust reconstruction of the transmitted video sequence. We then propose a sequential decoding algorithm based on this metric.

9.1.3 Group-Based Sequential Decoding

Notations

The symbols produced by a video coder before entropy coding are assumed to be generated by a source X, which has to satisfy some semantic rules. Consider a vector $\mathbf{x}_{1:k} = [x_1, \ldots, x_k]$ of k symbols generated by this source. The entropy coder associates a binary variable-length code word $\mathbf{b}_{x_\kappa}$ to each component x_κ of $\mathbf{x}_{1:k}$, $\kappa = 1, \ldots, k$, which is thus mapped onto a binary sequence $\mathbf{b} = [\mathbf{b}_{x_1}, \ldots, \mathbf{b}_{x_k}]$, with

$$\sum_{\kappa=1}^{k} \ell(\mathbf{b}_{x_\kappa}) = \ell(\mathbf{b}). \tag{9.1}$$

Then, $\mathbf{b}$ has to be compliant with the syntax of the variable-length code (VLC) and with the semantic rules of the source X.

At MAC layer, a header $\mathbf{h}$ is added at the beginning of $\mathbf{b}$, resulting in a concatenated vector $\mathbf{d} = [\mathbf{h}, \mathbf{b}]$. A CRC $\mathbf{c}$ is then computed from the data $\mathbf{d}$ and appended to the packet. This set of information is collected in a vector $\mathbf{p} = [\mathbf{h}, \mathbf{b}, \mathbf{c}] = [\mathbf{d}, \mathbf{c}]$, where $\mathbf{c} = \mathcal{F}(\mathbf{d})$, $\mathcal{F}$ being a generic encoding function.

The computation of $\mathbf{c}$ depends on some generator polynomial $g(z) = \sum_{i=0}^{\ell(\mathbf{c})} a_i z^i$, characterizing the CRC (Blahut, 1983). A systematic generator matrix $\mathbf{G} = [\mathbf{I}, \mathbf{\Pi}]$ may be associated with $g(z)$. Using $\mathbf{G}$, $\mathbf{c}$ may be determined by a recursive processing over the $\ell(\mathbf{d})$ bits of $\mathbf{d}$ as follows:

$$\mathbf{c}^{j+1} = \mathcal{F}(\mathbf{d}^{j+1}) = \mathbf{c}^j \oplus (d_{j+1} \cdot \boldsymbol{\pi}(d_{j+1})). \tag{9.2}$$

In (9.2), $\mathbf{d}^j = [d_1 \ldots d_j, 0 \ldots 0]$, $\boldsymbol{\pi}(d_j)$ is the jth row of $\mathbf{\Pi}$, i.e., the parity vector related to d_j and $\oplus$ represents the XOR operator. At initialization, $\mathbf{c}^0$ is set to $\mathbf{0}$. After $\ell(\mathbf{d})$ iterations, the vector $\mathbf{c}^{\ell(\mathbf{d})}$ contains the CRC value related to $\mathbf{d}$ (i.e., $\mathbf{c}^{\ell(\mathbf{d})} = \mathbf{c}$).

In our model, $\mathbf{p}$ is then binary phase-shift keying (BPSK)-modulated and transmitted over an additive white Gaussian noise (AWGN) channel that corrupts the modulated packets with a Gaussian noise of zero mean and variance σ^2. At the receiver, the observed vector is $\mathbf{y}_p = [\mathbf{y}_h, \mathbf{y}_b, \mathbf{y}_c]$, where $\mathbf{y}_h$, $\mathbf{y}_b$, and $\mathbf{y}_c$ are the observations of $\mathbf{h}$, $\mathbf{b}$, and $\mathbf{c}$, respectively. $\mathbf{y}_p$ contains the observations of $\mathbf{p}$ and represents a segment of the APL packet depicted in Figure 9.1. An overview of the transmission scheme is illustrated in Figure 9.2.

In practice, $\mathbf{b}$ is usually organized in groups of code words (e.g., texture information of a block or a macroblock), which are assumed to be encoded

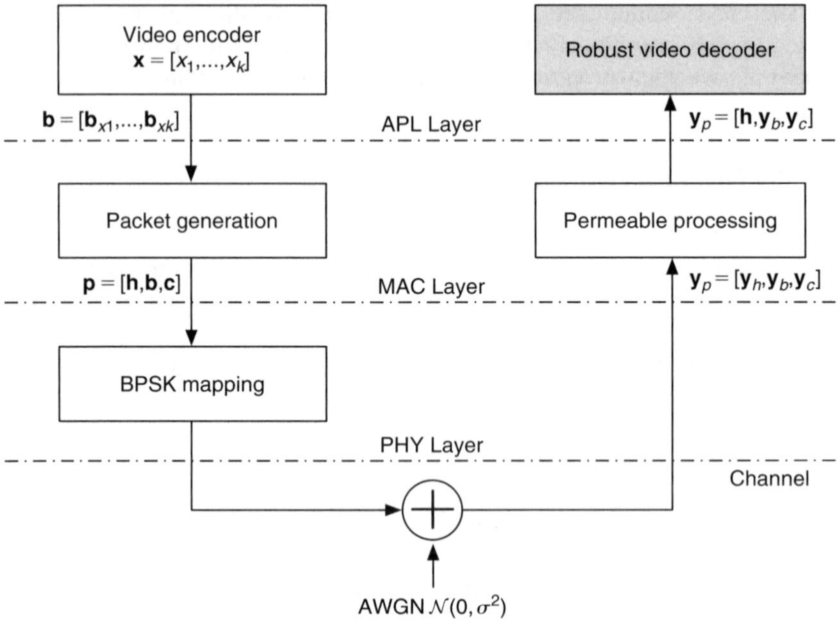

FIGURE 9.2 Overview of the transmission scheme.

independently.[1] Let $\mathbf{a}_1, \ldots, \mathbf{a}_e$ be the e groups of code words composing $\mathbf{b}$, i.e., $\mathbf{b} = [\mathbf{a}_1, \ldots, \mathbf{a}_e]$. The lengths $\ell(\mathbf{a}_\varepsilon)$, for $\varepsilon = 1, \ldots, e$, are supposed to be transmitted reliably as side information to the decoder. In the following, these lengths are called *position markers*. Using these markers, the decoding of each block may be performed separately by synchronizing the decoder over the corresponding portion in the received packet.

Decoding Algorithm

Assuming that the header $\mathbf{h}$ has been correctly received, the optimal MAP estimator $\widehat{\mathbf{a}}_\varepsilon$ for the εth group is given by

$$\widehat{\mathbf{a}}_\varepsilon = \arg\max_{\mathbf{a} \in \Omega_a^\varepsilon} P(\mathbf{a} | \mathbf{h}, \mathbf{y}_b, \mathbf{y}_c), \tag{9.3}$$

where Ω_a^ε is the set of valid combinations of $\mathbf{a}_\varepsilon$, i.e., compliant with the syntax of the VLC and the semantic of the source. When Ω_a^ε is not well structured, obtaining $\widehat{\mathbf{a}}_\varepsilon$ would require constructing the $2^{\ell(\mathbf{a}_\varepsilon)}$ possible combinations for

1. In realistic situations, the groups of code words belonging to a specific class of video coding parameters are correlated. However, we consider here that the existing dependencies are small and may be neglected.

$\mathbf{a}_\varepsilon$, keeping only the valid sequences (belonging to Ω_a^ε), and then evaluating $P(\mathbf{a}_\varepsilon|\mathbf{h},\mathbf{y}_b,\mathbf{y}_c)$ for each of them. When $\ell(\mathbf{a}_\varepsilon)$ is large (which is usually the case since this reduces the overhead due to the transmission of the side information), a sequential decoder is involved to reduce the decoding complexity [see Chapter 6 and Anderson and Mohan (1991b)].

Consider the nth step of the decoding of the εth group $\mathbf{a}_\varepsilon$. One may write

$$\mathbf{b}=[\mathbf{b}_{\varepsilon-1},\mathbf{u}_{\varepsilon,n},\mathbf{s}_{\varepsilon,n},\mathbf{r}_{\varepsilon,n}],$$

with

- $\mathbf{b}_{\varepsilon-1}=[\mathbf{a}_1,\ldots,\mathbf{a}_{\varepsilon-1}]$ are the bits of the first $\varepsilon-1$ groups. Note that for the decoding of $\mathbf{a}_\varepsilon$, $\mathbf{b}_{\varepsilon-1}$ is considered as a random vector and not as the decoded bitstream obtained previously. This prevents error propagation between successive group.
- $\mathbf{u}_{\varepsilon,n}$, the first bits of $\mathbf{a}_\varepsilon$ for which a set of valid combinations $\Omega_u^{\varepsilon,n}$ has been evaluated at step $n-1$ by the decoder.
- $\mathbf{s}_{\varepsilon,n}$, a vector for which, regardless of the syntax of the VLC and the semantic of the source, $2^{\ell(\mathbf{s}_{\varepsilon,n})}$ binary combinations are possible. Let $\Omega_s^{\varepsilon,n}$ be the set of these sequences.
- $\mathbf{r}_{\varepsilon,n}$, the $\ell(\mathbf{r}_{\varepsilon,n})$ remaining bits of $\mathbf{b}$. These bits have not yet been processed by the decoder, but they are involved in the computation of the CRC.

Figure 9.3 illustrates the considered structure of the packet. The observations associated to these four vectors are $\mathbf{y}_b^\varepsilon$, $\mathbf{y}_u^{\varepsilon,n}$, $\mathbf{y}_s^{\varepsilon,n}$, and $\mathbf{y}_r^{\varepsilon,n}$. Moreover, let $\Omega_{[u,s]}^{\varepsilon,n}\subset\Omega_u^{\varepsilon,n}\times\Omega_s^{\varepsilon,n}$ be the set of valid pairs $[\mathbf{u}_{\varepsilon,n},\mathbf{s}_{\varepsilon,n}]$.

At the nth step, the sequential decoding algorithm evaluates

$$P(\mathbf{u}_{\varepsilon,n},\mathbf{s}_{\varepsilon,n}|\mathbf{y}_b^{\varepsilon,n},\mathbf{y}_u^{\varepsilon,n},\mathbf{y}_s^{\varepsilon,n},\mathbf{y}_r^{\varepsilon,n},\mathbf{y}_c,\mathbf{h})\propto P(\mathbf{u}_{\varepsilon,n},\mathbf{s}_{\varepsilon,n},\mathbf{y}_b^{\varepsilon,n},\mathbf{y}_u^{\varepsilon,n},\mathbf{y}_s^{\varepsilon,n},\mathbf{y}_r^{\varepsilon,n},\mathbf{y}_c|\mathbf{h}), \tag{9.4}$$

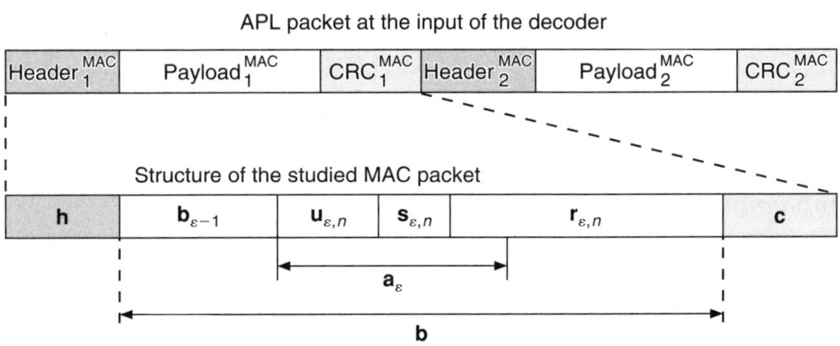

FIGURE 9.3 Partitioning of the received MAC packet at the nth iteration for the εth group.

for each $[\mathbf{u}_{\varepsilon,n}, \mathbf{s}_{\varepsilon,n}] \in \Omega_u^{\varepsilon,n} \times \Omega_s^{\varepsilon,n}$. In (9.4), one may write

$$P(\mathbf{u}_{\varepsilon,n}, \mathbf{s}_{\varepsilon,n}, \mathbf{y}_b^{\varepsilon,n}, \mathbf{y}_u^{\varepsilon,n}, \mathbf{y}_s^{\varepsilon,n}, \mathbf{y}_r^{\varepsilon,n}, \mathbf{y}_c|\mathbf{h}) = \sum_{\mathbf{b}_{\varepsilon-1}} \sum_{\mathbf{r}_{\varepsilon,n}} P(\mathbf{b}_{\varepsilon-1}, \mathbf{u}_{\varepsilon,n}, \mathbf{s}_{\varepsilon,n}, \mathbf{r}_{\varepsilon,n}, \mathbf{y}_b^{\varepsilon,n}, \mathbf{y}_u^{\varepsilon,n}, \mathbf{y}_s^{\varepsilon,n}, \mathbf{y}_r^{\varepsilon,n}, \mathbf{y}_c|\mathbf{h}). \tag{9.5}$$

Moreover,

$$P(\mathbf{b}_{\varepsilon-1}, \mathbf{u}_{\varepsilon,n}, \mathbf{s}_{\varepsilon,n}, \mathbf{r}_{\varepsilon,n}, \mathbf{y}_b^{\varepsilon,n}, \mathbf{y}_u^{\varepsilon,n}, \mathbf{y}_s^{\varepsilon,n}, \mathbf{y}_r^{\varepsilon,n}, \mathbf{y}_c|\mathbf{h}) = P(\mathbf{u}_{\varepsilon,n}, \mathbf{s}_{\varepsilon,n}|\mathbf{h}) \\ P(\mathbf{y}_u^{\varepsilon,n}|\mathbf{u}_{\varepsilon,n}, \mathbf{s}_{\varepsilon,n}, \mathbf{h}) P(\mathbf{y}_s^{\varepsilon,n}|\mathbf{y}_u^{\varepsilon,n}, \mathbf{u}_{\varepsilon,n}, \mathbf{s}_{\varepsilon,n}, \mathbf{h}) \\ P(\mathbf{b}_{\varepsilon}, \mathbf{r}_{\varepsilon,n}, \mathbf{y}_b^{\varepsilon}, \mathbf{y}_r^{\varepsilon,n}, \mathbf{y}_c|\mathbf{y}_u^{\varepsilon,n}, \mathbf{y}_s^{\varepsilon,n}, \mathbf{u}_{\varepsilon,n}, \mathbf{s}_{\varepsilon,n}, \mathbf{h}). \tag{9.6}$$

Using the fact that $\mathbf{u}_{\varepsilon,n}$ and $\mathbf{s}_{\varepsilon,n}$ do not depend on $\mathbf{h}$ and that the channel is memoryless, (9.6) becomes

$$P(\mathbf{b}_{\varepsilon-1}, \mathbf{u}_{\varepsilon,n}, \mathbf{s}_{\varepsilon,n}, \mathbf{r}_{\varepsilon,n}, \mathbf{y}_b^{\varepsilon,n}, \mathbf{y}_u^{\varepsilon,n}, \mathbf{y}_s^{\varepsilon,n}, \mathbf{y}_r^{\varepsilon,n}, \mathbf{y}_c|\mathbf{h}) = P(\mathbf{u}_{\varepsilon,n}, \mathbf{s}_{\varepsilon,n}) \\ \times P(\mathbf{y}_u^{\varepsilon,n}|\mathbf{u}_{\varepsilon,n}) P(\mathbf{y}_s^{\varepsilon,n}|\mathbf{s}_{\varepsilon,n}) P(\mathbf{b}_{\varepsilon-1}, \mathbf{r}_{\varepsilon,n}, \mathbf{y}_b^{\varepsilon}, \mathbf{y}_r^{\varepsilon,n}, \mathbf{y}_c|\mathbf{h}, \mathbf{u}_{\varepsilon,n}, \mathbf{s}_{\varepsilon,n}). \tag{9.7}$$

Now, combining (9.4), (9.5), and (9.7), one obtains

$$P(\mathbf{u}_{\varepsilon,n}, \mathbf{s}_{\varepsilon,n}|\mathbf{y}_b^{\varepsilon,n}, \mathbf{y}_u^{\varepsilon,n}, \mathbf{y}_s^{\varepsilon,n}, \mathbf{y}_r^{\varepsilon,n}, \mathbf{h}) \\ \propto P(\mathbf{u}_{\varepsilon,n}, \mathbf{s}_{\varepsilon,n}) P(\mathbf{y}_u^{\varepsilon,n}|\mathbf{u}_{\varepsilon,n}) P(\mathbf{y}_s^{\varepsilon,n}|\mathbf{s}_{\varepsilon,n}) \varphi(\mathbf{h}, \mathbf{u}_{\varepsilon,n}, \mathbf{s}_{\varepsilon,n}, \mathbf{y}_b^{\varepsilon}, \mathbf{y}_r^{\varepsilon,n}, \mathbf{y}_c), \tag{9.8}$$

with

$$\varphi(\mathbf{h}, \mathbf{u}_{\varepsilon,n}, \mathbf{s}_{\varepsilon,n}, \mathbf{y}_b^{\varepsilon}, \mathbf{y}_r^{\varepsilon,n}, \mathbf{y}_c) = \sum_{\mathbf{b}_{\varepsilon-1}, \mathbf{r}_{\varepsilon,n}} P(\mathbf{b}_{\varepsilon-1}, \mathbf{r}_{\varepsilon,n}, \mathbf{y}_b^{\varepsilon}, \mathbf{y}_r^{\varepsilon,n}, \mathbf{y}_c|\mathbf{h}, \mathbf{u}_{\varepsilon,n}, \mathbf{s}_{\varepsilon,n}). \tag{9.9}$$

In (9.8), $P(\mathbf{u}_{\varepsilon,n}, \mathbf{s}_{\varepsilon,n})$ represents the a priori probability of sequence $[\mathbf{u}_{\varepsilon,n}, \mathbf{s}_{\varepsilon,n}]$, which is null if $[\mathbf{u}_{\varepsilon,n}, \mathbf{s}_{\varepsilon,n}] \notin \Omega_{[u,s]}^{\varepsilon,n}$. Valid sequences are assumed to be equiprobable, i.e., $P(\mathbf{u}_{\varepsilon,n}, \mathbf{s}_{\varepsilon,n}) = 1/|\Omega_{[u,s]}^{\varepsilon,n}|$. Consequently, the metric $\mathcal{M}_\varepsilon$ associated to a valid sequence in group ε is given by

$$\mathcal{M}_\varepsilon([\mathbf{u}_{\varepsilon,n}, \mathbf{s}_{\varepsilon,n}] \in \Omega_{[u,s]}^{\varepsilon,n}|\mathbf{h}, \mathbf{y}_p) = P(\mathbf{y}_u^{\varepsilon,n}|\mathbf{u}_{\varepsilon,n}) P(\mathbf{y}_s^{\varepsilon,n}|\mathbf{s}_{\varepsilon,n}) \\ \times \varphi(\mathbf{h}, \mathbf{u}_{\varepsilon,n}, \mathbf{s}_{\varepsilon,n}, \mathbf{y}_b^{\varepsilon}, \mathbf{y}_r^{\varepsilon,n}, \mathbf{y}_c), \tag{9.10}$$

where $P(\mathbf{y}_u^{\varepsilon,n}|\mathbf{u}_{\varepsilon,n})$ and $P(\mathbf{y}_s^{\varepsilon,n}|\mathbf{s}_{\varepsilon,n})$ are the likelihoods of $\mathbf{u}_{\varepsilon,n}$ and $\mathbf{s}_{\varepsilon,n}$, respectively; their value being directly deduced from the channel model.

Implementation Issues and Complexity

In (9.9), $\varphi(\mathbf{h}, \mathbf{u}_{\varepsilon,n}, \mathbf{s}_{\varepsilon,n}, \mathbf{y}_b^{\varepsilon}, \mathbf{y}_r^{\varepsilon,n}, \mathbf{y}_c)$ is a sum involving $O(2^{\ell(\mathbf{b}_{\varepsilon-1})+\ell(\mathbf{r}_{\varepsilon,n})})$ operations. Consequently, the evaluation complexity of (9.4) for all $[\mathbf{u}_{\varepsilon,n}, \mathbf{s}_{\varepsilon,n}] \in \Omega_{[u,s]}^{\varepsilon,n}$ is $O\big(2^{\ell(\mathbf{b}_{\varepsilon-1})+\ell(\mathbf{r}_{\varepsilon,n})}|\Omega_u^{\varepsilon,n}||\Omega_s^{\varepsilon,n}|\big)$. $|\Omega_s^{\varepsilon,n}|$ depends on the number of bits taken into account at the nth steps and may thus be upper-bounded by a constant. One of the main difficulty comes from $|\Omega_u^{\varepsilon,n}|$, which is growing exponentially with n. To

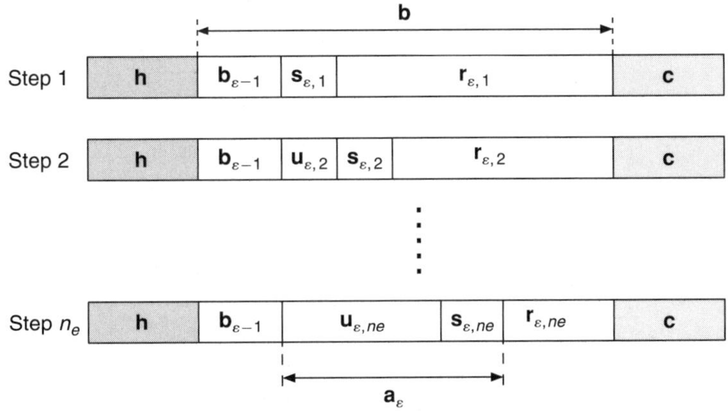

FIGURE 9.4 Evolution of the partitions through the sequential decoding steps for the eth group.

limit the complexity increase, an M-algorithm (see Section 6.4.2, page 177) is used. At each step, only the M most probable sequences belonging to $\Omega_{[u,s]}^{\varepsilon,n}$ are kept and stored in $\Omega_u^{\varepsilon,n+1}$. The parameter M allows to tune the tradeoff between complexity and efficiency.

Figure 9.4 illustrates the evolution of the parts $\mathbf{b}_{\varepsilon-1}$, $\mathbf{u}_{\varepsilon,n}$, $\mathbf{s}_{\varepsilon,n}$, and $\mathbf{r}_{\varepsilon,n}$ through the different steps. The flowchart of the decoding algorithm is depicted in Figure 9.5 and explanations are given in the following. Note that the metric $\mathcal{M}_\varepsilon([\mathbf{u}_{\varepsilon,n},\mathbf{s}_{\varepsilon,n}]|\mathbf{h},\mathbf{y}_p)$ is computed using (9.10). A each step, one obtains a sub-optimal algorithm, the complexity of which becomes $O(2^{\ell(\mathbf{b}_{\varepsilon-1})+\ell(\mathbf{r}_{\varepsilon,n})})$, mainly due to the evaluation of φ in (9.10).

Let n_ε be the number of steps necessary to reach the end of group ε. The number of bits $\ell(\mathbf{s}_{\varepsilon,n})$, for $n=1,\ldots,n_\varepsilon$, must, thus, be adjusted such that

$$\sum_{n=1}^{n_\varepsilon} \ell(\mathbf{s}_{\varepsilon,n}) = \ell(\mathbf{a}_\varepsilon), \tag{9.11}$$

for all $\varepsilon=1,\ldots,e$. In practice, the first $n_\varepsilon-1$ decoding depths are set to a constant value and the last one, i.e., $\ell(\mathbf{s}_{\varepsilon,n_\varepsilon})$, is chosen so that (9.11) is satisfied.

We now describe the complete sequential decoding algorithm for the εth group. At initialization ($n=1$), $\Omega_u^{\varepsilon,1}=\emptyset$. Afterwards, at each step $n>1$, the algorithm explores the new branches (on $\ell(\mathbf{s}_{\varepsilon,n})$ bits) of the decoding tree and only keeps the M most probable extended sequences $[\mathbf{u}_{\varepsilon,n},\mathbf{s}_{\varepsilon,n}]$. These M sequences are temporarily stored in a stack (corresponding to $\Omega_u^{\varepsilon,n+1}$), before being extended again at the next step.

This algorithm is applied to the decoding of H.264/AVC CAVLC sequences in the next section.

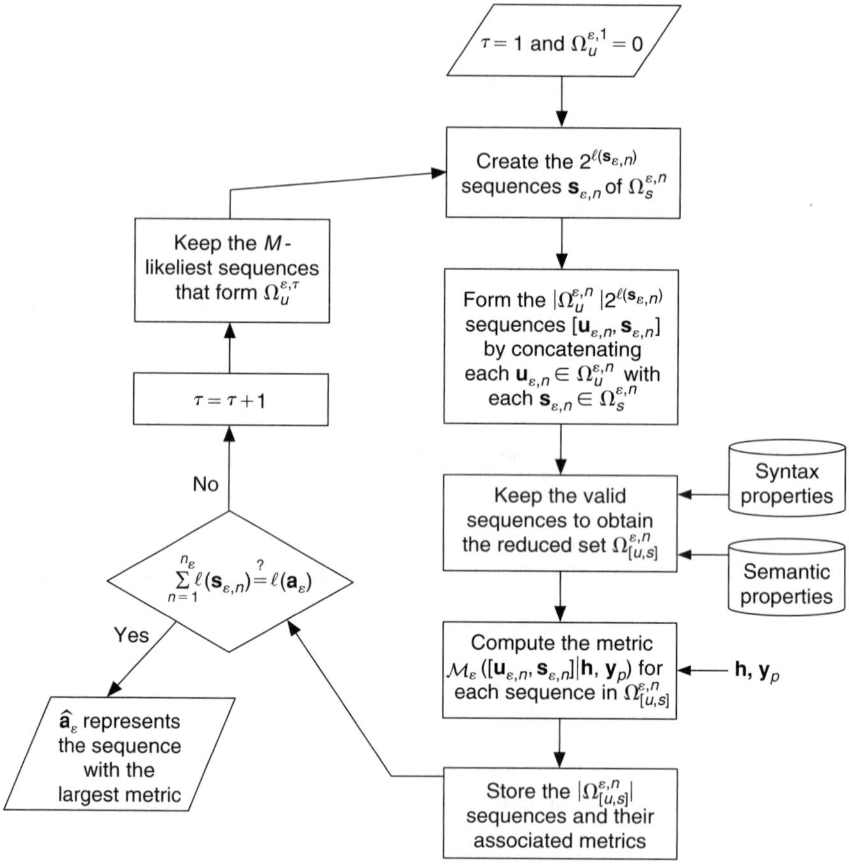

FIGURE 9.5 Proposed sequential decoding scheme.

9.1.4 Simulation Results

In the *extended* profile of H.264/AVC (ITU-T and ISO/IEC JTC 1, 2003), an error-resilience mode is provided. In this mode, the compressed picture data are classified according to their influence on the video quality. Three partitions are defined:

- Partition A contains the headers and the motion vectors of each encoded picture.
- Partition B consists of the texture coefficients of intracoded blocks.
- Partition C contains the texture coefficients of intercoded blocks.

This stream decomposition allows an adaptation of the protection to the sensitivity of the partition to be sent. After compression, each partition is encapsulated in a *Network Abstraction Layer Unit* (NALU) that is delivered to the RTP layer.

In our simulation, packets associated to the A partition are assumed heavily protected and correctly interpreted at the receiver. On the other hand, B and C packets are transmitted over a noisy channel and are corrupted by transmission errors. As previously mentioned, these packets contain the texture coefficients of the various 4×4 blocks of a picture, see Section 3.4.5. These blocks are encoded in CAVLC (Richardson, 2003).

We focus here on the decoding of the CAVLC sequences included in the B and C packets. Each CAVLC sequence is considered an independent group of code words that can be separated from the others by using the position markers, transmitted as side information (see Section 9.1.3). Consequently, the group-based sequential decoding method of Section 9.1.3 may be used for their estimation. Note that in H.264/AVC, the CAVLC sequences are not totally independent (adaptive context), but the existing dependencies are indeed small and are neglected here. The performance of the presented method has been evaluated by simulations and compared to that of two other decoding methods: a standard decoding method and a classical, robust decoding method (exploiting only the source properties).

The simulated system consists of a transmitter, a channel, and a receiver. The transmitter uses repeatedly the five first pictures of *Foreman.cif* with the IPPPP frame structure and generates the encoded partitions using the CAVLC H.264/AVC video coder. Video packets (partitions) are then processed by the protocol stack defined in Figure 7.15, page 207. At the MAC layer, IP packets are fragmented in several MAC packets of variable payload size. A CRC of 4 bytes, consistent with the 802.11 standard, is added at the end of each MAC fragment. At the PHY layer of the transmitter, the data are encoded by the convolutional channel coder of the 802.11*a* standard (IEEE, 1999). Next, the coded PHY packets are mapped onto BPSK symbols before being sent over the physical medium. To improve the decoding performance, the aforementioned position markers are sent as side information, indicating the location of each 4×4 encoded texture block in B and C packets. This side information is transmitted in a specific NALU, and the markers are compressed using the Exp-Golomb coding of H.264/AVC. The overhead due to the transmission of this redundancy represents approximately 30% of the total bitrate. The channel does not degrade the data contained in A packets nor the side information. On the other hand, it does add a white Gaussian noise to the other packets. At the receiver, the data are processed by a SISO channel decoder (BCJR algorithm) and are then delivered to the APL layer (following the permeable mechanism explained in Section 9.1.2). At the APL layer, three different decoders are considered:

1. A *standard* decoder performs hard decisions on the received soft data and makes usage of position markers to decode each block.
2. A *robust* decoder uses the source properties, the soft data, and the position markers but does not use the redundancy provided by the CRC. This decoder exploits the algorithm depicted in Section 9.1.3, but the metric in (9.10) does not include the term φ.

3. A *CRC-robust* decoder combines all previous sources of redundancy along with the CRC properties through the decoding method presented in Section 9.1.3.

Note that in our simulations, the two robust decoders use the same stack size $M = 20$ and the same default decoding depth $\ell(\mathbf{s}) = 4$ bits. The CRC-robust decoder uses the suboptimal method. For this purpose, the CRC is split into four blocks of 8 bits.

Figures 9.6 and 9.7 depict the *peak signal-to-noise ratio* (PSNR) of the decoded video as a function of the SNR for the three different decoders, with and without channel coding. In Figure 9.6, the channel coder/decoder at PHY layer was deactivated. In both figures, the standard, robust, and CRC-robust decoders are compared with a MAC payload size of 100 bytes. We can notice that the standard decoder is outperformed by the two robust decoders in both cases. Moreover, the two robust decoders are equivalent for low SNRs, and the CRC-robust decoder starts to outperform the classical robust decoder only beyond a given threshold. Above this threshold, the coding gain increases with the SNR. This behavior is specific to channel decoding performance. In our simulations, the threshold is approximately 8.5 dB in Figure 9.6 and 1.8 dB in Figure 9.7.

Figure 9.8 illustrates the 5th image of the *Foreman.cif* video sequence, along with its reproductions obtained after this image is transmitted and decoded by the standard, robust, and CRC-robust decoders. In this figure, the channel

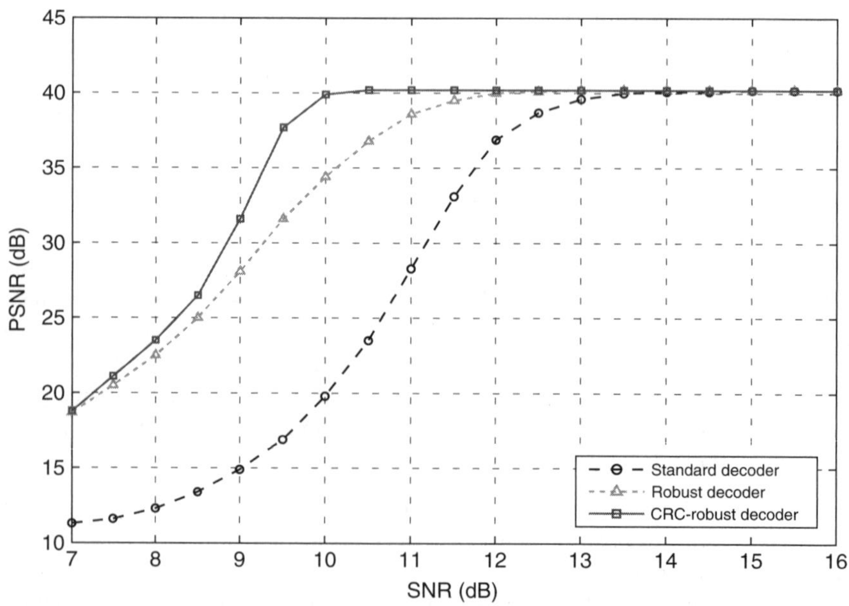

FIGURE 9.6 Decoded image quality (PSNR) versus SNR for the three types of decoders, with MAC payload size of 100 bytes and deactivated channel coder/decoder at PHY layer.

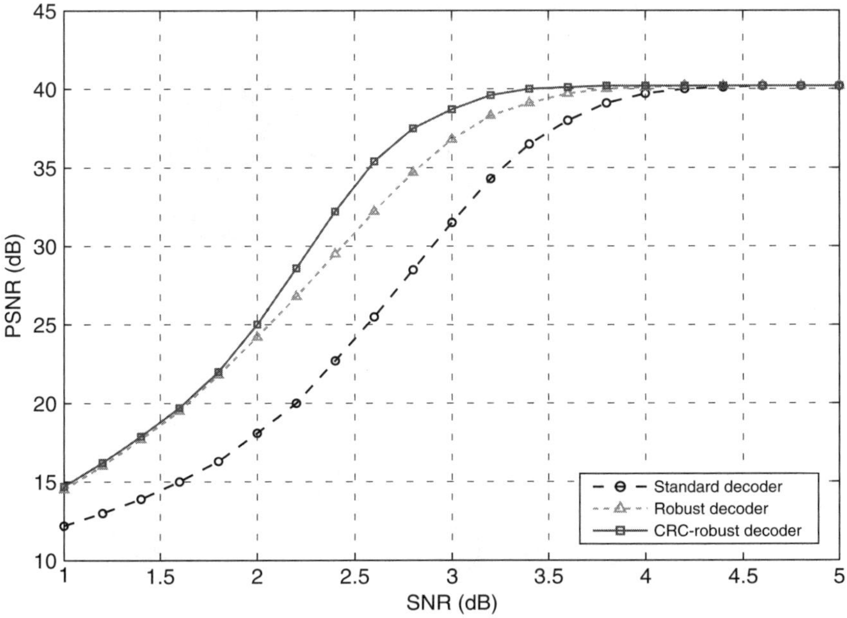

FIGURE 9.7 Decoded image quality (PSNR) versus SNR for the standard, robust, and CRC-robust decoders. In this case, the 802.11*a* channel coder/decoder is considered at PHY layer, and the MAC layer protocol of the transmitter generates 100-byte MAC payload.

coder/decoder is considered. This result was obtained with a payload size of 100 bytes and at an SNR of 2.8 dB for which the PSNR of the standard, robust, and CRC-robust decoders are 29, 35, and 38 dB, respectively (see Figure 9.7). Obviously, the image obtained with the standard decoder contains many artifacts and is of a very poor quality. On the other hand, the robust decoder strongly improves the quality even though some distortions are still visible. Finally, no visual difference may be noticed between the original image and the image obtained by the CRC-robust decoder.

9.1.5 Discussion

The proposed decoder shows that jointly exploiting the residual redundancy of the source coder along with the redundancy provided by the CRC at MAC layer may improve the decoding performance of JSCD schemes. The implementation of the MAP estimator combines a sequential decoding algorithm with a BCJR algorithm for evaluating the appropriate decoding metrics. We applied this method to H.264/AVC decoding of CAVLC sequences. Simulation results show that the information carried by the CRC does improve the decoding efficiency. More precisely, joint use of CRC and source properties becomes interesting above a certain threshold. It should be noted that the bitrate used for transmitting the side

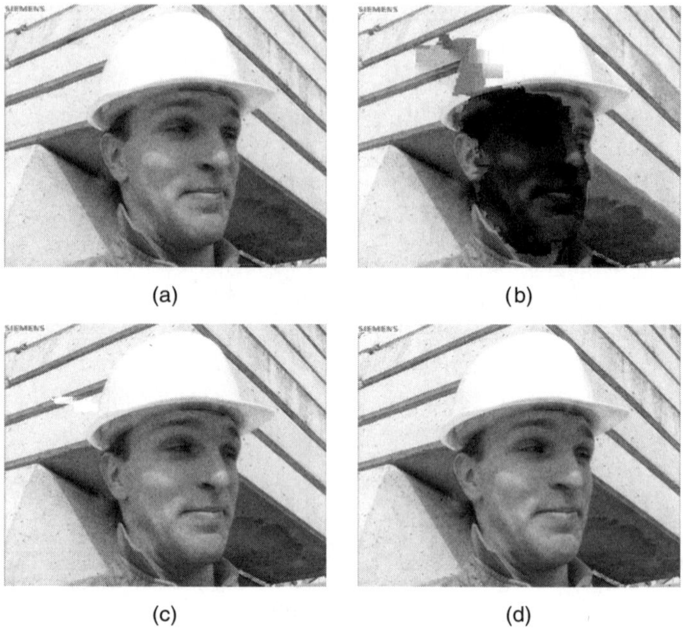

FIGURE 9.8 Fifth image of *Foreman.cif* obtained after (a) error-free decoding, (b) standard decoding, (c) robust decoding, and (d) CRC-robust decoding for a SNR of 2.8 dB and a MAC payload size of 100 bytes, with channel coding.

information is rather high (approximately 30%) in the presented experiments. The amount of side information may be significantly reduced, e.g., considering position markers indicating the location of each macroblock of 16×16 pixels.

Note that no iterative decoding was used in this example (see the next section), to maintain the complexity at a reasonable level. This is a good example of a situation where many tools explained in this book are combined to cope with (close to) practical situations.

9.2 ITERATIVE DECODING

In actual transmission chains, and especially those designed for wireless channels, the bits produced by the source coder are generally further encoded by a channel coder. The algorithms described in Chapters 5 and 6 bring some improvements to the source decoding, based on information taken from the PHY layer, but when a FEC device is used, the improvements provided by JSCD may be considerably reduced, see, e.g., the results of Section 9.1.4. It was shown in the above-mentioned chapters that efficient source decoding can be obtained by making use of an underlying state model, describing the feasible bitstreams. Channel coders are also usually described by state models, characterizing the set of all possible code words. When considering the conjunction of source coding and channel coding, the transmitted bitstream can, thus, be seen as produced by transitions on a global state model, product of the state models of the source and the channel coders. Optimal decoding can be performed on this product model (Lakovic et al., 1999; Murad and Fuja, 1999). However, the state-space dimension of this model explodes in practical cases.

Therefore, one relies on an iterative joint source-channel decoding system (Bauer and Hagenauer, 2000a,b, 2001a), in which iterations are performed between the two main parts of the receiver (source and channel decoders), just like it was proposed for turbo codes. Turbo decoding was later interpreted in terms of factor graphs (Forney, 2001) as being somewhat equivalent to running a belief propagation algorithm on a cyclic factor graph, which is an efficient procedure, provided the cycles are long enough. It was also observed that the simple introduction of an interleaver between two codes can make short cycles become long (Ge et al., 2001). An approximate iterative estimator can, thus, be designed, working alternately on the source coder model and the channel coder model, with significant gain in complexity. This turbo principle (Berrou et al., 1993) can be applied to many estimation problems beyond channel and source-channel decoding.

The presentation here follows the classical "turbo-decoding" presentation. Turbo decoding or iterative decoding is based on an iterative exchange of soft information between two SISO decoders. The global encoder considered in JSCD has the same structure as serially concatenated turbo codes (Benedetto et al., 1998). At the transmitter side, one of the encoders is the source encoder (outer code) and the other one is the channel encoder (FEC, inner code), separated by an interleaver for obtaining long cycles. Upon reception, SISO decoders have to be used, both for the channel decoder and the source decoder since they have to exchange soft information. This is depicted in Figure 9.9.

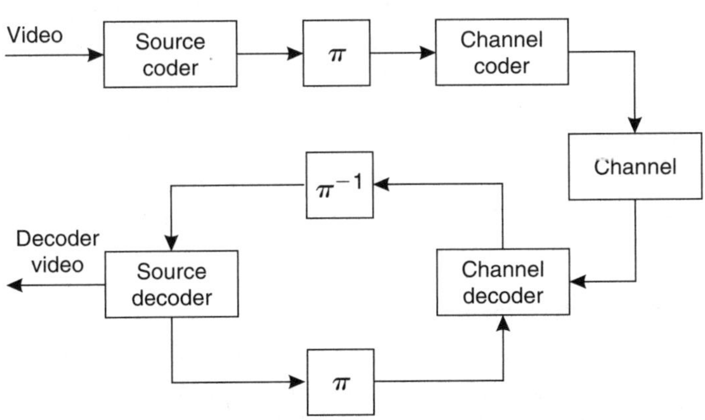

FIGURE 9.9 Iterative joint source-channel decoding of video transmitted on a noisy channel.

9.2.1 General Architecture

The soft-output values from the channel decoder are used by the source decoder for improving its decoding performance. Then, the soft-output values from the source decoder are used for improving the channel decoding. This loop is performed several times in an iterative process. Each SISO decoder uses its own redundancy (either implicit for the source or explicit for the channel), its own (channel or source) code structure, and also the soft information from the other SISO decoder, to decode the bitstream. In the case of the SISO source decoder, the redundancy remaining in source data, such as the one due to the syntax of the code and the source constraints, is exploited. This *iterative decoding process* is the core of the performance of the turbo codes. In our case, it is expected that the use of this algorithm in an iterative procedure brings further improvements because a small improvement on one decoder (here, the SISO source decoder) may be amplified by the iterative process. By doing so, one usually recovers most of the improvements brought by the robust source decoder when evaluated without FEC. There is a whole body of literature on the topic of iterative decoding, and we will not attempt to provide a constructive description of it. Instead, we provide inputs to the literature, an algorithmic view of the decoders, and the main characteristics that may impact their practical usage.

Most robust source decoders explained in this book can provide a posteriori probabilities for both the sequences of interest and for the individual bits, see Chapter 5. This is also the case for the BCJR (or forward-backward) algorithm applied to channel codes. However, similarly to the iterative decoding of turbo codes, information quantities that are exchanged between the SISO source decoder and the SISO channel decoder is the so-called *extrinsic information*. This quantity is used as an a priori probability in the other decoding block. The *extrinsic information* (or *extrinsic probability* when properly normalized) is defined as the ratio of the a posteriori probability evaluated by the given building block, divided by the a priori probability used at the same iteration by the same building block. In other words, it is also the ratio of the a posteriori probability evaluated by the given building block divided by the extrinsic probability provided by the other block as a result of its previous iteration.

This procedure is schematically depicted in Figure 9.10, where the a posteriori probabilities are denoted as $p_{\text{app},\nu}^{\text{s},k}$ for the APP of the νth bit provided by the source decoder at iteration τ, $p_{\text{app},\nu}^{\text{c},\tau}$ for the APP of the νth bit provided by the channel decoder at iteration τ, and $p_{\text{ext},\nu}^{\text{s},k}$ and $p_{\text{ext},\nu}^{\text{c},k}$, the corresponding extrinsic probabilities.

9.2.2 Convergence Issues

The convergence of iterative processes is always the subject of many questions. One has to determine whether the process converges. When this is the case, the convergence point has then to be determined. One may also be interested in the convergence speed.

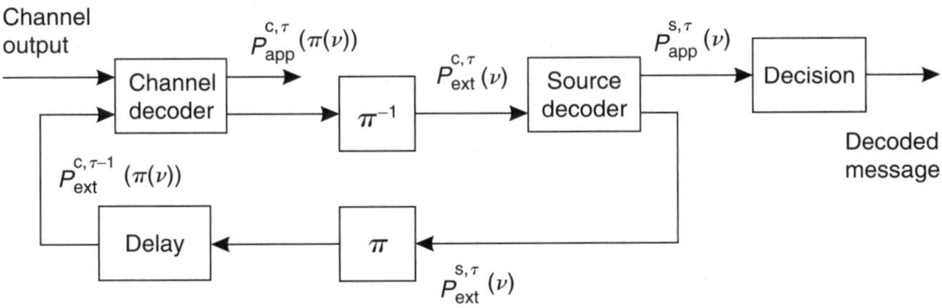

FIGURE 9.10 Iterations in the decoding process, $\pi(\nu)$ represents the index of the νth bit after interleaving.

The convergence of turbo decoding was analyzed in Ten Brink (2001) for turbo codes with the introduction of extrinsic information transfer (EXIT) charts, the principle of which is briefly introduced here in the case of the serial concatenation of a source and a channel code.

Assume that BPSK signaling is used after channel coding to get modulated bit $B_\nu = \pm 1$ sent onto an AWGN channel with variance σ^2. The quantity

$$a_\nu = \log\left(\frac{p(B_\nu = 1 | Y_\nu = y_\nu)}{p(B_\nu = -1 | Y_\nu = y_\nu)}\right) \tag{9.12}$$

obtained from the measurement y_ν at channel output may be viewed as a realization of the random variable

$$A_\nu = m_A B_\nu + N_A, \tag{9.13}$$

where N_A is Gaussian, with variance $\sigma_A^2 = 4/\sigma^2 = 2m_A$. Knowing that $B_\nu = b$, the probability distribution for A_ν is then,

$$p_A(\xi | B_\nu = b) = \frac{1}{\sqrt{2\pi\sigma_A^2}} \exp\left(-\frac{\left(\xi - (\sigma_A^2/2)\,b\right)^2}{2\sigma_A^2}\right). \tag{9.14}$$

This probability density function (pdf) is characterized by a single parameter σ_A^2. From (9.14), one may deduce the mutual information between B_ν and A_ν,

$$\begin{aligned} I(B_\nu, A_\nu) &= \frac{1}{2} \sum_{b=-1,1} \int_{-\infty}^{\infty} p_A(\xi | B_\nu = b) \log_2 \frac{2p_A(\xi | B_\nu = b)}{p_A(\xi | B_\nu = -1) + p_A(\xi | B_\nu = 1)} d\xi \\ &= 1 - \int_{-\infty}^{\infty} \frac{1}{\sqrt{2\pi\sigma_A^2}} \exp\left(-\frac{\left(\xi - (\sigma_A^2/2)\,b\right)^2}{2\sigma_A^2}\right) \log_2\left(1 + \exp(-\xi)\right) d\xi, \end{aligned} \tag{9.15}$$

assuming that $p(B_\nu = 1) = p(B_\nu = -1) = 1/2$, see Cover and Thomas (1991).

Ten Brink (2001) describes the a priori information at the input and the extrinsic information at the output of the constituent decoders, when expressed as log-likelihood ratios (9.12), are modeled as Gaussian distribution such as (9.14). To be reasonable, this assumption, justified by observation, requires the presence of long interleavers in the coding scheme.

Then, the mutual information $I(B_\nu, A_\nu)$ between B_ν and the a priori information A_ν and the mutual information

$$I(B_\nu, E_\nu) = 1 - \int_{-\infty}^{\infty} \frac{1}{\sqrt{2\pi\sigma_E^2}} \exp\left(-\frac{\left(\xi - (\sigma_E^2/2)\,b\right)^2}{2\sigma_E^2}\right) \log_2\left(1 + \exp(-\xi)\right) d\xi. \tag{9.16}$$

between B_ν and the extrinsic information E_ν are both written as in (9.15).

Under these assumptions, the constituent decoders may then be characterized by their information transfer characteristic defined as

$$I_E = T\left(I_A, E_b/N_0\right). \tag{9.17}$$

For a constant signal-to-noise ratio on the channel, one gets simply

$$I_E = T\left(I_A\right). \tag{9.18}$$

This function is monotonically increasing in I_A, $0 \leqslant I_A \leqslant 1$. An increased a priori information results in an increase of the extrinsic information. Thus, the inverse function

$$I_A = T^{-1}\left(I_E\right) \tag{9.19}$$

is defined over $T(0) \leqslant I_E \leqslant T(1)$.

If I_A and I_E are close to 1, this means that the estimation is very likely to correspond to the original signal B_ν, i.e., the estimation is good. Conversely, if I_A and I_E are close to 0, the estimation is bad (the estimate does not carry information about the true value).

There are very few decoders where information transfer characteristic may be explicitly obtained. Usually, one resorts to Monte–Carlo simulation to get estimates for $p_A(\xi | B_\nu = b)$ and $p_E(\xi | B_\nu = b)$ via histograms, from which estimates for (9.15), (9.16), and (9.17) are deduced.

Assuming that the information transfer characteristics T_c and T_s of the channel and source decoders are both known, the iterations between the decoders may be expressed as information transfers between decoders. At iteration $\tau = 0$, one starts with no a priori information $I_{A,\nu}^{c,0} = I\left(B_\nu, A_\nu^{c,0}\right) = 0$ at the input of channel decoder. At iteration τ, one gets

$$I_{E,\nu}^{c,\tau} = I\left(B_\nu, E_\nu^{c,\tau}\right) = T_c\left(I_{A,\nu}^{c,\tau-1}\right). \tag{9.20}$$

Then, $I_{E,\nu}^{c,\tau}$ is forwarded to the source decoder (interleaving does not change the mutual information) and becomes $I_{A,\nu}^{s,\tau}$. At the output of the source decoder, one gets

$$I_{E,\nu}^{s,\tau} = I\left(B_\nu, E_\nu^{s,\tau}\right) = T_s\left(I_{A,\nu}^{c,\tau}\right). \tag{9.21}$$

The extrinsic output of the source decoder is deinterleaved and fed back to the channel decoder as a priori information $I_{A,\nu}^{c,\tau} = I_{E,\nu}^{s,\tau}$ (up to a de-interleaving).

The iterations proceed as long as

$$I_{E,\nu}^{s,\tau+1} > I_{E,\nu}^{s,\tau}, \tag{9.22}$$

with

$$I_{E,\nu}^{s,\tau+1} = T_s\left(T_c\left(I_{E,\nu}^{s,\tau}\right)\right). \tag{9.23}$$

This may be reformulated as:

$$T_s^{-1}\left(I_{E,\nu}^{s,\tau+1}\right) > T_c\left(I_{E,\nu}^{s,\tau}\right). \tag{9.24}$$

Iterations stop when

$$T_s^{-1}\left(I_{E,\nu}^{s,\tau+1}\right) = T_c\left(I_{E,\nu}^{s,\tau}\right). \tag{9.25}$$

It can easily be shown that both sequences $I_{E,\nu}^{s,\tau}$ and $I_{E,\nu}^{c,\tau}$ have the same limit value at convergence.

Obviously, one is interested in systems such that the sequence

$$I_E^{c,0}, I_E^{s,0}, I_E^{c,1}, I_E^{s,1}, \ldots, I_E^{c,\tau}, I_E^{s,\tau}, \ldots$$

converges to some value as close to 1 as possible, with the fewest iterations as possible.

The analysis and optimization of joint source-channel turbo decoders have been carried out with EXIT charts, e.g., by Adrat and Vary (2005), Bauer and Hagenauer (2001a), Grangetto et al. (2004), Jaspar and Vandendorpe (2005b), Poulliat et al. (2005), and Thobaben and Kliewer (2003).

Intuitively, the first iteration of such decoders provide the same result as the classical ones, and the improvement brought by the iterations strongly depends on the EXIT characteristic of the constituent codes. In our case, the channel decoder remains identical, e.g., BCJR-based, and only the source decoder is now a soft decoder taking into account additional knowledge on the source bitstream. Therefore, the more efficient the source decoder is, i.e., if it makes good use of the source redundancies, the more efficient the iterations will be.

9.2.3 Situations in Which Iterative Decoding Is of Interest

In fact, many situations were studied in the literature. JSC turbo decoding was first proposed by Bauer and Hagenauer (2001b). The results obtained by testing different VLCs with a memoryless source, a convolutional code, and an AWGN channel was a good motivation for further work.

Other situations of interest were also studied:

- Sources with memory (Markov sources) (Garcia-Frias and Villasenor, 2001; Thobaben and Kliewer, 2003)
- Source semantics (Nguyen and Duhamel, 2004; Peng et al., 2000)
- VLCs were used at many places (Bauer and Hagenauer, 2001b; Guyader et al., 2001); FLCs were used by Adrat and Vary (2005); Gortz (2001)
- Resilient VLCs (Jaspar and Vandendorpe, 2005a)
- Arithmetic codes (Grangetto et al., 2004; Guionnet and Guillemot, 2003)

Obviously, also, the largest improvement is obtained by carefully adapting the source properties that are used to the type of channel code used as an inner code and of interleavers that are used to separate both constituents. Even if such a tuning was the topic of relatively few papers (with the exception of Poulliat et al., 2005), many types of situations were studied,

- turbo codes are considered by Guivarch et al. (2000), and Peng et al. (2000); sometimes without interleaver between the source code and the turbo code (unlike Figure 9.9).
- LDPCs are suggested by Poulliat et al. (2005).
- A parallel concatenation with a convolutional code is proposed by Kliewer and Thobaben (2003).

For more references, see Jaspar et al. (2007).

9.2.4 An Example: Video Transmitted on General Packet Radio Service (GPRS)

To evaluate the interplay between the amount of redundancy used by the soft source decoder, the channel decoder, and the final visual performance, we provide the following simulation, taken from Nguyen (2005).

The Source

The video sequence is a 12-second extract (from the 13th second to the 25th second) of the trailer of the movie *Matrix I*. The video is CIF (352×288 pixels) encoded at 12.5 frames per second. It is composed of one INTRA-encoded image followed by 149 INTER-encoded images. The video clip is characterized by a lot of motions, textures, and background changes. It is a typical action movie clip. Image blocks are extracted from this video sequence. The transmission of each of these image blocks on the enhanced GPRS (EGPRS) system is simulated.

The Transmission Chain

The GPRS system is the new generation of the Groupe Spécial Mobile (GSM), today's pan-European public land mobile system (Halonen et al., 2002). It supplements circuit-switched data and short message service by a packet-switched data system allowing file transfer, Web browsing, etc. However, the data rates provided by the GPRS system are quite small. Hence, an enhanced version of the GPRS system (EGPRS) was proposed, with a modified physical layer allowing higher data rates for the packet traffic channel. Instead of the four GPRS coding schemes (CS), nine modulation and coding schemes (MCS) are used in the EGPRS to increase the data rate per time slot (3GPP, 2009). The nine MCS are numbered in increasing order with the data rates and in decreasing order with the error-protection capacity. The GPRS CS are incorporated in the MCS-1 to MCS-4 of EGPRS using a GMSK modulation. The other five MCS (5–9) are based on an 8-PSK modulation. The channel coding can mainly be divided into three steps. The first is the rate $1/3$ convolutional coding, which multiplies by three, the amount of bits to be transmitted. This channel coding is defined by the polynomials

$$G_4 = D^6 + D^5 + D^3 + D^2 + 1$$
$$G_7 = D^6 + D^3 + D^2 + D + 1$$
$$G_5 = D^6 + D^4 + D + 1$$

Then, a puncturing operation depending on the selected MCS is performed. The third step is the interleaving of the selected data bits, so as to improve the quality of decoding. Table 9.1 summarizes the modulation type, the coding and puncturing rate, and the maximum bitrates for each MCS (3GPP, 2009).

The Simulation

Our intent here is not to provide a global simulation of a given application but rather to evaluate the impact of iterations on the received video, based on error patterns that are as close as possible to the ones generated by wireless networks. Therefore, the issue of the compatibility of available EGPRS bitrates with the throughput required by the video is not considered here. Moreover, the streaming server is the simplest one; it puts the bitstream into radio blocks, and the length

TABLE 9.1 The Various Modulation and Coding Schemes in EGPRS

CS	Modulation	Max Rate (kbps) with 1 Timeslot	Coding and Puncturing Rate	Max Rate (kbps) with 4 Timeslots
MCS-9	8-PSK	59.2	1	236.8
MCS-8		54.4	0.92	217.6
MCS-7		44.8	0.76	179.2
MCS-6		29.6	0.49	118.4
MCS-5		22.4	0.37	89.6
MCS-4	GMSK	17.6	1	70.4
MCS-3		14.8	0.85	59.2
MCS-2		11.2	0.66	44.8
MCS-1		8.8	0.53	35.2

in bits of each image block sequence is assumed to be known at the receiver. No header overheads are considered here. We assume that no retransmission (ARQ) is used.

The EGPRS error patterns and transmission chain have been obtained by a realistic EGPRS simulator made available to us by Alcatel–Lucent. The transmission is simulated for the MCS-4 scheme (see Table 9.1), and the main transmission parameters are: Typical Urban, 50 km/h (TU50) channel profile, carrier frequency 900 MHz, TDMA transmission mode, ideal frequency-hopping, no shadowing, no diversity.

Several decoders are used to decode each of the received image blocks.

1. An H.263 prefix-based decoder with a synchronization at each image block. Hard decision is taken on the output of the channel decoder, and the resulting bitstream is fed into a standard H.263 decoder. Synchronization is assumed to be based on the knowledge of the block length in bits so that errors do not propagate too much.
2. A SISO VLC syntax-based decoder iterated four times with the SISO channel decoder.
3. A SISO VLC syntax-based decoder with last constraint (see Section 4.3.2, page 112) iterated four times with the SISO channel decoder.
4. A SISO VLC syntax-based decoder with last and run constraints (see Section 4.3.2, page 112) iterated four times with the SISO channel decoder.

Since the proposed VLC decoder assumes that the boundaries between image blocks are known, the lengths of the image blocks are also supposed to be known by the H.263 prefix-based decoder 1 above. After being decoded by the decoders 2, 3, and 4, which correspond to increasing use of the residual redundancy, these image blocks are inserted back to the H.263 bitstream. The received video sequence is then played.

The first INTRA image extracted from the video sequence Matrix I, transmitted over EGPRS error-pattern channel (SNR = 18 dB) and decoded as described above is shown in Figure 9.11 for the prefix decoding (Decoder 1), in Figure 9.12 for an iterative decoding when the source decoder takes the VLC syntax only into account, in Figure 9.13 for an iterative decoding when the source decoder takes the VLC syntax and the *last* constraint into account, and in Figure 9.14 for an iterative decoding when the source decoder takes the VLC syntax and the *last* and *run* constraints into account.

Obviously, the hard decoding cannot produce an image of sufficient quality nor can the iterative scheme where the source decoder makes use of the VLC syntax, even if the iterations begin to improve the image, which is anyway far from acceptable. When some semantics of the bitstream are included in the process, the situation changes; taking the *last* constraint into account provides a useless image at the first iteration, but the additional redundancy is more effective in the iterations, and as a result, the image obtained at iteration 4 is almost correct. Finally, using more redundancy (last and run constraints) improves a lot the image

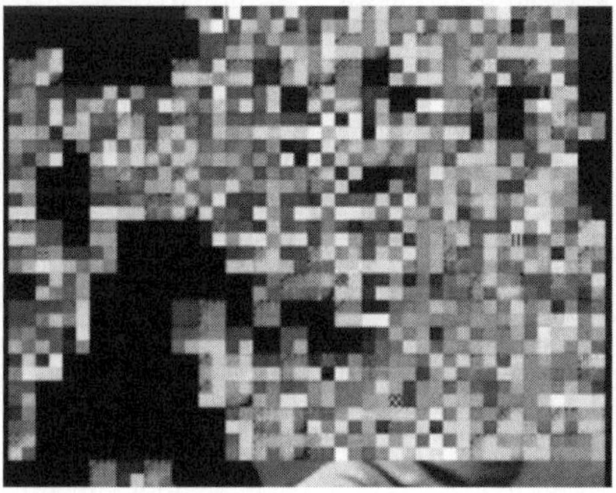

FIGURE 9.11 H.263 prefix-based decoder with known lengths of image blocks for the first INTRA image (SNR = 18 dB).

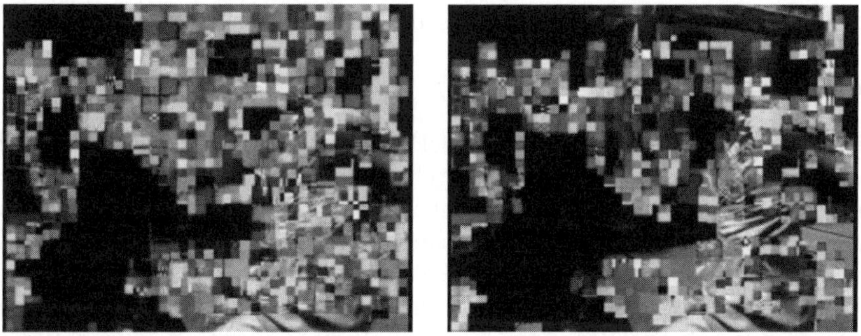

FIGURE 9.12 Image extracted from the video sequence Matrix I transmitted over EGPRS error pattern channel (SNR = 18 dB) and decoded by the VLC syntax-based decoder without iteration and after three iterations.

FIGURE 9.13 Image extracted from the video sequence Matrix I transmitted over EGPRS error pattern channel (SNR = 18 dB) and decoded by the VLC syntax-based decoder including the last constraint without iteration and after three iterations.

FIGURE 9.14 Image extracted from the video sequence Matrix I transmitted over EGPRS error pattern channel (SNR = 18 dB) and decoded by the VLC syntax-based decoder including the last and run constraints without iteration and after three iterations.

obtained at the first iteration, but the improvement at iteration 4 is not impressive compared to the simpler case of the last constraint.

9.3 DISCUSSION

As illustrated above, joint source-channel iterative decoding can improve a lot the resulting images. Moreover, this can be obtained quite often at a reasonable cost; if properly used in a more realistic context where ARQs are used, the iterations are performed only on the blocks that are erroneous.

However, the constraint is now the delay; the need for an additional interleaver between both constituent coders implies the need for performing source decoding and channel decoding on several blocks at a time, and this increases the processing delay.

EXIT charts have been briefly introduced as a tool to analyze the performance of iterative decoders. To evaluate the performance of individual components, *distance spectra* have been considered in the studies by Jaspar and Vandendorpe (2005a) for VLCs and by Ben-Jamaa et al. (2008) for quasi-arithmetic codes.

Chapter 10

Introduction to Joint Source-Channel Coding

It should be clear from the previous chapters that the general strategy behind joint source, protocol, and channel decoding is to take advantage of the redundancies in the signal sent onto the channel induced at any possible level: either left by the source coder, introduced deliberately by the channel coder for reducing the number of errors, or introduced by the network for other purposes.

Obviously, there are many possible underlying questions, such as:

- Should one really fight for obtaining no redundancy left by the source coder in the bitstream, since as a result some robustness is lost?
- Is it useful to ask the source coder to deliberately leave or introduce redundancy in the bitstream it generates?
- For a given global rate, what is the best possible allocation of the redundancy, since it may be introduced or left at many places in the communication chain?

Such questions are now of another nature than joint source-channel decoding (JSCD): one is now designing the transmitter side, and this strategy indeed corresponds to joint source-channel coding (JSCC).

This chapter first provides a short introduction to the classical way of performing JSCC, in which the entropy coding is usually removed (because the impact of transmission errors is difficult to model) and replaced by index assignment (IA). Then, one provides an outline of the work in this area that can be done using tools similar to the ones explained in the previous chapters. The strategy in this context is to deliberately introduce redundancy in the encoder, with a clear structure,

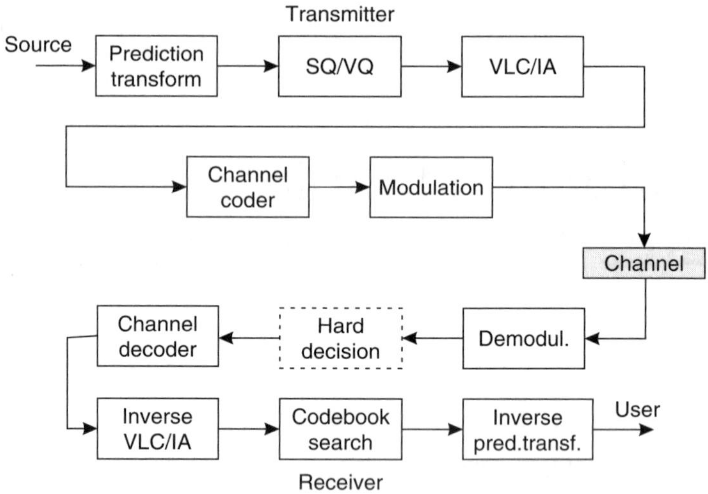

FIGURE 10.1 Block diagram of a transmission system (the network is assumed to be transparent).

so that its efficiency can be monitored. Within this framework, two main tools are described: error-correcting entropy codes (as a possible replacement for VLCs or ACs) and redundant descriptions, as possible replacements for the transformation/quantization building blocks. It should be noted that the improvements brought by entropy coders are so large that a complete elimination of this step does not seem to be competitive in our framework.

All methods are presented with reference to the communication scheme of Figure 10.1 in which all tasks that have to be completed in sequence are explicitly shown. Depending on the assumptions on which the JSCC methods rely, some blocks are merged and have to perform a more complex task which is then to be optimized for minimum distortion under noisy channel conditions. Other methods concentrate on one of these building blocks and propose possible replacements, able to deal with the possible transmission errors. In accordance with the traditional JSCC setting, the network layers are not taken into account in this section. They are assumed to be fully transparent, unless otherwise stated.

Communication Model

A general model of the transmission block diagram, forgetting the influence of the network layers, is depicted in Figure 10.1.

The message emitted from the source is first processed by some *prediction* and/or *transformation* blocks (for taking correlations into account). Source quantization is performed to eliminate less important data and allow efficient compression, see Chapter 3.

In most source coders, the next block is usually made of an entropy coder, which allows to reduce the length of the bitstream. However, the first generation of JSCC often did not consider this step, which was thought of as being incompatible with the existence of transmission errors. In contrast, it was realized that the existence of errors was making the IA more important for fixed-length codes. IA, also known as *Labeling*, aims at giving an appropriate bit pattern to each code vector. Note that if IA is used, optimized FLCs for minimum distortion are implicitly assumed at quantization step, either in a scalar (SQ) or in a vector form (VQ). If VLCs are used, the quantization is implicitly assumed to be uniform (this is the most efficient combination of quantization and coding as shown in Chapter 3).

The resulting bits are then protected by a channel encoder. The modulation shapes the signal before transmitting it to the physical channel, which introduces perturbations, according to some model.

A series of "inverse" operations is then performed. Demodulation, channel decoding, inverse IA (or VLC decoding), codebook search, and inverse transformation (and/or prediction) are applied to recover the original message.

Note that the hard decision is usually not implemented on the channel output, but at the output of the channel decoder. Moreover, if JSCD is used, hard decisions are avoided as much as possible.

This general model can be simplified in different ways. In fact, each method described below makes its own assumption on the model and combines some of the blocks in Figure 10.1 into a single block and/or easily omits some of the blocks. For example, a binary symmetric channel (BSC) simply models the modulator, channel noise, demodulator, and hard limiter gathered in a single block. Some methods make a single block from two or three other blocks and apply some optimization routines to it. As an extreme example, in Modulation ORganized Vector Quantization (MORVQ), see Section 10.1.3 below, vector quantization, IA, channel coding, and modulation are merged together and locally optimized.

10.1 TRADITIONAL VIEW OF JSCC

The methods that have been proposed for performing JSCC were initially performed on very simple models, such as those used in Chapter 2. Since then, more efficient algorithms were proposed, and we provide below a short review of the first-generation methods, pointing to the previous considerations relying on information theory and the generalized Lloyd algorithm when relevant.

10.1.1 Index Assignment

The IA step associates some code (*index*), which is transmitted to the channel, to the discrete real-valued quantizer outputs (*centroids*). In a perfect channel situation, if the source coder employs VQ optimized for minimum distortion, and if a fixed-length code is used, IA does not change the bitrate (nor the average distortion, obviously). Moreover, if optimized VQ or SQ is used, it is not usual to implement a variable-length code, the primary intent of optimized VQ being often precisely to avoid the use of VLCs, to avoid possible desynchronization or catastrophic decoding in presence of channel errors.

However, when the channel may introduce transmission errors, this assignment plays an important role in determining the overall performance, even if FLCs are used. Basically, the LBG algorithm, named after its authors, Lindé, Buzo, and Gray (Lindé et al., 1980), producing the optimized VQ, see Section 3.3.5, page 61, by itself does not provide any protection against channel noise because any change in a bit can redirect the initial code vector to any other one in the codebook, sometimes very far from the initial one in terms of quadratic distance. So, even a low *bit error rate* (BER) can heavily distort the signal if no specific IA strategy is used.

Once the origin of the problem is recognized, the task to be assigned to the IA is clear: channel errors should change the input index to other ones that are likely to be "close" to the initial one. It is shown by De Marca and Jayant (1987) that a nonnegligible reduction in distortion can be obtained through a well-designed IA rather than a random one. Hence, the problem is to find an IA for which the distance (to be chosen according to the channel model) between indexes is similar to the distance between centroids (Euclidian distance). This is called *pseudo-Gray coding* in the literature (Zeger and Gersho, 1990).

It must be noticed that designing an IA is a *nonpolynomial* (NP)-complete task. Let ℓ be the number of bits of the initial FLC; there are

$$\frac{(2^{\ell})!}{2^{\ell} \times \ell!} = \frac{(2^{\ell}-1)!}{\ell!} \tag{10.1}$$

possible distinct combinations to assign $J = 2^{\ell}$ code vectors to J code words. The 2^{ℓ} and the $\ell!$ factors in the denominator eliminate, respectively, the symmetric cases and the bit permutation cases. This results, e.g., in 8.3×10^{499} distinct possible combinations for $\ell = 8$ bits.

Of course, some strategies provide better initial estimates of the codebook than other ones. It has been soon recognized that the VQ codebook design strategy known as *splitting* (Lindé et al., 1980) training was efficient with that respect. In the splitting technique, the training begins with a few (possibly just one) code vectors. Each code vector is then divided into two sister code vectors each time with small perturbations, and these new centroids are updated. Farvardin (1990) has observed that when the splitting technique is used for VQ training, the resulting codebook has a natural ordering that can somehow protect the signal in the presence of channel errors. This is due to the splitting mechanism that makes sister code vectors behave similarly. However, this is not entirely efficient because if an error occurs on the first slitted bits, the resulting distortion can be much greater.

A solution to the IA problem is to perform the VQ design first and then permute the indexes in such a way that the resulting codebook becomes more robust against channel noise. Note that this strategy is not a simultaneous optimization of source and channel coding: only a "translation step" is performed. A more global strategy in which quantization, IA, and maybe channel coding are jointly optimized will be discussed in the next section.

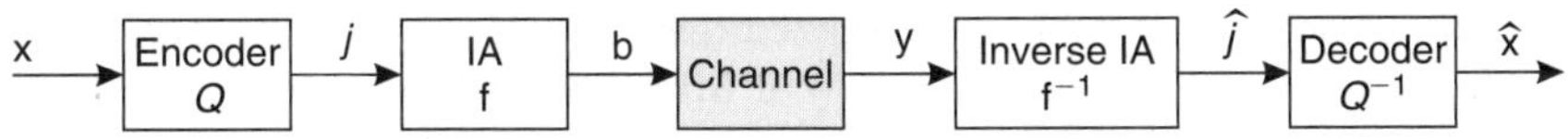

FIGURE 10.2 Block diagram of the VQ-based coding system used over a noisy channel.

Choice of the IA

The problem can be formulated simply as explained in Figure 10.2 where all building blocks between IA and Inverse IA are modeled through a BSC with crossover probability ε.

Simulated Annealing

Since IA is an NP-complete problem, Farvardin (1990) used *simulated annealing* (SA) to solve it. The SA algorithm can theoretically give the global optimum solution, unconditionally on the initial state, provided that the initial value of the parameter (named "temperature") and its schedule of decreasing are chosen appropriately. Unfortunately, this is difficult to achieve, and therefore, good optima from SA might be difficult to obtain in most practical cases.

As an example, Farvardin reported a signal-to-noise ratio (SNR) of about 8.95 dB for SA, compared to 8.87 dB for a naturally organized LBG with splitting. The test parameters were $\varepsilon = 10^{-2}$, the indexes had $\ell = 8$ bits, and the test signal was a first-order Gauss–Markov source with correlation coefficient $\rho = 0.9$.

Binary Switching Algorithm

Another algorithm for an optimum IA was proposed by Zeger and Gersho: *binary switching algorithm* (BSA). This algorithm changes the assignment progressively, starting by the code vectors that are most sensitive to error channels and trying to assign them the indexes that result in the least distortion due to channel errors, in the spirit of greedy algorithms. See Zeger and Gersho (1990) for more details.

Although a globally optimal IA is not necessarily obtained by BSA, good locally optimal solutions have been reported by Zeger and Gersho (1990). Simulation tests have been made with a first-order Gauss–Markov source, as well as with an i.i.d. source, and a sampled speech waveform. As an example, for $\varepsilon = 10^{-2}$, $\ell = 8$ bits, about 1.5 dB gain has been achieved compared to the initial randomly chosen, nonoptimized IA.

Conclusion on the IA

In our opinion, even if the IA problem cannot be forgotten, the search for optimal solutions is not really the panacea in this JSCC framework: it seems that in practical situations, the gains compared to reasonable solutions, e.g., by splitting techniques, is very moderate. Moreover, since it is usually not compatible with the use of entropy codes, there does not seem to be a large interest on this topic anymore. However, it should be realized that large Huffman codebooks such as those used in H.263 (ITU-T, 2005) or MPEG 4/AAC (ISO/IEC, 2005), for example, contain very large subsets of code words that are of the same length. The IA strategy could be used on these parts to reduce the impact of channel errors on the overall distortion.

10.1.2 Generalized Lloyd Algorithm

As explained in Chapter 2, information theory aims at maximizing the global rate, see Chapter 2. In actual situations, however, the overall system is generally constrained, and when the global architecture is decided, one tries to carefully tune its parameters to minimize the distortion D, for a given global rate.

This work is classically performed separately on the source and channel: First, minimize the distortion, assuming that the channel does not introduce any error, and then tune the error-correcting codes in such a way that the transmission errors are not "disturbing." However, this can be performed jointly by a procedure very similar to the Lloyd algorithm presented in Section 3.3, page 56, for classical quantization. This section is based on the paper by Zahir Azami et al. (1996), which describes this procedure in a very general setting. Section 10.1.3 makes the connection with previously proposed approaches.

The aim here is to minimize jointly the distortion introduced by the source coder and that resulting from channel errors. At this point, the results of Section 3.3 should be remembered: It was shown that minimizing the distortion without taking the rate into account was not optimal and that, for example, uniform quantization followed by entropy codes was more efficient than optimized quantization followed by an entropy code. It should be clear that, here again, even if one takes the distortion introduced by the channel into account, some kind of entropy-constrained optimization should be performed. To our knowledge, there is no reference in the literature on that topic, which remains for future research.

If one forgets the constraint on the rate, the criterion of interest is the total distortion, involving the quantization effects and the distortion introduced by the channel. Assume that $\mathbf{x} \in \mathbb{R}^k$ is generated by a source with pdf p, encoded using a coding function C (in the sense of information theory) to get a binary vector $\mathbf{b} = C(\mathbf{x})$, transmitted on a channel providing $\mathbf{y}$, from which $\widehat{\mathbf{x}} = F(\mathbf{y})$ is reconstructed at receiver side using a decoding function F, see Figure 10.3.

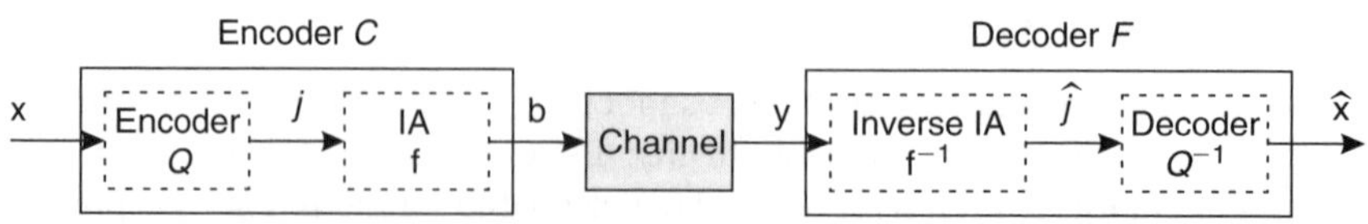

FIGURE 10.3 Block diagram of the VQ-based coding system used over a noisy channel. Here, the IA is included in the encoding process.

Considering a quadratic distortion measure, one may easily write

$$D = \frac{1}{k} E\left[||\mathbf{x} - \widehat{\mathbf{x}}||^2 \right] = \frac{1}{k} \sum\!\!\!\!\!\!\int_{\mathbf{x}} p(\mathbf{x}) \sum\!\!\!\!\!\!\int_{\mathbf{y}} p(\mathbf{y}|C(\mathbf{x})) ||\mathbf{x} - F(\mathbf{y})||^2, \tag{10.2}$$

where $p(\mathbf{y}|\mathbf{b})$ is the transition probability of the channel.

In a Lloyd algorithm, the minimization of D is performed by an iterative procedure in two steps.

1st Step (Generalized Centroid Condition)

The encoder C is fixed, and one optimizes the decoder F. This minimization is easily performed by first rewriting the criterion as

$$D = \frac{1}{k} \sum\!\!\!\!\!\!\int_{\mathbf{y}} \left[\sum\!\!\!\!\!\!\int_{\mathbf{x}} p(\mathbf{x}) p(\mathbf{y}|C(\mathbf{x})) ||\mathbf{x} - F(\mathbf{y})||^2 \right]. \tag{10.3}$$

The term between brackets is a distortion term, depending only on $\mathbf{y}$, say D_y, and the minimum D is the sum of the minimal contributions for each $\mathbf{y}$, as given by

$$D_{\min} = \sum\!\!\!\!\!\!\int_{\mathbf{y}} D_{y\min}. \tag{10.4}$$

Hence, the optimal decoder $\widehat{\mathbf{x}} = F^*(\mathbf{y})$ providing the minimum distortion is obtained by setting the derivative of D_y with respect to $\widehat{\mathbf{x}} = F(\mathbf{y})$ to zero

$$0 = \sum\!\!\!\!\!\!\int_{\mathbf{x}} p(\mathbf{x}) p(\mathbf{y}|C(\mathbf{x}))(\mathbf{x} - F^*(\mathbf{y})),$$

and we obtain

$$F^*(\mathbf{y}) = \frac{\sum\!\!\!\!\!\!\int_{\mathbf{x}} p(\mathbf{x}) p(\mathbf{y}|C(\mathbf{x}))\mathbf{x}}{\sum\!\!\!\!\!\!\int_{\mathbf{x}} p(\mathbf{x}) p(\mathbf{y}|C(\mathbf{x}))} \tag{10.5}$$

$$= \frac{\sum_{\mathbf{b}} p(\mathbf{y}|\mathbf{b}) \sum\!\!\!\!\!\!\int_{\mathbf{x}|C(\mathbf{x})=\mathbf{b}} p(\mathbf{x})\mathbf{x}}{\sum_{\mathbf{b}} p(\mathbf{y}|\mathbf{b}) \sum\!\!\!\!\!\!\int_{\mathbf{x}|C(\mathbf{x})=\mathbf{b}} p(\mathbf{x})} \tag{10.6}$$

$$= \frac{\sum_{\mathbf{b}} p(\mathbf{y}|\mathbf{b}) \sum\!\!\!\!\!\!\int_{\mathbf{x}|C(\mathbf{x})=\mathbf{b}} p(\mathbf{x})\mathbf{x}}{\sum_{\mathbf{b}} p(\mathbf{y}|\mathbf{b}) p(\mathbf{b})} \tag{10.7}$$

Here, (10.5) is true whatever the type of signals and channels, and (10.6) only holds when the channel symbols take discrete values. This formula shows a strong similarity with the classical centroid update in Section 3.3, but for the appearance of the weighting by $\sum_{\mathbf{b}} p(\mathbf{y}|\mathbf{b})$ which is only due to the channel model.

Note that, even if it is not demonstrated here, this formula corresponds to the classical result that this optimum decoder computes the conditional expectation of the source vector based on the channel output, i.e., $F(\mathbf{y}) = E[\mathbf{X}|\mathbf{Y} = \mathbf{y}]$.

2nd Step (Generalized Nearest Neighbor Condition)

Here, we optimize the encoder C for a given decoder F. First, rewrite the distortion as

$$D = \frac{1}{k} \sum\!\!\!\!\!\!\int_{\mathbf{x}} p(\mathbf{x}) \left[\sum\!\!\!\!\!\!\int_{\mathbf{y}} p(\mathbf{y}|C(\mathbf{x})) ||\mathbf{x} - F(\mathbf{y})||^2 \right], \tag{10.8}$$

and the optimal encoder is obtained as

$$C^*(\mathbf{x}) = \arg\min_{\mathbf{b}} \sum\!\!\!\!\!\!\int_{\mathbf{y}} p(\mathbf{y}|\mathbf{b}) ||\mathbf{x} - F(\mathbf{y})||^2,$$

which is the minimum of the "generalized distance" (taking into account the effect of the channel) between $\mathbf{x}$ and its reconstructed value $F(\mathbf{y})$. When there is no channel error, i.e., $p(\mathbf{y}|\mathbf{b}) = \delta(\mathbf{b}, \mathbf{y})$, this formula reduces to the classical one. Note that, when the channel is not very noisy, this is a reasonable approximation. Now, if $\widehat{\mathbf{x}}$ takes discrete values (depending on the channel model), this formula is easily rewritten as

$$C^*(\mathbf{x}) = \arg\min_{\mathbf{b}} \sum_{\widehat{\mathbf{x}}} ||\mathbf{x} - \widehat{\mathbf{x}}||^2 \sum\!\!\!\!\!\!\int_{\mathbf{y}|F(\mathbf{y}) = \widehat{\mathbf{x}}} p(\mathbf{y}|\mathbf{b}) \tag{10.9}$$

$$= \arg\min_{\mathbf{b}} \sum_{\widehat{\mathbf{x}}} ||\mathbf{x} - \widehat{\mathbf{x}}||^2 p(\widehat{\mathbf{x}}|\mathbf{b}). \tag{10.10}$$

The Algorithm

As classically done, the generalized Lloyd algorithm repeatedly performs both steps until the distortion (or the encoder and the decoder remain stable). Note that (10.7) and (10.10) can be practically evaluated in many situations.

10.1.3 Optimization of the Encoder and Channel Coder

In this generalized Lloyd procedure, one recognizes that the centroid values are adapted to the presence of a specific channel model. In this case, the full VQ is matched to the minimization of the global distortion. It turns out that several methods were proposed in the literature, which can be understood as special cases of this general procedure presented in the previous section. Some of them are described below.

Channel Optimized Vector Quantization

Farvardin proposed a joint optimization for the source and the channel coders (Farvardin and Vaishampayan, 1991a). It is in fact a straightforward application of the generalized Lloyd procedure in Section 10.1.2, applied on a BSC, with a quadratic distortion as a criterion to be minimized. First, rewrite the general distortion measure involving both the quantization error and the error due to channel perturbation (Ayanoglu and Gray, 1987) in the case of a BSC. Here, $C(\mathbf{x})$ denotes the index associated with emitted centroid and $F(\mathbf{y})$ denotes the centroid associated to the received index $\mathbf{y}$. The channel effects are represented by $p(\mathbf{y}|C(\mathbf{x}))$, the probability that some emitted index $C(\mathbf{x})$ is changed to index $\mathbf{y}$.

The resulting algorithm is very similar to the LBG algorithm and is named *channel optimized vector quantization* (COVQ). The cells, defined by C, are updated according to (10.7) and the centroids $F(\mathbf{y})$ according to (10.10).

In a few words, each input vector $\mathbf{x}$ is classified into the cell with the least expectation of distortion, while $F(\mathbf{y})$ represents the centroid of all input vectors that are decoded into $F(\mathbf{y})$, even if the received index, $\mathbf{y}$, is different from the emitted one $C(\mathbf{x})$. Of course, both equations can be simplified into the LBG learning equations by simply assuming that

$$p(\mathbf{y}|C(\mathbf{x})) = \begin{cases} 1 & : \quad \mathbf{y} = C(\mathbf{x}) \\ 0 & : \quad \mathbf{y} \neq C(\mathbf{x}). \end{cases} \tag{10.11}$$

This way, LBG can be regarded as a special case of COVQ when the parameter of the BSC is zero.

It can be shown that the obtained optimum encoding cells are convex polyhedrons and that some cells might vanish, thus, creating empty cells (Farvardin and Vaishampayan, 1991a). This means that the system trades quantization accuracy for less sensitivity to channel noise. Figure 10.4 shows an example of COVQ for a two-dimensional ($k = 2$), three-level ($J = 3$) VQ and a discrete memoryless channel (DMC) with the parameters as in the following Table 10.1.

Figure 10.4 illustrates that when the channel noise is large, there is a risk that some cells vanish. Assuming that there are J' nonempty encoding cells, $J' \leqslant J$, only J' code words need to be transmitted. Of course, any of J binary code words may be received and therefore the codebook must remain of size J. It is interesting to observe the analogy that exists between the presence of empty cells (code vectors with no corresponding input vector) and the added redundancy in channel coding.

Direct Modulation Organizing Scheme

Another possible source-channel configuration is the direct modulation organization. In this configuration, the encoder includes the modulator and benefits directly from the flexibility that is naturally present in a constellation. As shown in Figure 10.5, the channel is modeled as an additive white Gaussian noise (AWGN).

Several works have been done in this field. To mention some, we can indicate a competitive learning algorithm, which gives a direct mapping from input space to the signal space as presented by Skinnemoen (1994); the hierarchical modulation, in which the constellation points are located to minimize the

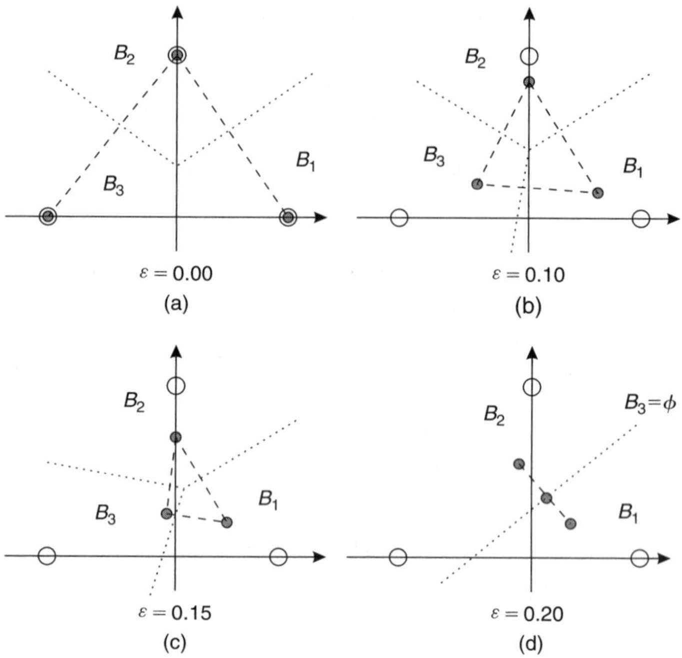

FIGURE 10.4 Evolution of the quantization cells when COVQ is used and with a simple discrete memoryless channel with $\varepsilon = 0.00, 0.10, 0.15,$ and 0.20; the code vectors get closer when ε increases and finally one of the cells, $\mathcal{B}_3$, vanishes for $\varepsilon = 0.20$. The circles are the reconstruction points for a nonnoisy environment.

TABLE 10.1 Transition Probability Matrix $P(i|j)$ in the DMC Example

$i\|j$	1	2	3
1	$1-2\varepsilon$	ε	ε
2	2ε	$1-4\varepsilon$	2ε
3	ε	ε	$1-2\varepsilon$

error expectation, is explained by Combelles et al. (1993) and Ramchandran et al. (1993). There exists some other works not detailed in this book: joint optimization of three blocks (source coder, channel coder, and modulator) (Vaishampayan and Farvardin, 1992) and trellis coding and Lattice coding, which are special kinds of covering the signal space by the constellation points (Laroia et al., 1994).

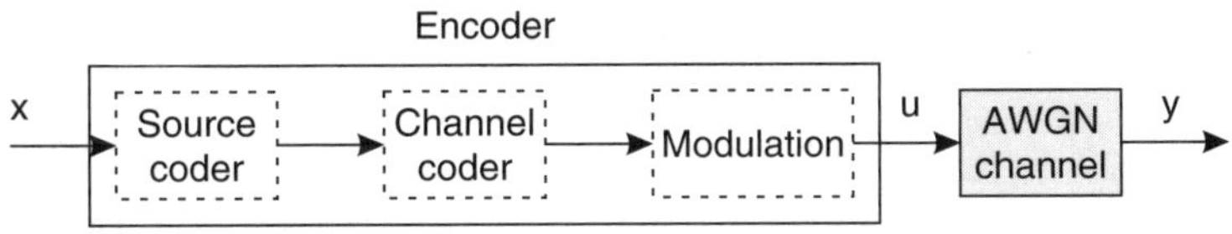

FIGURE 10.5 Block diagram of the direct modulation VQ-based coding system, used over an AWGN channel; the source coder, channel coder, and modulator are represented in one block; **u** belong to the set of constellation points.

Modulation Organized VQ

Withdrawing any binary representation, Skinnemoen proposed the MORVQ (Skinnemoen, 1994). This method uses a quantizer that maps the code vectors directly into the constellation plane. It makes efficient use of the Kohonen learning algorithm (Kohonen, 1990) to map the k-dimensional input space to the 2D signal space in such a manner that the close code vectors in the modulation space are assigned to the close points in the input space. This property is obtained by proper use of a neighborhood function (Kohonen, 1984, 1990), and the resulting codebook has some organized structure. Having this structure, most little changes due to channel noise make the output code vector to be one of the neighbors of the source vector, and so, the distortion will not be very important.

Skinnemoen observed a great difference between explicit error protection and the structure of a codebook. He states that any transmission system (with or without error protection) has a BER working threshold. Above that limit, the performance of the system breaks down. The role of MORVQ is to increase this threshold. However, in MORVQ, no more channel coding can be added since it does not produce any intermediate bit pattern which can be processed by classical channel coders. It was also observed that for MORVQ, the degradation curve by increasing channel noise is rather smooth, while for LBG there is a threshold above which the system performance drops rapidly.

Hierarchical Modulation

A multiresolution broadcast system has been proposed by Ramchandran et al. (1993). The main idea in this proposition consists in partitioning the information into two parts: the *coarse* information and the *refinement* or the *detail* information. This approach is intended to be used in conjunction with transformation-based source coding methods, like subband and wavelet coding, since they have a natural multiresolution interpretation. The coarse information is to be received correctly even in a very noisy transmission environment, while the detail information is mostly destinated to the receivers whose channels have better qualities (graceful degradation). This classification can even be made more precise, making several classes of importance.

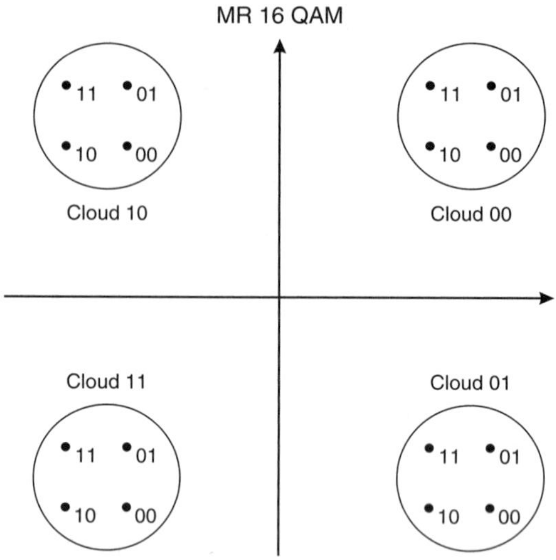

FIGURE 10.6 An example of multiresolution constellation. Each set of close points constitutes a cloud with four satellites points surrounding it. The detail information is presented in the satellites while the coarse information is represented in the clouds. So, there are 2 bits of coarse information and 2 bits for detail. Note also that the Gray code is used for numbering the satellites (and the clouds) in such a way that the code words with Hamming distance equal to 2 are far from each other. This is like the application of Karnaugh map in digital design and can be used for larger constellations too.

It has to be noted that this approach is naturally well suited to a broadcast situation rather than to a point-to-point link. Like the previous approach, the idea is to match the transmission constellation to the source-coding scheme, without merging both steps. Ramchandran et al. (1993) proposed the use of a multiresolution constellation as depicted in Figure 10.6. The coarse information is carried by the clouds; while inside each cloud, the miniconstellations or satellites provide the details. The loss of coarse information is associated with the receiver inability to decipher correctly which cloud was transmitted, while the loss of detail information occurs when the receiver confuses one intracloud signal point for another. This property is already present in any QAM constellation but is reinforced by the uneven localization of the points in the cloud.

Of course, many other configurations could be thought of yielding similar properties. The same idea has been used in conjunction with trellis modulation coding (*TMC*) and with embedded channel coding (Ramchandran et al., 1993).

The same idea of multiresolution coding has been proposed by Combelles et al. (1993), in conjunction with turbo codes, which were used to protect the coarse and detail information with 1/2 and 3/4 rates, respectively. They achieved 4 dB better performance for the coarse information while 2 dB degradation for the detail information, compared to a single-resolution system using turbo codes, with the same overall spectral efficiency to obtain the same error rate (10^{-4}), while for a Rayleigh fading channel their simulation shows 5 dB better performance for the coarse information and 3 dB degradation for the detail information.

In the study by Ramchandran et al. (1993) an example is given where with a multiresolution system, the broadcast coverage radius (64 km) of the coarse signal is much greater than for a single resolution system (45 km), while for the multiresolution system, the radius of full data availability is a little smaller (38 km).

Note that this hierarchical modulation was proposed as an option in the DVB-SH standard for video broadcasting (ETSI, 2007). To our knowledge, very little work is currently performed for evaluating its possible advantages.

10.2 DESIGN OF ROBUST ENTROPY CODES

Both types of entropy codes (VLCs and ACs) can be transformed in such a way as to be able to detect and correct transmission errors. Obviously, this is obtained at the cost of a reduced compression ratio, since some redundancy has been deliberately left in the output bitstream. While this strategy was known for a while for VLCs, it was generalized quite recently to robust arithmetic codes, see Section 10.2.3.

The redundancy left by VLCs and ACs, evaluated in Chapter 4, can indeed result in some error-correcting properties, as was seen in Chapters 5 and 6. Nevertheless, up to that point, these error-correcting properties were not characterized analytically in the same way as they are for channel coding. In this section, we introduce the quantities that allow to obtain bounds on the resulting error rates.

10.2.1 Free Distance and Distance Spectra

Free Distance

As for convolutional codes, the free distance (Blahut, 1983; Anderson, 2005) of a VLC, i.e., the minimum Hamming distance between two different equal-length sequences of code words, plays a very important role on error-correcting performance. Its definition is quite straightforward, but its evaluation is much more complex than for convolutional codes. The main difficulties here come from the fact that, intrinsically, the individual code words have different lengths and that the code is not linear, contrary to classical error-correcting codes. These two differences in the structure of codes has two main consequences: (i) one has to evaluate the free distance on a sequence of code words in terms of properties of the codebook and (ii) this evaluation, for VLCs, has to be performed on all *pairs* of sequences of the same length instead of using the all-zero code word as a reference, as done in linear codes.

The presentation here closely follows that adopted by Buttigieg (1995) and Lamy and Bergot (2003).

Let $\mathcal{C}=\{\mathbf{c}_1,\ldots,\mathbf{c}_J\}$ be a uniquely decodable VLC of cardinality J. Let w be the number of different code-word lengths (in bits) in $\mathcal{C}$. We denote these different lengths by $L_1, L_2, \ldots, L_w$, with $L_1 < L_2 < \cdots < L_w$.

Definition 10.1: *The* free distance

$$d_{\text{free}} = \min\left\{d_H(\mathbf{c}^1,\mathbf{c}^2) | (\mathbf{c}^1,\mathbf{c}^2) \in \mathcal{C}^*, \mathbf{c}^1 \neq \mathbf{c}^2,\ \ell(\mathbf{c}^1)=\ell(\mathbf{c}^2)\right\} \tag{10.12}$$

is the minimum Hamming distance in the set $\mathcal{C}^*$ of all arbitrary extension of the code $\mathcal{C}$, see Chapter 3, page 41. ■

Since d_{free} is much more difficult to evaluate for VLCs than for linear codes, as it involves pairs of sequences of code words, a lower bound for d_{free} has been introduced by Buttigieg (1995). This lower bound may be easily evaluated, as it involves several types of distances on individual code words which are defined in what follows.

Definition 10.2: *The* minimum block distance b_k *associated to the code-word length L_k of a code $\mathcal{C}$ is defined as the minimum Hamming distance between all distinct code words of $\mathcal{C}$ with length L_k. The* overall minimum block distance b_{min} *of $\mathcal{C}$ is the minimum block distance for every possible length L_k, i.e., $b_{\text{min}} = \min b_k$.* ■

Definition 10.3: *The* diverging *distance $d(\mathbf{c}_i,\mathbf{c}_j)$ (respectively, the* converging *distance $c(\mathbf{c}_i,\mathbf{c}_j)$) between two code words of different lengths $\ell(\mathbf{c}_i)$ and $\ell(\mathbf{c}_j)$ of a code $\mathcal{C}$ is defined as the Hamming distance between the prefixes (respectively, suffixes) of length $l = \min\{\ell(\mathbf{c}_i), \ell(\mathbf{c}_j)\}$ of the code words $\mathbf{c}_i$ and $\mathbf{c}_j$. The* minimum diverging *distance d_{min} (respectively,* minimum converging *distance c_{min}) of a code $\mathcal{C}$ is the minimum value for all diverging (respectively, converging) distances between every possible couple of code words in $\mathcal{C}$:*

$$d_{\text{min}} = \min_{\substack{(\mathbf{c}_i,\mathbf{c}_j)\in\mathcal{C}^2 \\ \ell(\mathbf{c}_i)\neq\ell(\mathbf{c}_i)}} d(\mathbf{c}_i,\mathbf{c}_j) \tag{10.13}$$

and

$$c_{\text{min}} = \min_{\substack{(\mathbf{c}_i,\mathbf{c}_j)\in\mathcal{C}^2 \\ \ell(\mathbf{c}_i)\neq\ell(\mathbf{c}_i)}} c(\mathbf{c}_i,\mathbf{c}_j). \tag{10.14}$$

■

The bound for the free distance variable-length error correcting (VLEC) proposed by Buttigieg (1995) is then

$$d_{\text{free}} \geqslant \min\{b_{\text{min}}, d_{\text{min}} + c_{\text{min}}\}. \tag{10.15}$$

Only a sketch of the proof is provided in what follows, see Buttigieg (1995) for more details. Chapter 5 has shown that succession of code words may be described by various trellises. Consider, for example, a bit-clock trellis. Each finite sequence of code words of $\mathcal{C}^*$ is associated to a path in this trellis, starting from the state 0 and ending at the same state 0. Consider two sequences of code words $(\mathbf{c}^1, \mathbf{c}^2)$ belonging to $\mathcal{C}^*$ such that $\ell(\mathbf{c}^1) = \ell(\mathbf{c}^2)$.

1. If the paths associated to $\mathbf{c}^1$ and $\mathbf{c}^2$ diverge at the beginning of a code word and converge again at the end of the code word, i.e., $\mathbf{c}^1$ and $\mathbf{c}^2$ differ only in a single code word. Since both sequences have the same length, the distance between these sequences is lower-bounded by b_{min}.
2. Now, assume that the paths associated to $\mathbf{c}^1$ and $\mathbf{c}^2$ diverge at some time ν and converge again at $\nu' > \nu + 1$, with $\mathbf{c}^1$ and $\mathbf{c}^2$ differing in more than one code word. In this case, the distance between sequences is at least the sum of the minimum diverging distance (since both paths diverge) and of the minimum converging distance (since the paths are converging again).

For catastrophic codes, i.e., codes for which there exist two semi-infinite input sequences differing in *infinitely* many positions that are encoded into two output sequences differing only in a *finite* number of positions, the bound (10.15) does not necessarily hold. Nevertheless, such codes are avoided in general when designing VLEC, due to their poor performance.

Distance Spectrum

While the free distance dominates the error correction capability of a code, a finer error analysis is possible by using the distance spectrum to evaluate an upper bound on error probability. The distance spectrum of a code is the sequence $\{A_d\}$ that counts the average number of paths at distance $d \geqslant d_{\text{free}}$ from the correct path. For a convolutional code, this is identical to its weight spectrum (i.e., A_d is the number of paths of weight d for $d \geqslant d_{\text{free}}$), due to linearity.

Let P_{er}^{b} be the bit error probability at any position in the code. Then, the following union upper bound holds (Lin and Costello, 1983, Ch. 11)

$$P_{\text{er}}^{\text{b}} \leqslant \sum_{d=d_{\text{free}}}^{\infty} dA_d P_d, \tag{10.16}$$

where P_d is the probability that the decoder selects an erroneous path at Hamming distance d instead of the correct path.

Without taking the code structure into account, the error probability for BPSK signaling over an AWGN channel is

$$P_d = \frac{1}{2} \operatorname{erfc} \sqrt{d \frac{E_{\text{b}}}{N_0}}, \tag{10.17}$$

where E_{b} is the energy per bit and $N_0/2$ is the variance of the zero-mean Gaussian channel noise (Viterbi and Omura, 1979).

Both bounds (10.16 and 10.17) characterize the bit error probability in the code domain, with and without the structural constraint of the code.

In practice, a more important figure of merit is the symbol error probability in the information domain. For convolutional encoders, this probability can be bounded using another spectrum, $\{B_d\}$, that counts the average number of nonzero information bits on paths of weight d. Then $P^{\mathrm{s}}_{\mathrm{er}}$, the symbol error probability at any source symbol position, is bounded by

$$P^{\mathrm{s}}_{\mathrm{er}} \leqslant \sum_{d=d_{\mathrm{free}}}^{\infty} B_d P_d. \tag{10.18}$$

The two distance spectra $\{A_d\}$ and $\{B_d\}$ and the resulting bounds (10.16) and (10.18) may be extended to a (nonlinear) time-invariant trellis code driven by a memoryless source. Buttigieg and Farrell (1995, 2000) carried this out for nonlinear VLEC trellises. The most important difference to convolutional codes is that it is no longer sufficient to consider path weights alone; hence, A_d has to be defined as the average number of converging path pairs at Hamming distance d, which can be computed assuming a stationary probability distribution on the trellis. Another difficulty arises from the fact that although the decoder compares converging code paths of equal length, these may be associated to source sequences of different lengths. Therefore the spectrum $\{B_d\}$ must be defined in terms of a distance measure that allows comparing unequal-length sequences; the most common choice is the Levenshtein distance d_{L} (Levenshtein, 1965), which is defined as the minimum number of deletions, insertions, or substitutions required to transform one sequence into the other. Hence, B_d is defined as the average Levenshtein distance between the input sequences of all converging pairs of paths whose output sequences are at Hamming distance d.

Extending the results for VLEC trellises to AC trellises (free distance, distance spectra) is straightforward, the proofs for the bounds (10.16) and (10.18) follow along the same lines as in Buttigieg and Farrell (1995, 2000), so that the only major difference is the computation of the distance spectra, see Ben-Jamaa et al. (2008) for more details.

Evaluating the Free Distance and Distance Spectra

The evaluation of d_{free} and of the distance spectra is carried out by an enumeration of the pairs of sequences of code words of the same length in C^*. This may be done using one of the two algorithms proposed by Buttigieg and Farrell (1995). Both algorithms perform an exhaustive search over all paths of the same length in bits in a bit clock. The first algorithm is fast, but requires a lot of memory, as all paths that have not converged are stored and compared. The search is stopped when some predetermined length $n_{\max}$ is reached. In the second algorithm, the required memory is limited, but the computations are slow. In that algorithm, only two paths are stored at a given time, but the same paths are repeatedly constructed, stored, and erased.

To reduce memory requirements and increase speed, the value of d for which A_d and B_d are computed may be limited, since the first few nonzero spectral components (with small d) dominate in the error bounds. In practice, also the length n is limited to $n_{\max}$, so only lower bounds on A_d and B_d are obtained.

This is more problematic with low-redundancy codes, which tend to have a huge number of short paths at $d_{\text{free}} = 1$. For higher redundancy, the first spectral components quickly approach their true value with growing n.

A polynomial-time evaluation of d_{free} and A_d is possible using iterative matrix-based algorithms, see Ben-Jamaa et al. (2008) and Weidmann and Kieffer (2009). Also, these algorithms evaluate a lower bound $A_d^{n_{\max}} \leqslant A_d$ by considering only path pairs of less than $n_{\max}$ bits. An upper bound on the approximation error is also obtained in similar fashion as for convolutional codes (Lassing et al., 2001), see Weidmann and Kieffer (2009).

10.2.2 VLEC Codes

This section aims at introducing two families of VLEC codes.

Reversible VLCs

Reversible variable-length codes (RVLCs) were proposed for facilitating the bidirectional decoding of a source-encoded bitstream, which mitigates, for example, the perceptual effects of transmission errors on multimedia data and are even able to provide error correction (Wen and Villasenor, 1997). They have been extensively studied particularly during the development of the video standards H.264 and MPEG-4 (Hanzo et al., 2007).

RVLCs are uniquely and instantaneously decodable from the beginning and the end of a sequence. This is due to the fact that not only do they have the prefix property (see Chapter 3), i.e., that no code word is the prefix of another one, but they are designed to meet also a suffix property, i.e., that no code word is the suffix of another one. Obviously, since there is an additional constraint, the average length of a RVLC is larger than that of the corresponding Huffman code.

Distance Properties of RVLCs

We consider the relationship between Huffman codes and RVLCs. Obviously, in the case of Huffman codes $d_{\text{free}} = b_{\min} = 1$, since a Huffman code is constructed so that the two longest code words differ only in the last bit, see Chapter 3.

In the case of RVLCs, things are slightly more involved. In Lakovic and Villasenor (2002), an adaptation of (10.15) to RVLC is introduced

$$d_{\text{free}}^{\text{RVLC}} \geqslant \min\{2, b_{\min}\}. \tag{10.19}$$

Thus, $b_{\min} \geqslant 2$ is a sufficient condition for $d_{\text{free}} \geqslant 2$.

Construction of RVLCs

Most RVLC constructions are based on a Huffman code designed for the source concerned. Then, the Huffman code words are replaced by identical length code words that satisfy both the prefix and suffix conditions. If the number of valid code words is insufficient, longer code words have to be assigned, resulting in an increased redundancy. By contrast, if there are more candidate code-words than necessary, different code word selection mechanisms may be applied. For example, the minimum repetition gap (Tsai and Wu, 2001) metric and the so-called affix index (Lakovic and Villasenor, 2003) metric may be invoked for designing asymmetric RVLCs. All the above code-word selection mechanisms attempt to match the code-word length distribution of the RVLC to that of the corresponding Huffman code. Since a RVLC has to satisfy the suffix condition in addition to the prefix condition, the desirable code-word length distribution of a Huffman code is usually not matched by that of the RVLC for the same source.

Error-Correcting VLCs

Obviously, when studying the distance properties of VLCs (see Buttigieg, 1995; Buttigieg and Farrell, 1994), one may have the temptation of designing VLEC codes by finding a variable-length code, which satisfies the specific distance requirements and additionally has a minimum average code-word length. Usually, one chooses $b_{\min} = d_{\min} + c_{\min} = d_{\text{free}}$ and $d_{\min} = \lceil d_{\text{free}}/2 \rceil$.

To our knowledge, there is no systematic algorithm for solving such a problem. Only heuristic construction methods were proposed by Buttigieg (1995) and Jin Wang et al. (2004).

10.2.3 Error-Correcting Arithmetic Codes

This lack of systematic design comes in contrast with the arithmetic codes, where it is rather easy to increase their robustness against transmission errors by introducing redundancy in the compressed bitstream.

As already explained in Chapter 3, arithmetic coders are to be implemented with finite precision to avoid computational accuracy problems. This process already introduces some redundancy in the bitstream. According to Howard and Vitter (1992), when considering integer probability spaces in $[0, T)$, the additional redundancy due to the integer approximation is at most $0.497/T + O(1/T^2)$ bits/symbol if correct probability estimates are used by the encoder. Limiting the value of variable f (for bits to follow), as defined in Chapter 3, may be another source of redundancy. Nevertheless, as the probability of having $f = f_{\max}$ decreases exponentially with $f_{\max}$, this additional redundancy remains small. Therefore, other techniques must be applied, usually in combination with finite precision implementation, since the resulting operations are, thus, described by a FSM.

A well-known JSC technique for AC is based on the introduction of a forbidden symbol (FS) in the source alphabet (Boyd et al., 1997). The FS is never emitted by the source, although a positive probability p_ε is assigned to it. In that way, decoding the FS indicates the occurrence of a transmission error. It can be shown that introducing a FS of probability p_ε adds a redundancy of $-\log(1-p_\varepsilon)$ bits/symbol to the coded bitstream (Chou and Ramchandran, 2000). Several FSs may be introduced, as suggested by Sayir (1999a,b), see Figure 10.7.

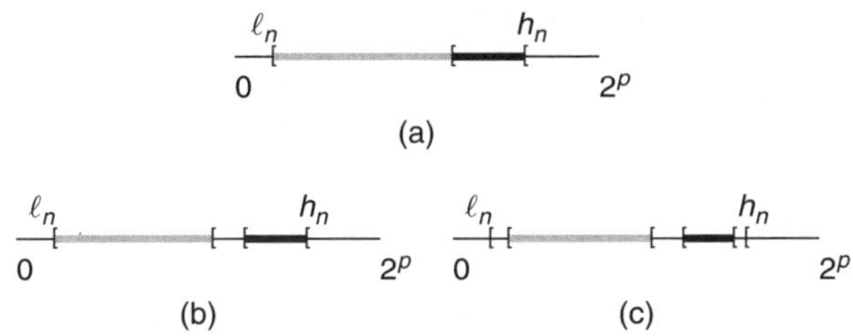

FIGURE 10.7 Partition of the coding interval with a classical binary arithmetic coder (a); Introduction of a forbidden symbol between the subintervals associated to 0 and 1 (b); Introduction of three forbidden subintervals (c).

The probability p_ε is distributed among the FSs, providing more design freedom to build good error-correcting arithmetic codes.

Obviously, when combined with finite precision implementation, one can derive FSMs describing the encoding and the decoding, which can in turn be used to derive trellises, and the techniques described previously in this book for MAP decoding can be applied.

Even if redundancy is easily evaluated, the error correction performance of such codes is not easily obtained. Ben-Jamaa et al. (2008) provide methods for evaluating the free distance of the "code" and even the corresponding distance spectrum. Based on these computations, some optimization to get the best possible location of the FSs can be undertaken.

10.3 OVERCOMPLETE REPRESENTATIONS

Compared to other ways of introducing redundancy in the bitstream, overcomplete representations have a strong advantage. Compare, for example, the possible usage of redundancy introduced by channel coding with that introduced by redundant transformations in situations where the channels may take very different qualities, such as broadcasting to mobile devices.

When the channel is good, the redundancy introduced by the channel is useless, since the corresponding code rate has been chosen for possibly very poor ones. The quality of the video which is received is the nominal one when the channel quality is above some threshold and decreases very quickly below that threshold. In contrast, source coders using redundant representations can be designed in such a way that the quality of the reconstructed video approximately matches the amount of information that is received. Therefore, if the channel is good, more "useful" information will reach the terminal, and the received quality matches the channel quality.

Obviously, this description is quite idealistic, but such considerations motivated a lot of work in this direction.

This section is only an introduction to the topic, which would deserve much more space for a comprehensive description. A good tutorial can be found in the study by Goyal (2001).

10.3.1 Multiple Description Coding

Multiple description coding (MDC) was introduced more than 20 years ago by the information theory community (Ozarow, 1980). It consists in representing the source by M independent streams, named *descriptions*. Each of these descriptions is sent over its own channel, characterized by the probability that the information is not delivered. The decoder aims at reconstructing the source with the best possible distortion, whatever the number of descriptions received. Obviously, this goal can be achieved (even for a small number of descriptions received) only if a certain amount of redundancy is found among all descriptions. Therefore, the main problem to be solved in this context is to find the best tradeoff between the total bitrate (sum of the bitrates among all descriptions) and the distortion after reconstruction. Note that in this setup, there is no hierarchy between the descriptions, since any subset of them should allow the recovery of the source with an acceptable distortion. This comes in contrast with hierarchical (or multiresolution) source coders where the main layers have to be received for the video to be correctly received.

A trivial solution would consist in using M identical distributions. Whatever the number of lost channels, the reconstructed signal would be of constant quality. However, the nominal bitrate would be very high, since this procedure would be exactly similar to some repetition coding. Moreover, this would not meet the explicit goal that quality should improve with the number of descriptions. Therefore, the solution consists in building descriptions that are individually good, but different enough, so that they can be combined in a constructive manner. The optimal tradeoff can be understood from the information theoretic approach, as described below in the specific case of Gaussian signals.

10.3.2 Theoretical Limits

This problem was studied in the early 1980s by several authors, see, for example, Gamal and Cover (1982) and Ozarow (1980). The problem was mainly studied in the case of two descriptions, as depicted in Figure 10.8. The source is assumed to be Gaussian, with unit variance. This source feeds two different channels, and a third receiver may receive both signals. Ozarow (1980) obtained an evaluation of the distortions (MSE) that each receiver can simultaneously achieve. This allows to evaluate the tradeoff mentioned in the introduction of this section between the necessary disparity of both descriptions for a good central receiver and the efficiency of the decoding using a single description.

The general setup for two descriptions is depicted in Figure 10.8. The message consists in k i.i.d. symbols $\mathbf{X} = (X_1, X_2, \ldots, X_k)$ drawn from a given alphabet with probability $p(X_\kappa)$. Let $\widehat{\mathbf{X}}^{(1)}$ and $\widehat{\mathbf{X}}^{(2)}$ be the corresponding estimates obtained from the outputs $\mathbf{Y}^{(1)}$ and $\mathbf{Y}^{(2)}$ of the individual channels, while estimate $\widehat{\mathbf{X}}^{(0)}$ is obtained through a decoder fed by both inputs. These estimates are constituted by k letters taken from the alphabet $\mathcal{A}$. The corresponding distortions D_0, D_1,

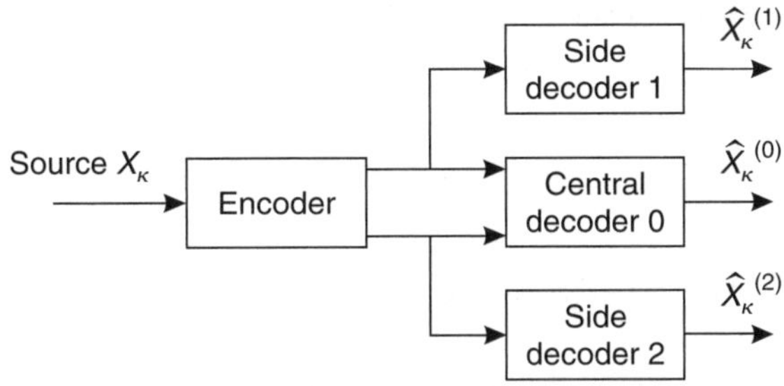

FIGURE 10.8 General setup for two descriptions feeding three receivers. The general case would contain M descriptions and $2^M - 1$ receivers.

and D_2 are obtained as

$$D_i = \frac{1}{k} E\left[d\left(\mathbf{X}^{(i)}, \widehat{\mathbf{X}}^{(i)}\right)\right],$$

where d is a real-valued distortion measure.

It has been recalled in Chapter 2 that the corresponding rate-distortion bound $R^{(i)}(D)$ is the smallest value that can be taken by the mutual information between $\mathbf{X}^{(i)}$ and $\widehat{\mathbf{X}}^{(i)}$. The problem, now, with reference to Figure 10.8, is to characterize the quintuplets $(R_1, R_2, D_0, D_1, D_2)$, as a means to evaluate the tradeoff between the efficiency of the reconstruction based on a single description versus that of the reconstruction based on both descriptions.

This problem, to our knowledge, has no solution in a general setting. However, Ozarow (1980) proposed a characterization of this quantity in the specific case of Gaussian signal and channels, the distortion measure being MSE. A trivial upper bound on D_i is obtained by applying the general formula (3.37), page 59, assuming that the source is of unit variance, which neglects the mutual influence of the encoders.

To be more precise, a separate application results in the following bounds:

$$D_1 \geqslant 2^{-2R_1} \tag{10.20}$$

$$D_2 \geqslant 2^{-2R_2} \tag{10.21}$$

$$D_0 \geqslant 2^{-2(R_1+R_2)}. \tag{10.22}$$

In a separate setting, the rate-distortion curve can be approached as closely as desired through the use of large block sizes. However, it has been shown by Ozarow (1980) that, in this case, since there are only two coders in the scheme rather than three as implicitly assumed in (10.22) above, this bound is not attainable, even with infinite block lengths. Instead, the actual attainable bounds are characterized by

$$D_1 \geqslant 2^{-2R_1} \tag{10.23}$$

$$D_2 \geqslant 2^{-2R_2} \tag{10.24}$$

$$D_0 \geqslant \frac{1}{1-\sqrt{(1-D_1)(1-D_2)}-\sqrt{D_1 D_2}} 2^{-2(R_1+R_2)}. \tag{10.25}$$

One can now evaluate the limits attainable by the reconstruction when only one link is available or when both are available to the central decoder. More precisely, one can see that, in order (10.25) to take values close to the unconstrained optimum (10.22), necessarily the individual distortions, i.e., when a single channel output is available, should be large. Conversely, if the qualities obtained by the single channel receivers are good (close to optimal), necessarily, the improvement obtained by the central receiver will not be impressive.

As usual, information theory provides useful information about the attainable performance and the corresponding tradeoffs, but provides little insight about practical implementations that would be close to optimal. Many articles addressed this topic, and we will only point out a few of them, only giving the underlying methodology for the main approaches.

10.3.3 Practical Implementations

Two main classes of MDC were proposed: the first one is based on quantization and is denoted as multiple description scalar quantization (MDSQ) (Vaishampayan, 1993).

Based on Quantization

The technique proposed by Vaishampayan (1993) is very similar to the general setup of MDCs, when the encoders are based on quantization. From a real-valued symbol x, one builds two descriptions $x^{(1)}$ and $x^{(2)}$ belonging to two dictionaries $\mathcal{A}^{(1)}$ and $\mathcal{A}^{(2)}$. If ever both descriptions arrive to the decoder, the reconstruction error D_0 should be smaller than when one of the descriptions is lost. Therefore, each description should be as reliable as possible, but both descriptions should not carry the same information.

Here again, a hierarchical quantization consisting in producing a coarse quantization first $x^{(1)}$ followed by a finer quantization of the residue (which would provide $x^{(2)}$) is not a good solution, since description $x^{(2)}$ would be useless without $x^{(1)}$. Another simple solution would be to use two quantizers of same step Δ, shifted by $\Delta/2$. Clearly, each description provides the same reconstruction quality, and the knowledge of both descriptions provides the knowledge of the signal with step $\Delta/2$. However, this is not either a good solution: if each description has rate R, the knowledge of both description costs $2R$ in terms of bitrate, while it provides $D(R+1)$ in terms of distortion.

A better solution to this problem requires some kind of Lloyd algorithm, as explained in Section 10.1.2. It can also be classified in the same categories as for the traditional quantization: scalar, vector, entropy constrained, etc. For example, entropy-constrained multiple description scalar quantization (ECMDSQ) has been proposed by Vaishampayan and Domaszewicz (1994).

Based on Transforms

The approach here can be seen as the converse of classical transform source coding: here, a linear transform correlates independent random variables. The amount of correlation is known to the decoder, which can then estimate the missing symbols, and therefore reduce the distortion arising when a single description is received. This technique, known as multiple description correlating transform (MDCT), was introduced by Wang et al. (1998) in the case of two variables. These results have been generalized to k variables in Goyal and Kovacevic (1998).

Comments on MDCs

It should be recognized that, implicitly, MDCs are working on a specific channel model, which is somewhat different to the one used in this book: the receiver knows exactly how many descriptions it received and which ones. This corresponds to a channel with erasures. This makes sense in a context where ACKs are used in the network, in which case the receiver knows exactly what it received and what is missing. Few articles considered the application of MDCs for channel with noise. In this case, descriptions are received, but they may be corrupted by transmission errors. JSCD techniques may then be combined with MDCs to recover the initial signal with the best quality (Guionnet et al., 2002; Larsen, 2006).

Moreover, many difficulties arise when trying to use such techniques in coders using prediction, since the images on which prediction is based can be of varying quality, which may result in a drift between the encoder and decoder. This has to be taken into account, and specific coding schemes have to be designed.

10.3.4 Representation Based on Frames

Given a signal, it is frequent, as seen in Chapter 3, that one has to represent it in another system, typically a basis, where its characteristics are more readily apparent in the transform coefficients. However, these representations are typically nonredundant, and thus, corruption or loss of transform coefficients can be serious. The idea of frames (Casazza, 2000) is that the representation should be chosen so that we can avoid those disasters. A frame is therefore the redundant counterpart of a basis.

As an example, consider a vector $\mathbf{x}$ of dimension k. Its decomposition on a frame ϕ_i, $i=1,\ldots,n$ with $n>k$ introduces redundancy. This redundancy is deterministic. Provided that the frame is well designed, any k coefficients among the total of n allow to recover the initial vector, see Goyal et al. (1998, 2001)

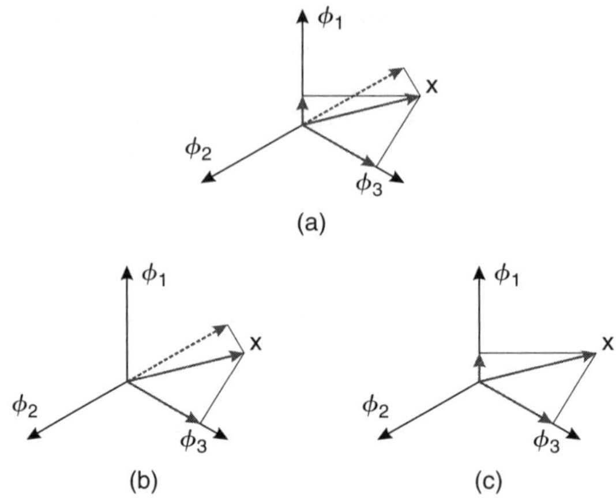

FIGURE 10.9 Expansion of **x** over a frame (a); Reconstruction of **x** from two projections only (b) and (c).

and Figure 10.9. Seen this way, the additional coefficients do not bring more information than a classical transform coefficient (in absence of erasure). In this respect, they would be very similar to block codes. Similarly, their infinite dimension version, based on filter banks, would be very similar to convolutive codes (Labeau et al., 2005). The drawback seems the same one: when the channel is good, all samples arrive to the destination, and the redundant ones are not useful in any respect, characterizing an inefficient use of the channel. In fact, the picture changes when taking into account the fact that the coefficients are quantized before being transmitted. In this case, the redundancy allows to reduce the contribution of quantization noise to the reconstructed signal (Goyal et al., 1998).

For a detailed and easy to read presentation of frames, see Kovacevic and Chebira (2007a,b).

Most of the work undertaken on frames deals with erasure channels, i.e., channels in which one knows whether the data arrived or not, even erroneous, with the exception of a few papers, such as a study by Labeau et al. (2005) where transmission errors are modeled as large errors, which can be detected and hopefully corrected. Few results are also available on the combination of this approach with VLCs (Marinkovic and Guillemot, 2005).

10.4 CONCLUSION

This chapter was only intended to provide a short introduction to JSCC, with the aim of showing the evolution of the research. Starting from a situation where VLCs were not taken into account, a lot of work was done on tools compatible with erasure channels, which are a coarse model of wireless channels. The recent tendency is to intentionally leave or reintroduce redundancy in building blocks that are found in efficient source coders. This is further discussed in Chapter 11. However, in light of what has been explained in this chapter, it should be clear that one of the main challenges is to build a coding system in which there is little reference to a nominal quality of the channel, therefore resulting in a reconstructed signal with a graceful degradation, following as closely as possible the $D(R)$ bound.

Chapter 11

Open Challenges

Obviously the topic joint source-channel coding/decoding (JSCC/D) is not fully mature yet, even if a lot of progress has been made in the recent years. Challenges are of various nature, and for most of them were already cited at appropriate places in the book. Let us summarize.

11.1 JOINT SOURCE-CHANNEL DECODING

Even if the techniques described in this book now seem to be applicable in many situations (and in this respect, broadcasting/multicasting of multimedia contents is the preferred situation), several important steps are still needed for a wide application of JSCD.

An extension which does not seem to be a major one is the reduction of the additional throughput which is sometimes needed for applying JSCD. At some places, in the previous chapters, one was needing some information which is usually not transmitted in the actual situation. A reduction of this additional information to a minimum without degrading the performance of the robust decoder is not fully available yet.

Another one is quite general and concerns the compatibility of robust decoding with the standard organization in layers of the protocol stack. In actual point-to-point communications, if one relaxes the requirement that a block at lower layer may be processed by higher layers without checking the Cyclic Redundancy Check (CRC), it may happen that erroneous blocks reach the source decoder, thus inducing a possibility of catastrophic failure. Some preliminary results can be found in Bouchireb et al. (2008), in which the authors replace the classical ACK/NACK decision based on a CRC check by an estimation of the fact that the robust decoder will provide acceptable results. More work in this direction is still needed. Generally, a better characterization of the performance of JSCD algorithms at various levels would allow to go from a "hard" CRC-check to some kind of "soft" CRC-check.

The control of the delay is still an open issue: in many cases, robustness is obtained at the cost of an additional delay in the decoding. In fact, basically, some part of a sequence is accepted if it fits well within a larger part of the video. Obviously, this strategy requires to take a decision after an additional delay compared with the classical strategy. This additional delay may not be compatible with some applications. Note, however, that depending on the layer where this decision takes place, the delay may vary by large quantities. The statistics of the wireless channels also have an impact there: in a mobile situation, errors arise by burst, which may correspond to a situation in which the robust decoder cannot provide an acceptable signal. This is typically the situation in which a cross-layer approach has to be adopted to finely tune the whole system. Such an approach has been adopted in the first results reported in Bouabdallah et al. (2007), but more work is still to be done to adopt a fully cross-layer approach of robust decoding, see Mastronarde and van der Schaar (2009), and Fu and van der Schaar (2009).

Computational complexity is obviously higher in the robust decoding methods, and this may prevent a general use of these techniques. The solution here is to take the benefit of all information available at all layers. Most packets (at many network levels) are indeed received without errors. This can be detected by CRC check. In this case, no robust processing is needed. The robust decoding will have to be implemented only in the small percentage of cases where the block is in error.

Finally, even if video is certainly the signal which will generate the largest increase in network throughput in the near future, other types of signals should not be forgotten. This book has already given a few examples of JSCD applied to parts of audio (MPEG4/AAC) bitstreams, first results were obtained on HTML files (Jaoua et al., 2008a,b). The available results, even if preliminary, seem to indicate that the procedure is of general applicability, if one takes into account all characteristics of the corresponding bitstream and source syntax.

11.2 JOINT SOURCE-CHANNEL CODING

The main strength of JSCD is that it ensures a backward compatibility, since the bitstream that is decoded is often exactly the same one as would be received by standard receivers. Therefore, the main advantage of JSCD is to obtain better receivers without changing the transmitter. Even if the performance improvements obtained by these JSCD algorithms can be quite large, it is likely that redesigning the transmitter could be more efficient. Such a JSCC strategy states a variety of unsolved problems.

11.2.1 Theoretical

Optimal strategies of JSCC as derived by information theory seem for most of them to be of analog nature. Is it really compulsory? How can we approach their performance through digital implementations?

When going to digital implementation, many more blocks are involved than in the analog model used by information theory, each one being devoted to small tasks. The complexity of the joint optimization of these blocks becomes huge. In such a context, it is unlikely that simulations will ever be able to provide reliable answers. Therefore, there is a strong need for analytical tools allowing to characterize each block in terms of JSCC/D possibilities: is robustness obtained at a loss of nominal quality and is this nominal quality sufficient?

Once such analytical tools are obtained, and cross-layer optimization used in conjunction with JSCC/D seems to be a necessity. Possible studies along these lines are outlined below.

- How to make the best use of the redundancy of the network layers to help the decoding of multimedia contents. What has been explained in this book is a very first approach of this study: it was explained that, by a careful study of some standard, some specific information could be obtained for a better source reception. However, in our opinion, the approach is much more general than it could seem at first sight: this redundant information is for providing the network with information that is needed for characterizing important quantities, such as sender address, receiver address, type of service, and source encoding, which will anyway have to be transmitted. Therefore, it is likely that this strategy will always be feasible, only its implementation will be different. Facilitating the use of this redundancy may still require more work.
- Redundancy in the headers was obviously noticed by many people, thus resulting in the use of Robust Header Compression (RoHC) (Bormann et al., 2001). The fact that the strategy just mentioned can still be used in presence of RoHC is not questionable, since, after all, this is exactly what is done when applying robust decoding to compressed video. However, the performance of robust header estimation in presence of RoHC will certainly decrease. To our knowledge, this has not been addressed yet.
- Some new computations could really facilitate the corresponding optimization, by trying at a given layer to predict whether other layers will be able to manage with the (possibly) remaining problems, unsolved at this layer. Reference Bouabdallah et al. (2007) is an example of such a study, which could take a variety of forms, see also Mastronarde and van der Schaar (2009), and Fu and van der Schaar (2009).

11.2.2 Practical

Some other problems that have to be solved before a wide application of JSCC are of a more practical nature, but should not be forgotten.

- Backward compatibility is often a need. When it is the case, standard transmission schemes and JSCC/D-based schemes should coexist.
- The proposed solutions should not (except, possibly, at both ends, in the source coders and decoders) be signal-dependent. At least, this dependence should be manageable by the network.
- A full JSCC treatment requires a deep knowledge of many layers, such as PHY, network, and source. The corresponding knowledge is usually not found in the same persons, not even in the same teams. This is true for the algorithm design and also for the practical implementation. Even if the processing in a terminal allows the layers to forward many more information than the classical implementation while remaining compatible with a given standard, a difficulty arises: the teams implementing the network layer part of the chips are usually not the same ones as for the PHY and the source.
- *Scalability* is a property of bitstreams which is really useful when broadcasting multimedia signals, since it allows to address terminals with different qualities at the same time, without increasing the throughput: the main layer is sent to all users and only those able to make use of higher quality will decode the additional part of the stream. This property of the bitstream to be prioritized brings many changes to the organization of the transmission system and may have an impact on the use of robust decoders.

Taken altogether, it seems that all these constraints promote some "decoding approach" of the coding, as outlined in the next section.

11.3 JOINT SOURCE-CHANNEL CODING/DECODING

Finally, it is obvious that both JSCC and JSCD should be used simultaneously.

In this case, we think that most results explained in this book demonstrate that the debate between leaving redundancy in source coding or fully compressing the data may change with the emergence of building blocks in which the remaining redundancy is not an effect of a lack of efficiency, but a deliberate introduction of structured redundancy (e.g., error correcting VLCs or ACs, and frames). As an example of this strategy let us quote Benedetto et al. (1998):

> Our intuition to use Variable Length Error Correcting Codes comes from the scheme of serial concatenation of convolutional codes (SCCC). In SCCCs, the characteristics of both constituent codes are important, hence both inner and outer codes may have an arbitrary share of the total rate. For example, given the total rate and the complexity of the constituent codes, we know that the free distance of the outer code determines the so-called interleaver gain.

Following these lines, a general application of this strategy would result in studying the essential problem of redundancy allocation: for a given total amount of redundancy, where should it be located? If the channel characteristics are well known, the complexity and delay unconstrained, the answer is known: all redundancy should be allocated at PHY layer to the channel coding. Current solutions implicitly allocated redundancy in the form of CRCs at higher layers, to cope with residual errors coming from the fact that the above assumptions are not really met. Note that the information contained in the headers will always be there, at least partly, since it is needed at the application level. Finally, in this book, we have illustrated that, even if the amount of redundancy left by source coders was very small, it could be of great usefulness in conjunction with the other ones, e.g., through iterative decoding, but not only.

Therefore, a crucial problem is the allocation of this redundancy along all processing blocks, so that the quality of the signal reaching the source decoder matches the quality of the actual channel. Such a problem cannot really be addressed without the use of many analytical tools to predict the performance of large systems, encompassing the effects of redundancy at all layers. This is, in our opinion, the most important task to be performed in this area.

Appendix A

Format of 802.11 Packets

This appendix aims at describing the format of packets used at various layers, from the physical (PHY) to the application (APL) layer, for a transmission of multimedia contents using the 802.11 standard (IEEE, 1999). In Section 8.2.4, page 223, dedicated to Joint protocol-channel decoding, the content of many fields is assumed to be known or predictable. This content is specified here in many cases, the assumptions being those considered in Section 8.2.4.

A.1 PHY PACKETS FORMAT

The direct-sequence spread spectrum (DSSS) protocol allows the transmission of binary data coming from the upper layers onto the wireless channel between the access point and the terminal. These data are allocated to packets, inside the payload field. A header is added to each packet and contains transmission parameters associated to the content. A *preamble* is added in front of each packet and allows synchronization between transmitter and receiver. The detailed format of packets issued by the PHY layer is provided in Figure A.1.

In the DSSS protocol, the preamble and header of the PHY packets are always transmitted at 1 Mbps and make use of the more robust modulation scheme (BPSK). This allows the receiver to synchronize efficiently with respect to the incoming sequence and to maximize the probability to receive the parameters associated with the transmission of the rest of the packet (payload). The payload itself is transmitted at a variable bitrate of 1 or 2 Mbps. The various fields of the PHY packets are described below.

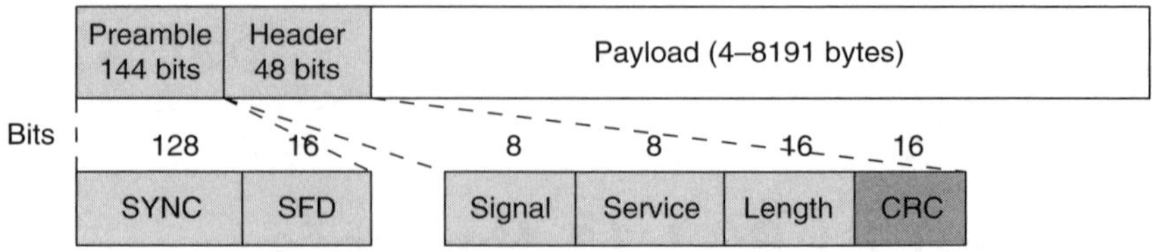

FIGURE A.1 Format of packets for the 802.11 PHY layer.

- The SYNC field is constant and consists of 128 bits set to 1. It allows the receiver to detect the beginning of a message (by detecting the energy coming from the channel) to synchronize on the carrier frequency and to adjust the input amplifier gain.
- The SFD field is constant, of length 2 bytes, set to $F3A0_{16}$. It is used by the receiver to perform time synchronization on the bitstream.
- The Signal field is 1 byte, specifying the bitrate used for transmitting the payload. This value is $0A_{16}$ for the BPSK modulation and 14_{16} for the QPSK modulation.
- The Service field (1 byte) is reserved for future recommendations. It is set to 00_{16} in the 802.11 standard.
- The Length field specifies the time (in microseconds) necessary for transmitting the payload. Its value (2 bytes) lies between 16 and $2^{16}-1$. It depends on two parameters: the size and the bitrate of the payload.
- The Signal, Service, and Length fields are protected by a 2-byte cyclic redundancy check (CRC), which corresponds to a CCITT CRC-16, with generator polynomial $g(x)=x^{16}+x^{12}+x^5+1$. This allows the receiver to detect packets with erroneous headers. In the case of erroneous header, the packet is erased.
- The payload contains the data to be transmitted, provided by the Media Access Control (MAC) layer. This field can contain from 4 to 8191 bytes. Its size is easily obtained from the Signal and Length fields.

A.2 FORMAT OF THE MAC PACKETS ASSOCIATED TO THE DCF PROTOCOL

The distributed coordination function DCF protocol allows to transmit the IP packets from the access point to the terminals by means of a random access to the wireless channel. The PHY layer then inserts each MAC packet in the PHY payload. Channel reservation is obtained through exchange of RTS and CTS frames between the transmitter and receiver. MAC fragments incorporating IP data are then transmitted to the receiver that acknowledges each of them separately. This section defines the format of packets involved in the communication between one access point and a specific terminal. First, the format of payload fragments is described, followed by control packets (RTS, CTS, ACK).

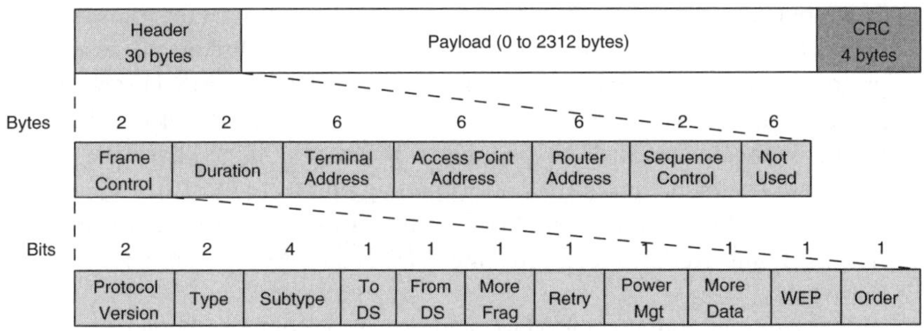

FIGURE A.2 MAC fragment format.

The MAC fragments contain binary data corresponding to segments of IP packets, which are encapsulated in the payload. A header containing parameters necessary to the link layer of the receiver is added at the beginning of each fragment, and a CRC is appended for checking data integrity. The format of the MAC packets is illustrated in Figure A.2, and the various fields are described below.

- Protocol Version (2 bits) specify the version of the Wi-Fi standard that is used. It is set to 10_2 for the 802.11 standard.
- The Type and Subtype fields of 2 and 4 bits, respectively, inform about the packet functionality. For a data fragment in DCF mode, Type is set to 10_2 and Subtype to 0000_2.
- The two 1-bit fields To DS and From DS characterize the network architecture and the direction of transmission. In our case, we are in a downlink transmission, on an extended network: data come from the Internet and are distributed by the access point to the terminals. In this case, To DS is set to 0 and From DS to 1.
- The More Frag field (1 bit) indicates whether the current fragment is the last one (0) or not (1).
- The 1-bit Retry field specifies if the current fragment is the retransmission of a previous one, in which case its value is set to 1. Otherwise, it is set to 0.
- The Power Mgt field (1 bit) specifies if the transmitter is in energy savings mode. This is not considered here, and in our simulations, it is set to 0.
- The More Data field (1 bit) is useful in PCF mode and indicates to a base station in energy savings mode if the access point has other information to be transmitted. More Data is set to 0 in our simulations.
- The WEP field (1 bit) specifies whether the payload is encrypted or not. In our case, we do not consider encryption, and *WEP* is set to 0.
- The Order field (1 bit) indicates if the fragments are transmitted in the right order. In the simulations considered in Section 8.2.4, transmission is ordered, and Order is set to 1.

- The 2-byte Duration field specifies the estimated duration of the transmission, expressed in microseconds of the future fragment, including the associated protocol information (SIFS, ACK) that are attached to it, see (8.11) and (8.12). Its value lies between 0 and $2^{15}-1$ in the DCF mode and allows the other terminals to adjust their NAV parameter. It can be obtained from the size of the future fragment and of the corresponding bitrate.
- Terminal Address contains the MAC address of the receiver (6 bytes).
- Access Point Address contains the MAC address of the access point (6 bytes).
- Router Address contains the MAC address of the router (6 bytes).
- The Sequence Control field (2 bytes) contains two parameters: a sequence number, coded on 12 bits, which is an IP packet counter, and a MAC fragment counter, on 4 bits. These counters are used to rearrange the received fragments and packets in the correct order in the terminal.
- The last field of the header is composed of 6 bytes and is used for local network transmissions. In our study, they are set to 0.
- The payload can contain from 0 to 2312 bytes.
- A 4-byte CRC protects the header and payload. It is defined by the following generator polynomial $g(x) = x^{32}+x^{26}+x^{23}+x^{22}+x^{16}+x^{12}+x^{11}+x^{10}+x^{8}+x^{7}+x^{5}+x^{4}+x^{2}+x+1$. It allows the receiver to detect erroneous packets and to ask for their retransmission.

In contrast with the data fragments, the control packets of the MAC layer do not contain any payload and are made only of a header and a CRC. The RTS, CTS, and ACK frames are described in what follows.

An RTS control frame is sent by the access point for sending information to a specific terminal. It contains the MAC address of the access point and the MAC address of the target terminal. The full format is described in Figure A.3.

The terminal answers the access point by means of a CTS control frame. This packet is shorter than the RTS frame, since it only contains the access point address. The corresponding format is illustrated in Figure A.4.

When the terminal has correctly received some MAC fragment, an ACK frame is sent to the access point. This frame is of same size as a CTS frame, and its precise format is represented in Figure A.5.

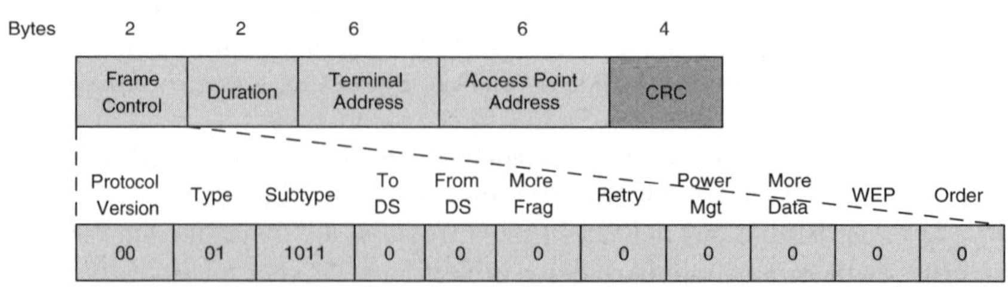

FIGURE A.3 Format of a RTS packet.

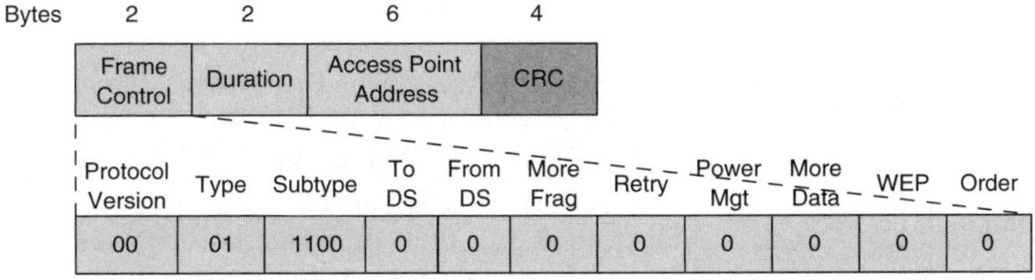

FIGURE A.4 Format of a CTS frame.

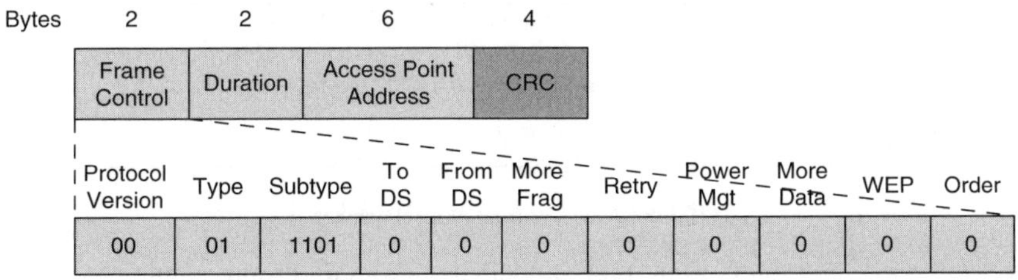

FIGURE A.5 Format of an ACK packet.

Remark A.1: *In fact, the Link layer is made of two successive layers: the MAC layer (adapted to the PHY layer) and the* Logical Link Control *(LLC) layer (IEEE 802.2 standard, 1998), which follows the* 802.2 *standard. This LLC layer establishes the logical link between the MAC layer and the network layer. When the MAC layer of the IEEE* 802.11 *standard is used, the LLC layer only specifies the protocol to be used at the upper layer. The current trend is toward the use of an IP protocol, and the LLC (sub) layer becomes progressively obsolete. In this book, we consider that the information provided by this layer is redundant, and the LLC layer is not taken into account.*

A.3 FORMAT OF IP PACKETS

UDP packets are encapsulated as the payload of IP packets. A header is added in front of the packet, containing all parameters required by the network layer to process the data. The IP packet format is illustrated in Figure A.6, and the content of the fields is as follows:

- The Version field (4 bits) specifies the IP protocol version. It is set to 4 for the IPv4 protocol.
- The Header Size field, of 4 bits, specifies the size of the header in terms of 32-bits word. In our situation, its value is 5.
- The Type of Service field (1 byte) is concerned with the characteristics of the transmission (bitrate, delay, reliability) depending on the type of data contained in the packet. All our simulations are run with default mode, in which case this parameter is set to 00_{16}.

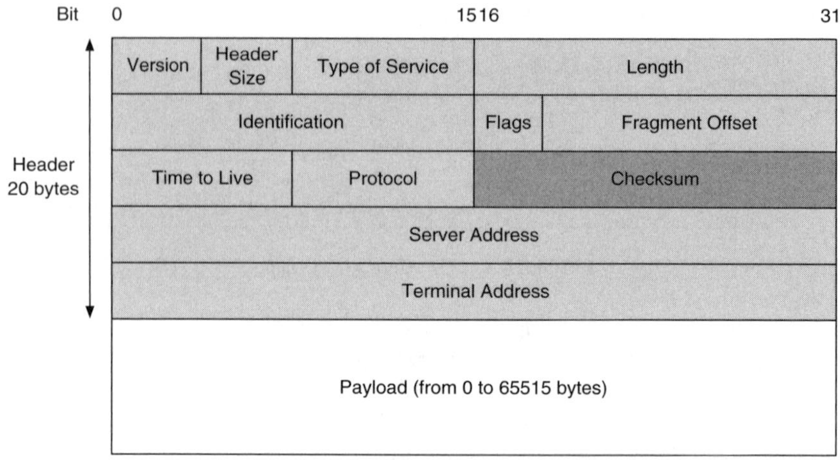

FIGURE A.6 Format of an IP packet.

- The 2 bytes of the Length field indicate the number of bytes of the IP packet, including the header. This field can take values between 20 and 65535.
- The Identification field (2 bytes) is an identifier characterizing the IP packet. This parameter is randomly generated, its value changes with each packet.
- The fields Flags and Fragment Offset of 3 and 13 bits, respectively, contain fragmentation parameters.
- The field Time to Live (1 byte) is the maximum number of routers the IP packet can meet. At server level, a standard initialization value is 255. Each router decrements the parameter by 1 when processing the packet. When the value goes to zero, the packet is erased. This allows to kill packets that could be lost in the network.
- The Protocol field (1 byte) specifies the protocol used at the upper layer. UDP corresponds to the value 17.
- A checksum of 2 bytes protects the header. In case of erroneous packets, the whole packet is erased.
- The 4 bytes of the Server Address field contain the IP address of the server.
- The 4 bytes of the Terminal Address field contain the IP address of the terminal.
- Finally, the Payload encapsulates the data to be transmitted (UDP packet). This field may contain 0 to 65515 bytes.

A.4 THE TRANSPORT LAYER (UDP/RTP)

A.4.1 UDP Packets Format

The UDP protocol handles packet multiplexing by managing ports. Ports allow the devices to distinguish between users and processes. The UDP mode is nonconnected and does not ensure communication reliability. However, it is able to detect erroneous packets.

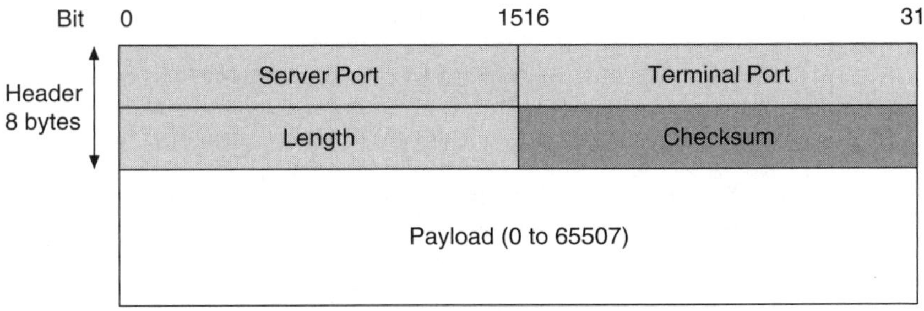

FIGURE A.7 Format of a UDP packet.

The payload of a UDP packet is made of data issued by the RTP packets. A header contains the parameters used by the UDP layer. The format of UDP packets is depicted in Figure A.7 and the various fields are described below.

- The field Server Port (2 bytes) gives the number of the server port. If the terminal answers the server, it should be through this port.
- The field Terminal Port (2 bytes) corresponds to the reception port of the terminal.
- The 2 bytes of the field Length contain the length (in bytes) of the UDP packet, including the header. Its value can be between 0 and 65515 (this takes into account the header of the IP layer).
- A checksum of 2 bytes protects all data in the packet (header plus payload). If ever an error is detected, the packet is erased.
- The Payload contains all data to be transmitted (RTP packet). This field can contain up to 65507 bytes.

A.4.2 RTP Packets Format

The RTP protocol provides the required services for transmitting IP data with real-time constraints. This protocol is directly associated to the current application, and some of its parameters are even used in the upper layer. It allows to rearrange the data in the correct order and to synchronize the decoded streams, thanks to temporal markers.

RTP packets carry data generated by the application layer. In our case, these data consist of NAL packets provided by the H.264/AVC coder, which are placed in the payload. A header is added at the beginning of each packet, containing useful information at the RTP level, as described below. The format of a RTP packet is illustrated in Figure A.8.

- the 2 bits of the field V (or Version) indicates the version of the RTP protocol. This value is always equal to 2.
- The field P (or Padding) is 1 bit long and specifies if padding bytes are used at the end of the packet. In our case, this parameter is 0.

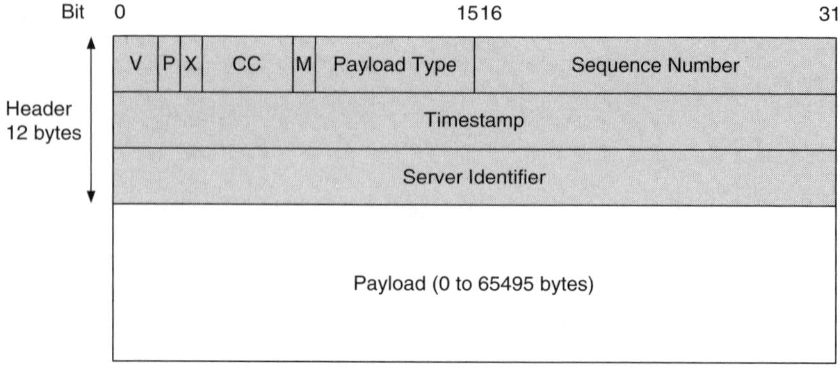

FIGURE A.8 Format of an RTP packet.

- The field X (or Extension) indicates if extensions are added in the packet header. In our case, this parameter is set to 0.
- When several users communicate altogether (such as in audio conference), an identifier (CSRC) characterizes each source (user). Each user sends its data to a central point (mixer), which is in charge of mixing them and to transmit these data to the various users. Each RTP packet transmitted by the central point contains all contributed data (CSRC). The identifiers are inserted after the header. The 4 bits of the CC (or CSRC Count) indicate their number. In our simulations, there is only one transmitter and one receiver, and *CC* is set to 0.
- The 1-bit field M (or Marker) is used by the decoder to separate the images. It is set to 1 in an RTP packet that contains the last slice (group of macroblocks) of an image and to 0 for the intermediate slices.
- The field Payload Type (7 bits) identifies the type of data that are transmitted within the RTP packet. It is set to 105 for the H.264/AVC standard.
- The 2 bytes of the field Sequence Number contain the RTP packet counter. It is incremented by 1 for each new transmitted packet. It is used for reordering the packets upon reception.
- The field Timestamp (4 bytes) characterizes the instant when the image must be displayed and is used by the decoder to synchronize the image flow. In the H.264/AVC standard, its value depends on the cycles of a 90-kHz clock. In the first image of a video stream, the field is initialized to 0, and it is constant in all RTP packets containing slices of the same image.
- The 4 bytes of the field Server Identifier correspond to a unique identifier of the server.
- The Payload contains the data to be transmitted (NAL packet), issued by the video encoder of the application layer. This field can contain from 0 to 65495 bytes (including all IP, UDP, and RTP headers).

Bibliography

3GPP, 2009. Technical specification group gsm/edge radio access network; channel coding (release 8). Technical Report TS 45.003 V8.2.0, 3rd Generation Partnership Project.

802.11, 2005. Wireless Networks: The Definitive Guide, second ed. O'Reilly, Sebastopol, Canada.

Adrat, M., Vary, P., 2005. Iterative source-channel decoding: Improved system design using EXIT charts. Eurasip J. Appl. Signal Process. 2005 (6), 928–947.

Ahmed, N., Natarajan, T., Rao, K.R., 1974. Discrete cosine transform. IEEE Trans. Computer 23 (1), 90–93.

Alajaji, F., Phamdo, N., Fuja, T., 1996. Channel codes that exploit the residual redundancy in CELP-encoded speech. IEEE Trans. Speech Audio Process. 4 (5), 325–336.

Anderson, J.B., 2005. Digital Transmission Ingineering, second ed. IEEE Press, Piscataway, NJ.

Anderson, J.B., Mohan, S., 1991. Source and Channel Coding: An Algorithmic Approach. Kluwer Academic Publishers, Norwell, MA.

Andrews, J.G., Ghosh, A., Muhamed, R., 2007. Fundamentals of WiMAX: Understanding Broadband Wireless Networking. Pearson Education, Upper Saddle River, NJ.

Antonini, M., Barlaud, M., Mathieu, P., Daubechies, I., 1992. Image coding using wavelet transform. IEEE Trans. Image Process. 1 (2), 205–220.

Ayanoglu, E., Gray, R.M., 1987. The design of joint source and channel trellis waveform coders. IEEE Trans. Inf. Theory 33 (6), 855–865.

Bahl, L.R., Cocke, J., Jelinek, F., Raviv, J., 1974. Optimal decoding of linear codes for minimizing symbol error rate. IEEE Trans. Info. Theory 20 (2), 284–287.

Bates, R.J., 2002. GPRS: General Packet Radio Service. McGraw-Hill Professional, New York.

Bauer, R., Hagenauer, J., 1998. On variable length codes for iterative source/channel decoding. In: Proceedings of the IEEE Data Compression Conference, Snowbird, UT, pp. 272–282.

Bauer, R., Hagenauer, J., 2000a. Iterative source/channel decoding using reversible variable length codes. In: Proceedings of the IEEE IData Compression Conference (DCC '00), pp. 93–102.

Bauer, R., Hagenauer, J., 2000b. Iterative source/channel decoding based on a trellis representation for variable length codes. In: Proceedings of the IEEE International Symposium on Information Theory, pp. 238.

Bauer, R., Hagenauer, J., 2000c. Symbol-by-symbol MAP decoding of variable length codes. In: Proceedings of the 3rd ITG Conference Source and Channel Coding, München, pp. 111–116.

Bauer, R., Hagenauer, J., 2001a. On variable length codes for iterative source/channel decoding. Proceedings of DCC, Snowbird, Utah, USA, pp. 273–282.

Bauer, R., Hagenauer, J., 2001b. The turbo principle in joint source channel decoding of variable length codes. In: Proceedings IEEE Information Theory Workshop, pp. 128–130.

Ben-Jamaa, S., Kieffer, M., Duhamel, P., 2009. Efficient MAP estimation of CABAC encoded data with objective adjustment of the complexity/efficiency tradeoff. IEEE Trans. Commun. 57 (7), 2014–2023.

Ben-Jamaa, S., Weidmann, C., Kieffer, M., 2008. Analytical tools for optimizing the error correction performance of arithmetic codes. IEEE Trans. Commun. 56 (9), 1458–1468.

Benedetto, S., Divsalar, D., Montorsi, G., Pollara, F., 1998. Serial concatenation of interleaved codes: Performance analysis, design and iterative decoding. IEEE Trans. Inf. Theory 44 (3), 909–926.

Berger, T., 1971. Rate Distortion Theory: A Mathematical Basis for Data Compression. Prentice-Hall, Englewood Cliffs, NJ.

Berger, T., 1972. Optimum quantizers and permutation codes. IEEE Trans. Inf. Theory 18 (6), 759–765.

Berrou, C., Glavieux, A., Thitimajhima, P., 1993. Near shannon limit error correcting coding and decoding: Turbo-codes. Proceedings of ICC, Geneva, pp. 1064–1070.

Bi, D., Hoffman, W., Sayood, K., 2006. State machine interpretation of arithmetic codes for joint source and channel coding. Proceedings of DCC, Snowbird, Utah, USA, pp. 143–152.

Blahut, R.E., 1983. Theory and Practice of Error Control Codes. Addison-Wesley, Reading, MA.

Blahut, R.E., 1987. Principles and Practice of Information Theory. Addison Wesley, Reading, MA.

Bormann, C., Burmeister, C., Degermark, M., Fukushima, H., Hannu, H., Jonsson, L.-E., et al., 2001. Robust header compression (ROHC): framework and four profiles. Technical Report RFC 3095.

Bouabdallah, A., Kieffer, M., Lacan, J., Sabeva, G., Scot, G., Bazile, C. et al., 2007. Evaluation of cross-layer reliability mechanisms for satellite digital multimedia broadcast. IEEE Trans. Broadcast. 53 (1), 391–404.

Bouchireb, K., Marin, C., Duhamel, P., Kieffer, M., 2008. Improved retransmission scheme for video communication systems. In: IEEE International Symposium on Personal, Indoor and Mobile Radio Communications (PIMRC '08), Cannes, France, pp. 1–5.

Boyd, C., Cleary, J., Irvine, I., Rinsma-Melchert, I., Witten, I., 1997. Integrating error detection into arithmetic coding. IEEE Trans. Comm. 45 (1), 1–3.

Busacker, R.G., 1965. Finite Graphs and Networks. McGraw-Hill, New York.

Buttigieg, V., 1995. Variable-Length Error-correcting Codes. Ph.D. thesis, University of Manchester.

Buttigieg, V., Farrell, P.G., 1994. On variable-length error-correcting codes. In: Proceedings of the IEEE International Symposium on Information Theory (ISIT '94), p. 507.

Buttigieg, V., Farrell, P.G., 1995. A MAP decoding algorithm for variable-length error-correcting codes. In: Codes and Cyphers: Cryptography and Coding IV, pp. 103–119, Essex, England. The Inst. of Mathematics and its Appl.

Buttigieg, V., Farrell, P.G., 2000. Variable-length error-correcting codes. IEE Proc. Commun. 147 (4), 211–215.

Cain, J.B., Clark, Jr., G.C., Geist, J.M., 1979. Punctured convolutional codes of rate $(n-1)/n$ and simplified maximum likelihood decoding. IEEE Trans. Inf. Theory 25 (1), 97–100.

Caire, G., Narayanan, K., 2007. On the SNR exponent of hybrid digitalanalog space time coding. IEEE Trans. Inf. Theory 53 (8), 2867–2878.

Cantillo, J., Lacan, J., Buret, I., Arnal, F., 2008. Procédé et dispositif de délinéation d'un flux de données et système de communication comprenant ledit dispositif. French patent application FR0800968.

Capocelli, R., de Santis, A., 1989. Tight upper bounds on the redundancy of Huffman codes. IEEE Trans. Inf. Theory 35 (5), 1084–1091.

Casazza, P.G., 2000. The art of frame theory. Taiwanese J. Math. 4 (2), 129–202.

Chen, Q., Subbalakshmi, K.P., 2003. Joint source-channel decoding for MPEG-4 video transmission over wireless channels. IEEE J. Sel. Areas Commun. 21 (10), 1780–1789.

Chiani, M., Martini, M.G., 2007. Analysis of optimum frame synchronization based on periodically embedded sync words. IEEE Trans. Commun. 55 (11), 2056–2060.

Choi, D., 1990. Frame alignment in a digital carrier system – a tutorial. IEEE Commun. Mag. 28 (2), 47–54.

Chou, J., Ramchandran, K., 2000. Arithmetic coding-based continuous error detection for efficient ARQ-based image transmission. IEEE J. Sel. Areas Commun. 18 (6), 861–867.

Combelles, P., Castelain, D., Berrou, C., 1993. Système de codage canal hiérarchique pour la télévision numérique. In: Quatorzième Colloque GRETSI, pp. 615–618.

Cover, T.M., Thomas, J.M., 1991. Elements of Information Theory. Wiley, New York.

Dahlman, E., Parkvall, S., Skold, J., Beming, P., 2007. 3G Evolution: HSPA and LTE for Mobile Broadband. Academic Press, Oxford.

Daubechies, I., 1988. Orthonormal bases of compactly supported wavelets. Commun. Pure Appl. Mathematics. 11 (1), 909–996.

Daubechies, I., 1992. Ten Lectures on Wavelets. SIAM, Philadelphia, PA.

Daubechies, I., Sweldens, W., 1998. Factoring wavelet transform into lifting steps. J. Fourier Anal. Appl. 4 (3), 247–269.

Deutsch, L.P., 1996. GZIP compressed data format specification. Technical Report RFC 1952, The Internet Society.

De Marca, J.R.B., Jayant, N.S., 1987. An algorithm for assigning binary indices to the codevectors of a multidimensional quantizer. In: Proceedings of the IEEE International Conference on Communications, pp. 1128–1132.

Drmota, M., Hwang, H.K., Szpankowski, W., 2002. Precise average redundancy of an idealized arithmetic coding. In: Proceedings of the Data Compression Conference, Snowbird, pp. 222–231.

ETSI. Digital video broadcasting (DVB); transmission system for handheld terminals (DVBH). Technical report, EN 302 304 V1.1.1.

ETSI, 2007. Framing structure, channel coding and modulation for satellite services to handheld devices (SH) below 3 GHz. Technical report, EN 302 583 V1.1.1.

El Gamal, A.A., Cover, T.M., 1982. Achievable rate for multiple description. IEEE Trans. Inf. Theory 28 (6), 851–857.

Elmasry, G.F., 1999. Joint lossless-source and channel coding using automatic repeat request. IEEE Trans. Commun. 47 (7), 953–955.

Equitz, W.H.R., Cover, T.M., 1991. Successive refinement of information. IEEE Trans. Inf. Theory 37 (3), 460–473.

Fano, R., 1963. A heuristic discussion of probabilistic decoding. IEEE Trans. Info. Theory 9 (2), 64–74.

Farvardin, N., 1990. A study of vector quantization for noisy channels. IEEE Trans. Inf. Theory 36 (4), 799–809.

Farvardin, N., Modestino, J., 1984. Optimum quantizer performance for a class of non-gaussian memoryless sources. IEEE Trans. Inf. Theory 30 (3), 485–497.

Farvardin, N., Vaishampayan, V., 1991a. On the performance and complexity of channel optimized vector quantizers. IEEE Trans. Inf. Theory 37 (1), 155–160.

Farvardin, N., Vaishampayan, V., 1991b. Optimal quantizer design for noisy channels: an approach to combined source-channel coding. IEEE Trans. Inf. Theory 33 (6), 827–837.

Fingscheidt, T., Vary, P., 2001. Softbit speech decoding: A new approach to error concealment. IEEE Trans. Speech Audio Proc. 9 (3), 240–251.

Forney, G.D., 1973. The Viterbi algorithm. Proceedings of the IEEE 61 (3), 268–278.

Forney, G.D., 2001. Codes on graphs: normal realizations. IEEE Trans. Inf. Theory 47 (2), 520–548.

Fu, F., van der Schaar, M., 2009. A new systematic framework for autonomous cross-layer optimization. IEEE Trans. Vehicular Technol. 58 (4), 1887–1903.

Gallager, R., 1978. Variations on a theme by Huffman. IEEE Trans. Inf. Theory 24 (6), 668–674.

Gallager, R.G., 1968. Information Theory and Reliable Communication. Wiley, New York.

Garcia-Frias, J., Villasenor, J.D., 1997. Combining hidden Markov source models and parallel concatenated codes. IEEE Commun. Lett. 1 (4), 111–113.

Garcia-Frias, J., Villasenor, J.D., 2001. Joint turbo decoding and estimation of hidden Markov sources. IEEE J. Sel. Areas Commun. 19 (9), 1671–1679.

Gatspar, M., Rimoldi, B., Vetterli, M., 2003. To code or not to code: Lossy source-channel communication revisited. IEEE Trans. Inf. Theory 49 (5), 1147–1158.

Ge, X., Eppstein, D., Smyth, P., 2001. The distribution of loop lengths in graphical models for turbo decoding. IEEE Trans. Inf. Theory 47 (6), 2549–2553.

Gormish, M.J., Allen, J.D., 1993. Finite state machine binary entropy coding. In: Storer, J.A., Cohn, M. (Eds.), Proceedings of the Data Compression Conference, Snowbird, Utah, p. 449.

Gortz, N., 2001. On the iterative approximation of optimal joint source-channel decoding. IEEE J. Sel. Areas Commun. 19 (9), 1662–1670.

Goyal, V.K., 2001. Multiple description coding: compression meets the network. IEEE Signal Process. Mag.

Goyal, V.K., Kovacevic, J., 1998. Optimal multiple description transform coding of gaussian vectors. Proc. Data Compression Conference, pp. 388–397, Snowbird.

Goyal, V.K., Kovacevic, J., Kelner, J.A., 2001. Quantized frame expansions with erasures. J. Appl. Comput. Harmonic Anal. 10 (3), 203–233.

Goyal, V.K., Kovacević, J., Vetterli, M., 1998. Multiple description transform coding: Robustness to erasures using tight frame expansion. In: Proceedings of the IEEE International Symposium on Information Theory, Boston, MA, p. 408.

Goyal, V.K., Vetterli, M., Thao, N.T., 1998. Quantized overcomplete expansions in $\mathbb{R}^n$: analysis, synthesis, and algorithms. IEEE Trans. Inf. Theory 44 (1), 16–31.

Grangetto, M., Cosman, P., Olmo, G., 2005. Joint source/channel coding and MAP decoding of arithmetic codes. IEEE Trans. Comm. 53 (6), 1007–1016.

Grangetto, M., Scanavino, B., Olmo, G., 2004. Joint source channel iterative decoding of arithmetic codes. In: Proceedings of the IEEE ICC, pp. 886–890.

Gray, R.M., Neuhoff, D.L., 1998. Quantization. IEEE Trans. Inf. Theory 44, 2325–2383.

Guionnet, T., Guillemot, C., 2003. Soft decoding and synchronization of arithmetic codes: Application to image transmission over noisy channels. IEEE Trans. Image Process. 12 (12), 1599–1609.

Guionnet, T., Guillemot, C., 2004. Soft and joint source-channel decoding of quasi-arithmetic codes. EURASIP J. Appl. Signal Process. 2004 (3), 393–411.

Guionnet, T., Guillemot, C., Fabre, E., 2002. Soft decoding of multiple descriptions. In: Multimedia and Expo, 2002. ICME '02. Proceedings. 2002 IEEE International Conference, vol. 2, pp. 601–604.

Guivarch, L., Carlach, J.-C., Siohan, P., 2000. Joint source channel soft decoding of Huffman codes with turbo-codes. In: Proceedings of the IEEE Data Compression Conference, p. 88.

Guivarch, L., Siohan, P., Carlach, J.C., 2000. Low complexity soft decoding of Huffman encoded Markov sources using turbo-codes. In: Proceedings of ICT, Acapulco, pp. 872–876.

Guyader, A., Fabre, F., Guillemot, C., Robert, M., 2001. Joint source-channel turbo decoding of entropy coded sources. IEEE J. Sel. Areas Commun. 19 (9), 1680–1695.

Haberle, H., 1969. Frame synchronization. Electr. Commun. 44 (4), 208–287.

Haccoun, D., Ferguson, M.J., 1975. Generalized stack algorithms for decoding convolutional codes. IEEE Trans. Inf. Theory 21 (6), 638–651.

Hagenauer, J., Hoeher, P., 1989. A Viterbi algorithm with soft-decision outputs and its applications. In: Proceedings of the Globecom, Dallas, TX, pp. 1680–1686.

Halonen, T., Romero, J., Melero, J., 2002. GSM, GPRS and EDGE Performance: Evolution Toward 3G/UMTS. John Wiley & Sons, Chichester.

Han, T.S., 2002. Information-Spectrum Methods in Information Theory. Springer, Berlin.

Hanzo, L., Cherriman, P., Streit, J., 2007. Video Compression and Communications: From Basics to H.261, H.263, H.264, Mpeg4 for DVB and HSDPA-Style Adaptive Turbo-Transceivers, second ed. John Wiley & Sons Ltd, Hoboken.

Hanzo, L., Wong, C.H., 2002. Adaptive Wireless Transceivers: Turbo-Coded, Turbo-Equalised and Space-Time Coded TDMA, CDMA, MC-CDMA and OFDM Systems. Wiley, New York.

Herre, J., Shulz, D., 1998. Extending the mpeg-4 aac codec by perceptual noise substitution. In: Proceedings of the 104th International Conference of the Audio Engineering Society. preprint #4720.

Hong, M.C., Schwab, H., Kondi, L.P., Katsaggelos, A.K., 1999. Error concealment algorithms for compressed video. Signal Process. Image Commun. 14, 473–492.

Hotelling, H., 1933. Analysis of a complex of statistical variables into principal components. J. Educ. Psychol. 24, 417–441, 498–520.

Howard, P.G., Vitter, J.S., 1992. Practical implementation of arithmetic coding. In: Storer, J.A. (Ed.), Image and Text Compression. Kluwer Academic Publisher, Boston.

Howard, P.G., Vitter, J.S., 1994. Arithmetic coding for data compression. Proceedings of IEEE 82 (6), 857–865.

Hsiang, S.T., Woods, J.W., 2000. Embedded image coding using zeroblocks of subband/wavelet coefficients and context modeling. In: Proceedings of IEEE Int. Symp. On Circuits and Systems, Genova, pp. 662–665.

Huber, R., Kollmeier, B., 2006. PEMO-Q, a new method for objective audio quality assessment using a model of auditory perception. IEEE Trans. Audio, Speech, Lang. Process. 14 (6), 1902–1911.

Huffman, D.A., 1952. A method for the construction of minimum-redundancy codes. Proc. Inst. Radio Eng. 40 (9), 1098–1101.

Huusko, J., Vehkapera, J., Amon, P., Lamy-Bergot, C., Panza, G., Peltola, J., Martini, M.G., 2007. Cross-layer architecture for scalable video transmission in wireless network. Image Commun. 22 (3), 317–330.

IEEE 802.16e Amendment, 2005. Physical and medium access control layers for combined fixed and mobile operation in license bands. Technical report, IEEE.

IEEE, 1999. 802.11, part 11: Wireless LAN Medium Access Control (MAC) and Physical layer (PHY) specifications. Technical report.

IEEE, 2004. 802.16: IEEE standard for local and metropolitan area networks, air interface for fixed broadband wireless access systems. Technical report.

IEEE, 2006. IEEE standard for local and metropolitan area networks part 16: Air interface for fixed and mobile broadband wireless access systems amendment 2: Physical and medium access control layers for combined fixed and mobile operation in licensed bands and corrigendum 1. Technical Report IEEE 802.16e, IEEE.

IEEE 802.16 Standard, 2004. Local and metropolitan area networks part 16: Air interface for fixed broadband wireless access systems amendment for physical and medium access control layers for combined fixed and mobile in license bands. Technical report, IEEE.

IEEE 802.2 Standard, 1998. Local and metropolitan area networks part 2: Logical link control. Technical report, IEEE.

ISO/IEC JTC 1, 1993. Coding of moving pictures and associated audio for digital storage media at up to about 1.5 Mbis/s – part 2: Video. Technical report, ISO/IEC 11172-2 (MPEG-1).

ISO/IEC JTC 1, 1999. Coding of audio-visual objects – part 2: Video. Technical report, ISO/IEC 14496-2 (MPEG-4 visual version 1).

ISO/IEC JTC 1, 2000. JPEG 2000 part i: Final draft international standard (ISO/IEC FDIS15444-1). Technical report, ISO/IEC JTC1/SC29/WG1 N1855.

ISO/IEC JTC 1/SC 29, 1996. Generic coding of moving pictures and associated audio information: Systems. Technical report, ISO/IEC 13818-1 (MPEG-2 Part 1).

ISO/IEC, 1997. MPEG-2 advanced audio coding, AAC. Technical Report 13818-7, International Organization for Standardization.

ISO/IEC, 2005. MPEG-4 advanced audio coding, AAC. Technical Report 14496-3, International Organization for Standardization.

ITU, 1992. Recommendation IT-81, information technology – digital compression and coding of continuous-tone still images – requirements and guidelines (JPEG). Technical report, ITU – CCITT.

ITU-T, 1990. Video codec for audiovisual services at px64 kbits/s. Technical report, ITU-T Rec. H.261, version 1.

ITU-T, 2005. H263 – video coding for low bitrate communications. Technical report, ITU-T Rec. H.263 (01/2005).

ITU-T and ISO/IEC JTC 1, 2003. Advanced video coding for generic audiovisual services. Technical report, ITU-T Rec. H.264, and ISO/IEC 14496-10 AVC.

ITU-T/SG16/VCEG, ISO/JTC1/SC29/WG11, 2002. H264 Joint Committee Draft, Document JVT-C167. 3rd meeting: Fairfax, Virginia, USA.

Itoh, Y., Cheung, N.M., 2000. Universal variable length code for DCT coding. In: Proceedings International Conference on Image Processing, vol. 1, pp. 940–943.

Jain, A.K., 1989. Fundamentals of Digital Image Processing. Prentice-Hall, Englewood Cliffs, NJ.

Jaoua, Z., Mokraoui-Zergainoh, A., Duhamel, P., 2008a. Robust transmission of HTML files: iterative joint source-channel decoding of deflate codes. In: European Signal Processing Conference (EUSIPCO), Lausanne.

Jaoua, Z., Mokraoui-Zergainoh, A., Duhamel, P., 2008b. Robust transmission of HTML files: Iterative joint source-channel decoding of Lempel Ziv–77 codes. In: IEEE International Conference on Acoustics, Speech, and Signal Processing (ICASSP '08), Las Vegas, pp. 2993–2996.

Jaspar, X., Vandendorpe, L., 2005a. Design and performance analysis of joint source channel turbo schemes with variable length codes. In: Proceedings of the IEEE ICC, vol. 3, pp. 526–530.

Jaspar, X., Vandendorpe, L., 2005b. Performance and convergence analysis of joint source channel turbo schemes with variable length codes. In: Proceedings of the IEEE ICASSP, vol. 3, pp. 485–488.

Jaspar, X., Guillemot, Ch., Vandendorpe, L., 2007. Joint source-channel turbo techniques for discrete-valued sources: From theory to practice. Proc. IEEE 95 (6), 1345–1361.

Jeanne, M., Carlach, J.-C., Siohan, P., 2005. Joint source channel decoding of variable length codes for convolutional codes and turbo codes. IEEE Trans. Commun. 53 (1), 10–15.

Jegou, H., Malinowski, S., Guillemot, C., 2005. Trellis state aggregation for soft decoding of variable length codes. In: Proceedings of the IEEE Workshop on Signal Processing Systems, pp. 603–608.

Jelinek, F., 1969. Fast sequential decoding algorithm using a stack. IBM J. Res. Develop. 13 (6), 675–685.

Jenkac, H., Stockhammer, T., Xu, W., 2005. Permeable-layer receiver for reliable multicast transmission in wireless systems. In: Proceedings of the IEEE Wireless Communications and Networking Conference, vol. 3, pp. 1805–1811.

Jin Wang, Lie-Liang Yang, Hanzo, L., 2004. Iterative construction of reversible variable-length codes and variable-length error-correcting codes. IEEE Commun. Lett. 8 (11), 671–673.

Hopcroft John E., Ullman Jeffrey D., 1979. Introduction to Automata Theory, Languages and Computation. Addison-Wesley, Reading, MA, 1979.

Kaiser, S., Bystrom, M., 2000. Soft decoding of variable-length codes. In: Proceedings of the IEEE ICC, vol. 3, New Orleans, pp. 1203–1207.

Karhunen, H., 1947. Über Lineare Methoden in der Wahrscheinlichkeitsrechnung. Ann. Acad. Scien. Fennicae, Series A 37, 7–79.

Kliewer, J., Thobaben, R., 2003. Parallel concatenated joint source channel coding. Electron. Lett. 39 (23), 1664–1666.

Koch, W., Baier, A., 1990. Optimum and suboptimum detection of coded data disturbed by time-varying intersymbol interference. In: Proceedings of the IEEE Globecom, vol. 3, pp. 1679–1684.

Kohonen, T., 1984. Self Organization and Associative Memory. Springer-Verlag, Berlin.

Kohonen, T., 1990. The self-organizing map. Proc. IEEE 78 (9), 1468–1480.

Kopansky, A., Bystrom, M., 1999. Sequential decoding of MPEG-4 coded bitstreams for error resilience. In: Proceedings of the 33rd Annual Conference on Information Sciences and Systems.

Korhonen, J., Wang, Y., 2003. Schemes for error resilient streaming of perceptually coded audio. In: Proceedings of the International Conference on Multimedia and Expo (ICME '03), vol. 3, pp. 165–168.

Kovacevic, J., Chebira, A., 2007a. Life beyond bases: The advent of frames (part ii). IEEE Signal Process. Mag. 24 (5), 115–125.

Kovacevic, J., Chebira, A., 2007b. Life beyond bases: The advent of frames (part i). IEEE Signal Process. Mag. 24 (4), 86–104.

Krichevsky, R., Trofimov, V., 1981. The performance of universal encoding. IEEE Trans. Inf. Theory 27 (2), 199–207.

Kullback, S., Leibler, R.A., 1951. On information and sufficiency. Ann. Math. Stat. 22 (1), 79–86.

Kung, W.-Y., Kim, C.-S., Kuo, C.-C.J., 2006. Spatial and temporal error concealment techniques for video transmission over noisy channels. IEEE Trans. Circuits Syst. Video Technol. 16 (7), 789–802.

Kurose, J.F., Ross, K.W., 2005. Computer Networking: A Top-Down Approach Featuring the Internet, third ed. Addison Wesley, Boston.

Labeau, F., Chiang, J.C., Kieffer, M., Duhamel, P., Vandendorpe, L., Mack, B., 2005. Oversampled filter banks as error correcting codes: theory and impulse correction. IEEE Trans. Signal Process. 53 (12), 4619–4630.

Lakovic, K., Villasenor, J., 2002. On design of error-correcting reversible variable length codes. IEEE Commun. Lett. 6 (8), 337–339.

Lakovic, K., Villasenor, J., 2003. An algorithm for construction of efficient fix-free codes. IEEE Commun. Lett. 7 (8), 391–393.

Lakovic, K., Villasenor, J., Wesel, R., 1999. Robust joint Huffman and convolutional decoding. In: Proc. IEEE Vehicular Technol. Conf., vol. 5, pp. 2551–2555.

Lamy, C., Bergot, F.X., 2003. Lower bounds on the existence of binary error-correcting variable-length codes. In: Information Theory Workshop, Paris, pp. 300–303.

Lamy, C., Merigeault, S., 2002. Procédé de correction d'une trame erronée par un récepteur. French patent no. 0206501.

Lamy, C., Perros-Meilhac, L., 2003. Low complexity iterative decoding of variable length codes. In: Proceedings of the Picture Coding Symposium, St Malo, France, pp. 275–280.

Lamy, C., Pothier, O., 2001. Reduced complexity maximum a posteriori decoding of variable-length codes. In: Proceedings of the IEEE Global Telecommunications Conference, San Antonio, USA, pp. 1410–1413.

Laroia, R., Farvardin, N., Tretter, S.A., 1994. On optimal shaping of multidimensional constellation. IEEE Trans. Inf. Theory 40 (3), 1044–1056.

Larsen, M.H., Weidmann, C., Kieffer, M., 2006. Iterative decoding of entropy-constrained multiple description trellis-coded quantization. In: Proceedings of GLOBECOM. (submitted).

Larzon, L.A., Degermark, M., Jonsson, L.E., Fairhurst, G., 2004. The lightweight user datagram protocol (UDP-Lite). Technical Report RFC 3828, The Internet Society.

Lassing, J., Ottosson, T., Strom, E., 2001. On the union bound applied to convolutional codes. In: Proceedings of the 54th IEEE Vehicular Technology Conference (VTC 2001 Fall), vol. 4, pp. 2429–2433.

Lee, C.M., Kieffer, M., Duhamel, P., 2005. Soft decoding of VLC encoded data for robust transmission of packetized video. In: Proc. ICASSP, pp. 737–740.

Lescuyer, P., Bott, F., 2004. UMTS: Origins, Architecture and the Standard. Springer-Verlag, New York.

Levenshtein, V.I., 1965. Binary codes with correction of deletions, insertions and substitution of symbols. Dokl. Akad. Nank. SSSR 163 (4), 845–848.

Lin, S., Costello, D.J., 1983. Error Control Coding: Fundamentals and Applications. Prentice-Hall, Englewood Cliffs.

Lindé, Y., Buzo, A., Gray, R., 1980. An algorithm for vector quantizer design. IEEE Trans. Commun. 28 (1), 84–94.

Lloyd, S.P., 1957. Least squares quantization in PCM. Bell Telephone Laboratories Paper.

Lloyd, S.P., 1982. Least squares quantization in pcm. IEEE Trans. Inf. Theory 28 (2), 129–137.

Loeliger, H.A., Lustenberger, F., Helfenstein, M., Tarkoy, F., 1998. Probability propagation and decoding in analog VLSI. In: Proceedings of the IEEE International Symposium on Information Theory (ISIT '94), p. 146.

Loève, M., 1948. Fonctions aléatoires du second ordre. In: P. Lévy (Ed.), Processus Stochastiques et Mouvement Brownien. Gauthier-Villars, Paris.

Ma, S., Gao, W., Lu, Y., 2005. Rate-distortion analysis for H.264/AVC video coding and its application to rate control. IEEE Trans. Circuits Syst. Video Technol. 15 (12), 1533–1544.

MacKay, D.J.C., 2003. Information Theory, Inference, and Learning Algorithms. Cambridge University Press, Cambridge.

Malinowski, S., Jegou, H., Guillemot, C., 2007. Synchronization recovery and state model reduction for soft decoding of variable length codes. IEEE Trans. Inf. Theory 53 (1), 368–377.

Mannoni, V., Siohan, P., Jeanne, M., 2006. A simple on-line turbo estimation of source statistics for joint source-channel VLC decoders: Application to MPEG-4 video. In: Proceedings of the 4th International Symposium on Turbo Codes & Related Topics, Munich, Germany, 2006.

Manstetten, D., 1992. Tight bounds on the redundancy of Huffman codes. IEEE Trans. Inf. Theory 38 (1), 144–151.

Marin, C., Duhamel, P., Bouchireb, K., Kieffer, M., 2007. Robust video decoding through simultaneous usage of residual source information and mac layer crc redundancy. In: Proceedings of Globecom, pp. 2070–2074.

Marin, C., Leprovost, Y., Kieffer, M., Duhamel, P., 2008. Robust header recovery based enhanced permeable protocol layer mechanism. In: Proceedings of the IEEE Workshop on Signal Processing Advances in Wireless Communications, pp. 91–95.

Marinkovic, S., Guillemot, C., 2005. Joint source-channel coding by means of oversampled filter banks codes. Eurasip J. Appl. Signal Process. (4), 510–524.

Marpe, D., Schwarz, H., Weigand, T., 2003. Context based adaptative binary arithmetic coding in the H.264/AVC video compression standard. IEEE Trans. Circuits Syst. Video Technol. 13 (7), 620–636.

Martini, M.G., Chiani, M., 2009. Optimum metric for frame synchronization with gaussian noise and unequally distributed data symbols. In: Proceedings of the IEEE International Workshop on Signal Processing Advances for Wireless Communications, Perugia, Italy.

Martini, M.G., Mazzotti, M., Lamy-Bergot, C., Huusko, J., Amon, P., 2007. Content-adaptive network aware joint optimisation of wireless video transmission. IEEE Commun. Mag. 45 (1), 84–90.

Massey, J.L., 1972. Variable-length codes and the Fano metric. IEEE Trans. Inf. Theory 18 (1), 196–198.

Mastronarde, N.H., van der Schaar, M., 2009. Towards a general framework for cross-layer decision making in multimedia systems. IEEE Trans. Circuits Syst. Video Technol. 19 (5), 719–732.

Max, J., 1960. Quantizing for minimum distorsion. IRE Trans. Inf. Theory 6, 7–12.

Maxted, J., Robinson, J., 1985. Error recovery for variables length codes. IEEE Trans. Inform. Theory 31 (6), 794–801.

McEliece, R., 2002. The Theory of Information and Coding. Cambridge University Press, Cambridge.

Mealy, G.H., 1955. A method for synthesizing sequential circuits. Bell System Tech. J. 34, 1045–1079.

Mittal, U., Phamdo, N., 2002. Hybrid digital-analog (HDA) joint source-channel codes for broadcasting and robust communication. IEEE Trans. Inform. Theory 48 (5), 1082–1102.

Mohammad-Khani, G.R., Lee, C.M., Kieffer, M., Duhamel, P., 2006. Simplification of VLC tables with application to ML and MAP decoding algorithms. IEEE Trans. Commun. 54 (10), 1835–1844.

Morrow, R., 2002. Bluetooth: Operation and Use. McGraw-Hill Professional, New York.

Murad, A.H., Fuja, T.E., 1999. Robust transmission of variable-length encoded sources. In: Proceedings of the IEEE Wireless Communications and Networking Conference, vol. 2, New Orleans, LA, pp. 968–972.

Narasimha, M.J., Peterson, A.M., 1978. On the computation of the discrete cosine transform. IEEE Trans. Commun. 26 (6), 934–936.

Nguyen, H., 2005. Décodage source-canal conjoint utilisant la sémantique source pour la transmission robuste de flux vidéo sur liens mobiles. PhD thesis, Univ Paris-Sud.

Nguyen, H., Duhamel, P., 2003. Compressed image and video redundancy for joint source-channel decoding. In: Proceedings of the Globecom, vol. 4, pp. 2198–2202.

Nguyen, H., Duhamel, P., 2004. Iterative joint source channel decoding of variable length encoded video sequences exploiting source semantics. In: Proceedings of the IEEE ICIP, pp. 3221–3224.

Nieto, J.W., Furman, W.N., 2007. Cyclic redundancy check (CRC) based error method and device. US Patent US 2007/0192667 A1.

Noll, P., Zelinski, R., 1978. Bounds on quantizer performance in the low bit-rate region. IEEE Trans. Commun. 26 (2), 300–304.

Nuaymi, L., Bouida, N., Lahbil, N., Godlewski, P., 2007. Headers overhead estimation, header suppression and header compression in wimax. In: Proceedings of the IEEE International Conference on Wireless and Mobile Computing, Networking and Communication, Los Alamitos, CA, USA, p. 17.

Omura, J.K., 1969. On the Viterbi decoding algorithm. IEEE Trans. Inform. Theory 15 (1), 177–179.

Osorio, R.R., Bruguera, D., 2004. Arithmetic coding architecture for H.264/AVC CABAC compression system. Proceedings of the EUROMICRO Systems on Digital Dystem Design, pp. 62–69.

Ozarow, L., 1980. On a source-coding problem with two channels and three receivers. Bell Syst. Techn. J. 59 (10), 1909–1921.

Panza, G., Balatti, E., Vavassori, G., Lamy-Bergot, C., Sidoti, F., 2005. Supporting network transparency in 4G networks. In: Proceedings of the IST Mobile and Wireless Communication Summit.

Park, M., Miller, D.J., 1998. Decoding entropy-coded symbols over noisy channels by MAP sequence estimation for asynchronous hmms. In: Proceedings of the 32nd Annual Conference on Information Sciences and Systems (CISS '98), pp. 477–482.

Park, M., Miller, D.J., 1999. Joint source-channel decoding for variable-length encoded data by exact and approximate map sequence estimation. In: Proceedings of the IEEE International Conference Acoustics, Speech, Signal Processing (ICASSP '99), vol. 5, pp. 2451–2454.

Park, M., Miller, D.J., 2000. Joint source-channel decoding for variable length encoded data by exact and approximate MAP sequence estimation. IEEE Trans. Commun. 48 (1), 1–6.

Peng, Z., Huang, Y.-F., Costello, D.J., 2000. Turbo codes for image transmission: a joint channel and source decoding approach. IEEE J. Sel. Areas Commun. 18 (6), 868–879.

Perros-Meilhac, L., Lamy, C., 2002. Huffman tree based metric derivation for a low-complexity soft VLC decoding. Proc. IEEE ICC. 2, 783–787.

Pettijohn, B.D., Hoffman, W., Sayood, K., 2001. Joint source/channel coding using arithmetic codes. IEEE Trans. Commun. 49 (5), 826–836.

Phamdo, N., Farvardin, N., 1994. Optimal detection of discrete Markov sources over discrete memoryless channels, applications to combined source-channel coding. IEEE Trans. Inform. Theory 40 (1), 186–193.

Plotnik, E., Weinberger, M.J., Ziv, J., 1992. Upper bounds on the probability of sequences emitted by finite-state sources and on the redundancy of the lempel-ziv algorithm. IEEE Trans. Inf. Theory 38 (1), 66–72.

Postel, J., 1980. User datagram protocol. Technical Report RFC 768, Information Sciences Institute.

Postel, J., 1981a. Internet protocol. Technical Report RFC 791, Information Sciences Institute.

Postel, J., 1981b. Transport control protocol. Technical Report RFC 793, Information Sciences Institute.

Poulliat, C., Declercq, D., Lamy-Bergot, C., 2005. Analysis and optimization of irregular LDPC codes for joint source-channel decoding. IEEE Commun. Lett. 9 (12), 1064–1066.

Proakis, J.G., 2001. Digital Communications. Maidenhead, Berkshire, UK.

Ramchandran, K., Ortega, A., Uz, K.M., Vetterli, M., 1993. Multiresolution broadcast for digital HDTV using joint source/channel coding. EEE Trans. Sel. Areas Commun. 11 (1), 3–14.

Reimers, U., 2005. DVB: The Family of International Standards for Digital Video Broadcasting. Springer, Berlin.

Richardson, I., 2003. H.264 and MPEG-4 Video Compression: Video Coding for Next-Generation Multimedia. John Wiley and Sons, Chichester.

Richardson, T., Urbanke, U., 2008. Modern Coding Theory. Cambridge University Press, Cambridge.

Rimoldi, B., 1994. Successive refinement of information: Characterization of the achievable rates. IEEE Trans. Inf. Theory 40 (1), 253–259.

Rioul, O., 2007. Théorie de l'information et du codage. Hermes Science – Lavoisier.

Rissanen, J., 1976. Generalised Kraft inequality and arithmetic coding. IBM J. Res. Dev. 20 (3), 198–203.

Rissanen, J., 1986. Complexity of strings in the class of Markov sources. IEEE Trans. Inf. Theory 32 (4), 526–532.

Rodeh, M., 1982. A fast test for unique decipherability based on suffix trees. IEEE Trans. Inf. Theory 28 (4), 648–651.

Sabeva, G., Ben-Jamaa, S., Kieffer, M., Duhamel, P., 2006. Robust decoding of H.264/AVC encoded video transmitted over wireless channels. In: Proceedings of MMSP, Victoria, Canada, pp. 9–13.

Said, A., Pearlman, W.A., 1996. A new fast and efficient image codec based on set partitioning in hierarchical trees. IEEE Trans. Circuits Syst. Video Technol. 6 (3), 243–250.

Sardinas, A.A., Patterson, C.W., 1953. A necessary sufficient condition for the unique decomposition of coded messages. In: IRE International Convention Record, vol. 8, pp. 104–108.

Savari, S.A., 1997a. Redundancy of the Lempel-Ziv incremental parsing rule. IEEE Trans. Inf. Theory 43 (1), 9–21.

Savari, S.A., 1997b. Redundancy of the Lempel-Ziv-Welch code. In: Proceedings of Data Compression Conference, DCC '97, pp. 191–200, 25–27.

Sayir, J., 1999a. Arithmetic coding for noisy channels. Proceedings of the IEEE Information Theory Workshop, pp. 69–71.

Sayir, J., 1999b. On Coding by Probability Transformation. ph.D. Thesis Nr. 13099, EE Department, ETH Zurich, Switzerland.

Sayood, K., 2005. Introduction to Data Compression, second ed. Morgan Kufmann, San Francisco.

Sayood, K., Borkenhagen, J.C., 1991. Use of residual redundancy in the design of joint source/channel coders. IEEE Trans. Commun. 39 (6), 838–846.

Schulzrinne, H., Casner, S., Frederick, R., Jacobson, V., 1996. A transport protocol for real-time applications. Technical Report RFC 1889, Network Working Group.

Schwartz, M., 1980. Information, Transmission, Modulation, and Noise. McGraw Hill, NY.

Shannon, C.E., 1948. A mathematical theory of communication. Bell Syst. Tech. J. 27, 379–423, 623–656.

Shapiro, J.M., 1993. Embedded image coding using zerotrees of wavelet coefficients. IEEE Trans. Signal Process. 41 (12), 3445–3462.

Skinnemoen, P.H., 1994. Robust Communication with Modulation Organized Vector Quantization. PhD thesis, Norvegian Institute of Technology.

Stubley, P.R., 1994. On the redundancy of optimum fixed-to-variable length codes. In: Proceedings of Data Compression Conference, DCC '94, pp. 90–97.

Sullivan, G.J., Wiegand, T., 2005. Video compression – from concepts to the H.264/AVC standard. Proc. IEEE, 93 (1), 18–31.

Swaszek, P.F., DiCicco, P., 1995 More on the error recovery for variable length codes. IEEE Trans. Inform. Theory 41 (6), 2064–2071.

Szpankowski, W., 2000. Asymptotic average redundancy of Huffman (and other) block codes. IEEE Trans. Inf. Theory 46 (7), 2434–2443.

Takishima, Y., Wada, M., Murakami, H., 1995. Reversible variable length codes. IEEE Trans. Commun. 43 (2–4), 158–162.

Tanenbaum, A.S., 2002. Computer Networks. Prentice Hall, Upper Saddle River.

Taubman, D., 2000. High performance scalable image compression with EBCOT. IEEE Trans. Image Process. 9 (7), 1158–1170.

Ten Brink, S., 2001. Convergence behavior of iteratively decoded parallel concatenated codes. IEEE Trans. Commun. 49 (10), 1727–1737.

Thobaben, R., Kliewer, J., 2003. On iterative source-channel decoding for variable-length encoded Markov sources using a bit-level trellis. In: Proceedings of the IEEE SPAWC, pp. 50–54.

Thobanen, R., Kliewer, J., 2003. Robust decoding of variable-length encoded Markov sources using a three-dimensional trellis. IEEE Commun. Lett. 7 (7), 320–322.

Tian, C., Steiner, A., Shamai, Sh., Diggavi, S.N., 2008. Successive refinement via broadcast: Optimizing expected distortion of a gaussian source over a gaussian fading channel. IEEE Trans. Inf. Theory 54 (7), 2903–2918.

Tsai, C.W., Wu, J.L., 2001. On constructing the Huffman-code based reversible variable length codes. IEEE Trans. Commun. 49 (9), 1506–1509.

Ueda, H., Yamaguchi, U., Miki, N., Watanabe, R., 2003. A method of improving misframe for hec-based variable length frame suitable for ip services. Electron. Commun. Japan (Part I: Communications), 86 (6), 46–58.

Ueda, U., Yamaguchi, H., Watanabe, R., 2001. Reducing misframe frequency for hec-based variable length framesuitable for ip services. In: IEEE International Conference on Communications, ICC 2001, vol. 4, pp. 1196–1200.

Usevitch, B.E., 2001. A tutorial on modern lossy wavelet image compression: Foundations of JPEG 2000. IEEE Signal Process. Mag. 18 (5), 22–35.

Vaishampayan, V.A., 1993. Design of multiple description scalar quantizers. IEEE Trans. Inf. Theory 39 (3), 821–834.

Vaishampayan, V., Domaszewicz, J., 1994. Design of entropy-constrainted multiple description scalar quantizers. IEEE Trans. Inf. Theory 40 (1), 245–250.

Vaishampayan, V.A., Farvardin, N., 1992. Joint design of block source codes and modulation signal sets. IEEE Trans. Inf. Theory 38 (4), 1230–1248.

Van der Schaar, M., Shankar, S., 2005. Cross-layer wireless multimedia transmission: Challenges, principles, and new paradigms. IEEE Wireless Commun. Mag. 12 (4), 50–58.

Vembu, S., Verdu, S., Steinberg, Y., 1995. The source-channel separation theorem revisited. IEEE Trans. Inf. Theory 41 (1), 44–54.

Vetterli, M., Herley, C., 1992. Wavelets and filter banks: Theory and design. IEEE Trans. Signal Proc. 40 (9), 2207–2231.

Vetterli, M., Kovačević, J., 1995. Wavelets and Subband Coding. Prentice Hall, Englewood Cliffs.

Viterbi, A.J., Omura, J.K., 1979. Principles of Digital Communication and Coding. McGraw-Hill, New York.

Vitter, J.S., 1987. Design and analysis of dynamic Huffman codes. J. ACM 34 (4), 825–845.

Vucetic, B., Yuan, J., 2000. Turbo Codes – Principles and Applications. Kluwer, Dordrecht.

Walsh, J.L., 1923. A closed set of orthogonal functions. Am. J. Math. 55, 5–24.

Wang, J., Venkatachalam, M., Fang, Y., 2007. System architecture and cross-layer optimization of video broadcast over WiMAX. IEEE J. Sel. Areas Commun. 25 (4), 712–721.

Wang, Y., Orchard, M.T., Reibman, A., 1998. Optimal pairwise correlating transforms for multiple description coding. In: Proceedings of the International Conference on Image Processing, vol. 1, pp. 679–683.

Weidmann, C., Kieffer, M., 2009. Evaluation of the distance spectrum of variable-length finite-state codes. IEEE Trans. Commun., in press.

Welch, T.A., 1984. A technique for high-performance data compression. IEEE Comput. 17 (6), 8–19.

Wen, J., Villasenor, J., 1997. A class of reversible variable length codes for robust image and video coding. In: Proceedings of the IEEE International Conference on Image Processing, vol. 2, pp. 65–68.

Wen, J., Villasenor, J.D., 1999. Utilizing soft information in decoding of variable length codes. In: Proceedings of the IEEE Data Compression Conference (DCC '99), pp. 131–139.

Wen, J., Villasenor, J.D., 2002. Soft-input soft-output decoding of variable length codes. IEEE Trans. Commun. 50 (5), 689–692.

Wiegand, T., Sullivan, G.J., Bjøntegaard, G., Luthra, A., 2003. Overview of the H.264/AVC video coding standard. IEEE Trans. Circuits Syst. Video Technol. 13 (7), 560–576.

Winkler, G., 2005. Image Analysis, Random Fields and Dynamic Monte Carlo Methods: A Mathematical Introduction, second ed. Number 27 in Applications of Mathematics. Springer, Berlin.

Witten, I.H., Neal, R.M., Cleary, J.G., 1987. Arithmetic coding for data compression. Commun. ACM 30 (6), 520–540.

Wolf, J.K., 1978. Efficient maximum-likelihood decoding of linear block codes using a trellis. IEEE Trans. Inform. Theory 24 (1), 76–80.

Wong, K.K.Y., 2006. The Soft-Output M-Algorithm and its Applications. PhD thesis, Queen's University, Kingston, Canada.

Yang, E.H., Song, L., Shamir, G.I., Kieffer, J.C., 2005. On the pointwise redundancy of the LZ78 algorithm. In: Proceedings of International Symposium on Information Theory, pp. 495–499.

Yeung, R.W., 2002. A First Course in Information Theory. Springer, New York.

Zahir Azami, S. B., Duhamel, P., Rioul, O., 1996. Joint source channel coding : Panorama of methods. Proceedings of CNES workshop on Data Compression.

Zeger, K., Gersho, A., 1990. Pseudo-Gray coding. IEEE Trans. Commun. 38 (12), 2147–2158.

Zehavi, E., Wolf, J.K., 1988. On runlength codes. IEEE Trans. Inf. Theory 34 (1), 45–54.

Zhou, G., Zhang, Z., 2002. Synchronization recovery of variable-length codes. IEEE Trans. Inf. Theory 48 (1), 219–227.

Zigangirov, K., 1966. Some sequential decoding procedures. Probl. Peredach. Inform. 2 (4), 13–15.

Ziv, J., Lempel, A., 1977. A universal algorithm for sequential data compression. IEEE Trans. Inf. Theory 23 (3), 337–343.

Ziv, J., Lempel, A., 1978. Compression of individual sequences via variable-rate coding. IEEE Trans. Inform. Theory 24 (5), 530–536.

Index

E

F

G

H

I

J

K

L

M